FOREWORD

Machine Milking, the forerunner of this book, was published in 1959 by Her Majesty's Stationery Office, for the Ministry of Agriculture, Fisheries & Food as Bulletin No. 177. No authoritative and comprehensive work of the same nature has succeeded it and a replacement is overdue. The Institute is grateful to HMSO and MAFF, who willingly agreed to the proposal that the NIRD should write and publish a new book to replace Bulletin 177, and to use material from it.

The same style and treatment of subject have been retained as well as the title, and most of the original authors have contributed. The enlarged subject matter is now divided into 12 chapters, each substantially complete in itself. They are written at varying technical levels to suit the nature of the material. International milking machine terminology and S.I. units are used throughout, equivalents in Imperial units usually being given. This time, not all the authors are members of the staff of the NIRD. Dr G.A. Mein is with the State Department of Agriculture, Victoria, Australia, and Mr J.G. Anderson and Mr L.J. Coombs are with the Avon Rubber Co.

I am most grateful to Mr H.S. Hall for initiating this work, and especially to Drs C.C. Thiel and F.H. Dodd, who completed the work on Mr Hall's retirement. The NIRD also acknowledges the help of many members of staff and others who assisted in the preparation of the book. We are particularly indebted to manufacturers who provided much information on their products, to Mr M. Marshall who prepared the line drawings, and to Mr B. Bone, the NIRD Librarian, for the index. It is also a pleasure to thank Mr J. Armstrong and his staff of the Printing Division of the College of Estate Management at the University of Reading for their interest and care in printing the volume.

I believe that this work will become established as the standard reference book on this important topic.

B.G.F. Weitz
Director, 1967-1978

The first impression of this book has sold well and it has become necessary to produce a reprint. The text remains up to date and changes made are corrections and minor revision. We are indebted to the authors, and to Mr C. Line, Mrs J. Geens and Mr M.F.H. Shearn of the NIRD for their assistance and care in the preparation of this reprint.

C.C. Theil and F.H. Dodd
May 1979

MACHINE MILKING

EDITORS :— C.C. Thiel and F.H. Dodd

Technical Bulletin 1

The National Institute for Research in Dairying, Reading, England
The Hannah Research Institute, Ayr, Scotland

First published 1977
Reprinted with minor revision, 1979

© NIRD 1979
ISBN 0 7084 0116 3

For titles of other publications in the series,
and ordering, see inside back cover

Printed at the College of Estate Management, Reading

637.124028
M259t

83-1732

12/14/83 - P

CONTENTS

Page

Foreword .... *B.G.F. Weitz*

Chapter I	History and Development .... *H.S. Hall*	1
	The establishment of principles	
	Comparative trials of milking machines 1913 and 1916	
	Vacuum level and pulsation rate	
	Major developments in the 1920s	
	The pattern of development	
	Growth of milking machines	
Chapter II	Basic Mechanics .... *H.S. Hall*	23
	The concept of pressure and vacuum	
	Fluid flow in milking machine installations	
	The mechanics of vacuum production and control	
Chapter III	Description and Performance of Components .... *D.N. Akam*	37
	Vacuum pumps and associated components	
	Pulsation equipment	
	Milking units	
	Milk transport systems	
	Stand-by equipment	
	Miscellaneous components	
Chapter IV	Maintenance and Mechanical Testing .... *H.S. Hall*	102
	Maintenance	
	Mechanical tests	
Chapter V	Action of the Cluster During Milking . . *C.C. Thiel & G.A. Mein*	116
	Patterns of flowrate during a milking	
	Patterns of flowrate within a single pulsation cycle	
	Vacuum conditions in the cluster and liner wall movement	
	Action of the liner in relation to the teat	
	Vacuum stability	
	Summary	
Chapter VI	Anatomy and Physiology of the Udder .... *A.T. Cowie*	156
	Anatomy of the udder	
	Physiology of lactation	
Chapter VII	Milking Routines .... *F.H. Dodd & T.K. Griffin*	179
	Milking methods and rate of milk secretion	
	Milking routines	
	Rate of milking and lactation	
	Summary	

Chapter VIII   Machine Milking in Cowsheds & Milking Parlours   *P.A. Clough*   201
   Cowsheds
   Milking parlours
   Machine milking performance
   Selection of a milking parlour

Chapter IX   Machine Milking and Mastitis
   R.G. Kingwill, F.H. Dodd & F.K. Neave   231
   Udder infection and mastitis
   Mode of infection
   Hygiene
   Milking machine factors
   Machine milking practice
   Mastitis control
   Conclusions

Chapter X   Cleaning and Disinfection in Milk Production
   Christina M. Cousins & C.H. McKinnon   286
   Sources of contamination of milk
   Preparation of the cow for milking
   Methods for milking equipment
   Cleaning farm bulk milk tanks
   Monitoring and assessment of cleaning and disinfection
   Cleaning and disinfecting agents
   Legal requirements

Chapter XI   Rubber and the Milking Machine
   J.G. Anderson & L.J. Coombs   330
   Nature and properties of rubber
   Manufacture of rubber articles
   Rubber parts in milking machines
   Conclusion

Chapter XII   Milk Cooling Equipment . . . . *J.B. Hoyle*   353
   The background of bulk collection
   Refrigerated farm milk tanks
   Tank cleaning
   Energy saving
   Possibilities for the future

Terminology and definitions   378

Conversion factors   384

Suppliers mentioned in the text   385

Index   386

Chapter I

HISTORY AND DEVELOPMENT

H S Hall

Ideas for milking cows by machine, to replace the centuries-old practice of milking by hand, first became apparent about 150 years ago. The earliest devices were metal tubes (cannulae) inserted into the teat canal so that milk in the udder and teat sinuses could flow out by the forces of gravity and intramammary pressure. It seems likely that the use of straws for this purpose was already well-known. Probably the first patent for such a metal device was taken out by Blurton in 1836. In this, the four cannulae were connected to a pail suspended from the cow. Although the risk of teat damage must have been evident, others followed his example and the idea was developed commercially, persisting for many years (Fig I 1).

THE ESTABLISHMENT OF PRINCIPLES

The simple physics of the problem led to the idea of using vacuum and two British inventors, Hodges and Brockedon, in 1851, were probably the first to use this. Their patent[1] was primarily concerned with extracting gunshot from wounds, and other medical procedures but they visualized the possible application of their invention to milking cows. For this a vessel having a 'shield' or upper plate with apertures for the teats would be pressed upwards against the udder by a 'band fastened over the cow's loins'. The vessel would be 'exhausted of air to the extent required' and would form the milk receiver or be connected to a separate vessel under vacuum. L.O. Colvin, of Cincinnatus USA, developed this idea and took out a British Patent[2] in 1860 (Fig I 2). This machine had individual 'teat cups' (probably the first use of this term) connected to hand-operated diaphragm vacuum pumps by short rubber tubes to facilitate

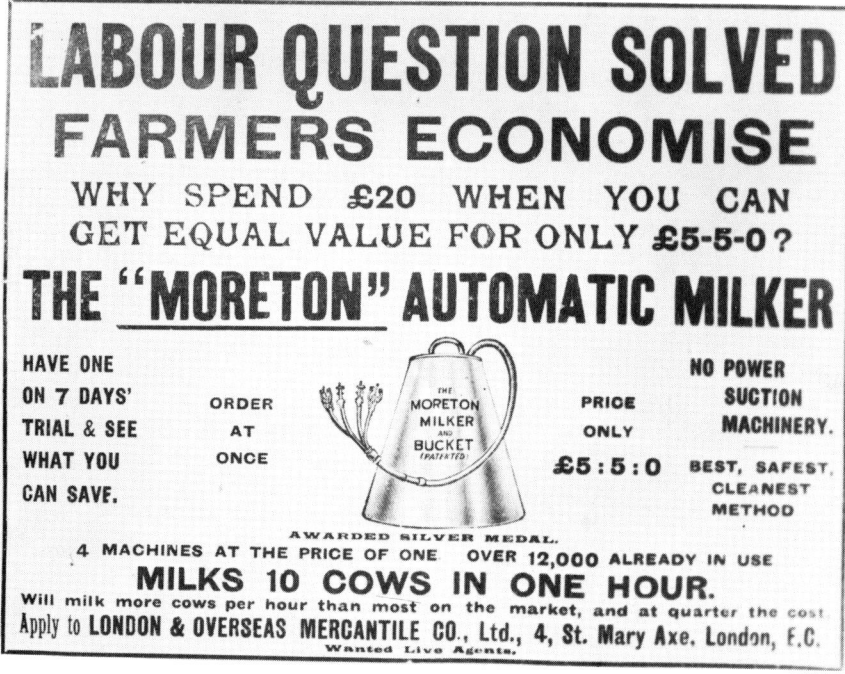

Fig I 1. Copy of an advertisement from the *Mark Lane Express Agricultural Journal*, March 27, 1916. This device was similar to the Blurton design of 1836 using metal cannulae.

adjustment. The teatcups, presumably metal, had internal rings to improve the grip on the teat. In 1863 Louis Grasset, a French tinman, invented a vacuum milking machine which had some elements of hygienic construction[3]. The teatcups (called 'funnels'), claw and long milk tube were all made from caoutchouc (rubber) and the milk tube was connected to a free-standing bucket on which a hand-operated piston-type vacuum pump was fitted. The bucket had a window with a graduated scale to measure the contents and a large access port for manual cleaning. The teatcups were intended to be washed by plunging in warm water which was then drawn into the bucket by vacuum to clean the claw and long milk tube.

For the next 15 years development of vacuum milking machines appears to have made little or no progress. In 1878 the Royal Agricultural Society of England (RASE) created a special class for milking machines in its Trial of Dairy

Implements at the Royal Show. No entries were received in spite of the offer of a prize of £50. Perhaps for this reason, and certainly because of the lack of development of vacuum milking machines, inventors turned their attention to mechanical pressure-type machines, sometimes called lactators. From 1878 onwards a spate of designs came forward from various countries in Europe and North America. These machines sought to imitate hand milking by applying pressure to the outside of the teat rather than by applying vacuum. Some were crude in the extreme but there were many which were amazing examples of mechanical ingenuity using plates, bars, rollers or belts, manually operated through mechanical, hydraulic or pneumatic transmission. One of the early

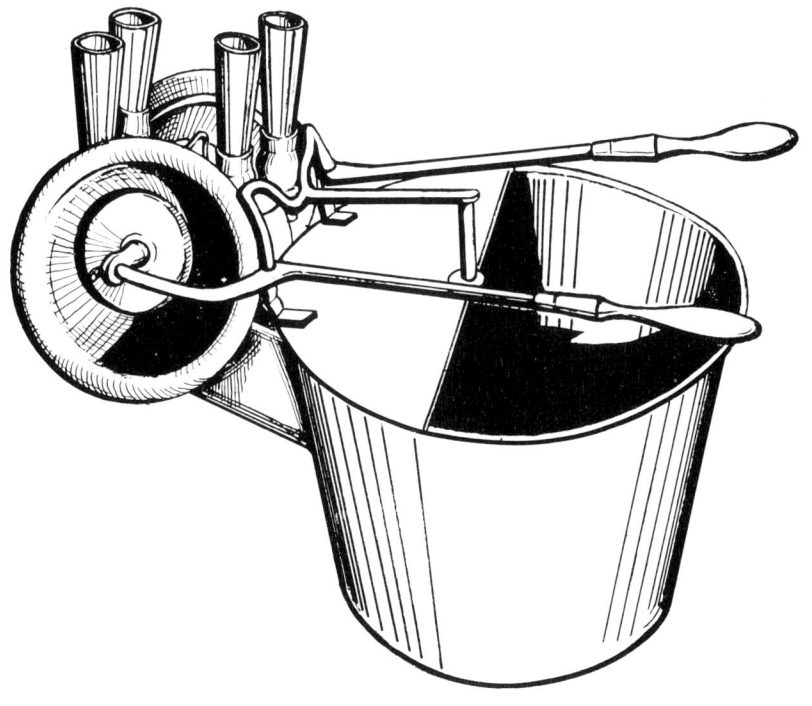

Fig I 2. The Colvin hand-operated vacuum milker (1860). The handles were moved to and fro to operate the diaphragm vacuum pumps.

designs was the subject for patent by A.B. Crees[4], a bank manager from Ilminster. This machine (Fig I 3) had two pairs of travelling chains driven through gears from a hand wheel. Each chain carried two rubber rollers which were synchronized with the rollers on the second chain of a pair so applying a downward squeeze to two teats simultaneously. This squeezing apparatus was mounted on top of an open bucket and carried by spring-loaded legs to provide vertical adjustment. A few of these pressure-type machines were manufactured commercially and three makes survived to compete in trials held by the RASE in

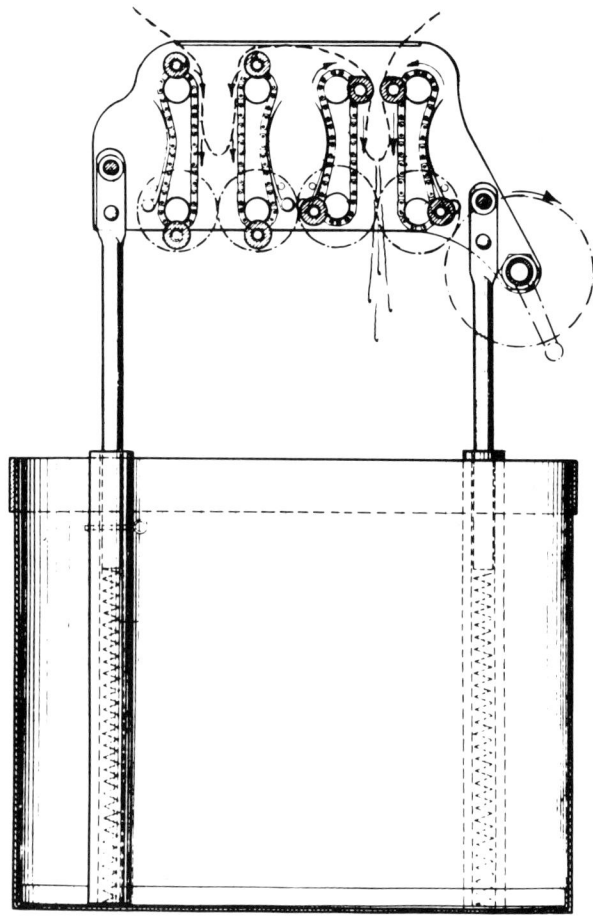

Fig I 3. The Crees lactator (1881). The rollers on the travelling chains squeezed milk from pairs of teats alternately.

1913 (see below). However, the inherent disadvantages of cumbersome mechanisms close to the cow, difficulties in cleaning and, probably, teat damage were responsible for the disappearance of this type of machine before 1920 and it has never been revived. The RASE repeated its offer of a prize in 1882 but again no entries were received. However, this encouragement, and probably the competition from inventors of pressure-type machines, stimulated further development of the idea of milking by vacuum. In 1889 William Murchland, a Kilmarnock plumber, produced the first successful commercial milking machine[5]. This was a great step forward as it was not merely a machine for milking a cow but a milking machine installation designed to operate with a central vacuum system and a number of milking units (Fig I 4).

Murchland's machine used a hand-operated vacuum pump which exhausted a 1 in (25 mm) pipe system extending overhead round the cowshed and provided with a stall cock between each pair of cows. A vacuum of 11 in of mercury (37 kPa) regulated by a water column was used, and it was stated that one boy could maintain the necessary vacuum for three girls to operate two or three units each. The bucket was suspended from the cow and connected to the

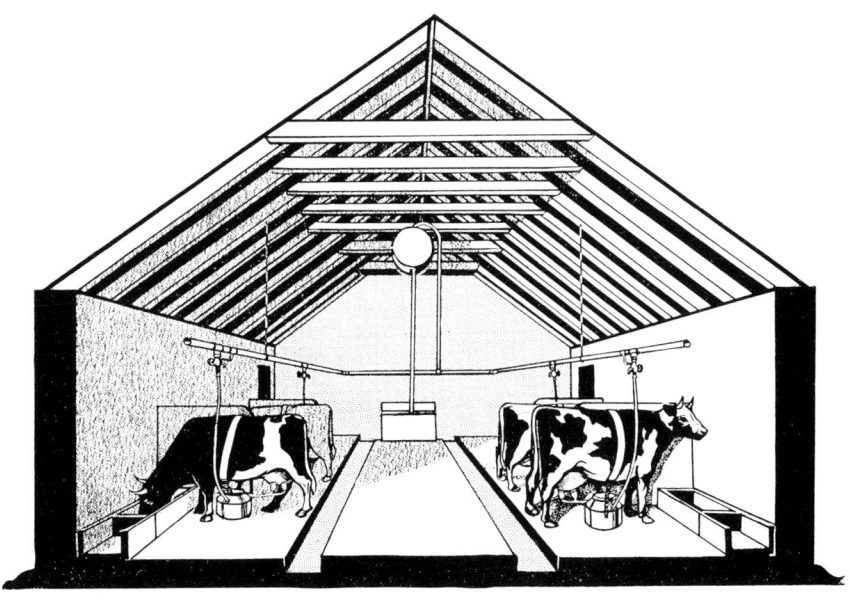

Fig I 4. The Murchland milking installation (1889). The upper and lower tanks at the rear are associated with control of vacuum level by height of a water column.

stall cock by a rubber hose. The vacuum applied to the teat was continuous and the teatcups were specifically designed to keep the teat surrounded with milk. Originally the teatcup had a rubber mouthpiece, but later this was extended to form an open lining (Fig I 5). Murchland had ideas of piping the milk under vacuum to a central collecting point.

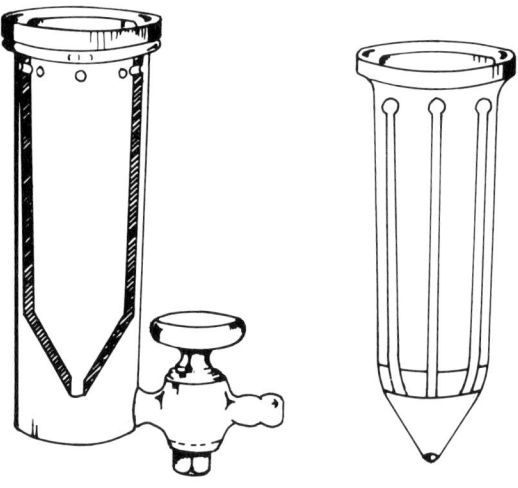

Fig I 5. The Murchland teatcup and liner.

South-west Scotland was, and still is, a predominantly dairying area with large herds. Murchland's machine must therefore have created considerable interest and acted as a stimulus to other inventors and manufacturers in the area. One of these was the firm of Nicholson & Gray of Stranraer. The RASE again offered prizes for competition at the Doncaster Royal Show in 1891. The Nicholson & Gray machine[6] was the only entry but it gained the prize of £20 and a silver medal after three working installations had been inspected. It used a hand or power-driven reciprocating vacuum pump connected to a pipeline laid on the floor behind the cows and provided with a short vertical branch at each stall division. To this was connected a rubber tube communicating with a pail which stood beside the cow. The teatcups were made from cow's horns and each was fitted with a rubber mouthpiece. The vacuum was nominally 15 in of mercury (51 kPa), but fluctuated with the strokes of the pump. This pulsation was reported to cause udder troubles; however, it prevented a continuous flow of milk, and on this count it was criticized by the judges. In the original plan it was intended that the vacuum pipeline should be glass-lined so that the milk from each cow could be drawn into a central reservoir: the buckets were introduced for individual cow recording and were provided with a graduated scale for this purpose.

Building on Nicholson and Gray's experience, Dr Shiels of Glasgow took out patents in 1895 for a device which he called a "pulsator". He had formed the Thistle Mechanical Milking Machine Company and exhibited his machine at the Darlington Royal Show in 1895, gaining a silver medal (Fig I 6). His idea was to relieve the teats of constant suction and in fact, to develop the very feature for which the Nicholson and Gray machine had been criticized. A mechanically operated valve on the vacuum pump admitted air to the pipeline with a regular frequency so that the vacuum applied to the teat varied rhythmically or "pulsated" between 15 in and 4½ in of mercury (51 and 15 kPa). A non-return valve in a chamber on the lid of the bucket allowed milk to collect therein at constant vacuum. The teatcups were made of thick rubber with thin areas on opposite sides and at the top moulded in so that under the influence of the pulsating vacuum inside, and the constant atmospheric pressure outside, they squeezed the teats to some extent at pulsation frequency.

The continuous and intermittent vacuum principles, exemplified by the Murchland and Thistle machines, continued to battle for honours and came into open competition at a trial organized by the Highland and Agricultural Society of Scotland in 1897[8]. The need for hygiene was beginning to be realized, and the trial was purposely arranged for the end of May to obtain warm weather conditions. Three Murchland and four Thistle plants were inspected, milk samples were taken for keeping quality tests and the milking and cleaning procedures compared. The Murchland gained the award because it produced milk with a keeping quality equal to hand drawn milk — generally exceeding two days. The Thistle gave keeping qualities of less than one day and caused bad flavours, even when, as in one plant, steam was used for cleaning the pipeline. The judges also took exception to the intricate construction of the Thistle pulsator. This is probably why the Thistle cost almost five times as much as the Murchland.

The Thistle Company faded out but one of the directors, Robert Kennedy, went into partnership with William Lawrence, a Glasgow engineer, and produced the Lawrence-Kennedy machine in 1897. They retained Shiels' principle of a pulsating vacuum, but this was now obtained from a piston-type pulsator situated on the pail lid. Like its predecessor it came into competition with the Murchland and both machines were entered in the RASE trials at York in 1900[9]. The Murchland milked more quickly and more completely than the Lawrence-Kennedy, but evidently the judges set great store on hygiene and on this occasion failed the Murchland on this account. Neither machine effectively fulfilled the requirements, however, and no award was made. The Murchland had been exported to New Zealand and with the Brookside, a machine similar to the

History and Development

Fig 1 6. The Thistle milking machine (1895). The massive vacuum pump was driven by a steam engine (not shown).

Nicholson and Gray, was on sale there in 1893. In general, however, progress up to 1900 in other countries lagged behind that in Great Britain, though there was no lack of inventions, particularly in the USA.

The next advance in the principle of mechanical milking came in 1902 from Hulbert and Park of New York. They introduced the new idea of applying pressure to the outside of the teat in addition to applying vacuum to the teat orifice. They did this by fitting a rubber lining to the teatcup so creating an annular chamber which could be supplied with pulses of compressed air, so squeezing the teat. There were three such annular chambers, arranged one immediately above the other, and the compressed air supply ports were opened in succession from the top downwards by a spring-loaded slide valve. Thus a three-stage downward squeeze was produced. It is not known if this device reached the stage of commercial production but clearly it would have suffered from mechanical and hygienic problems. However, it was quickly overtaken by a series of inventions by Alexander Gillies, a dairyman from Terang, Australia, who devised the basic principle of the teatcup which is in universal use today.

Gillies filed a British Patent in December 1902[10] describing a teatcup consisting of a rigid rubber casing having a thin flexible lining. The interior of the lining was in communication with the milk receiver and therefore under constant vacuum. The annular space between the lining and the casing was in communication with a pulsating vacuum produced, for example, by a Lawrence-Kennedy pulsator. This idea was elaborated in a further patent in June 1903[11]. The teatcup illustrated in this patent (Fig I 7) showed distinct improvement in that it could be dismantled easily and the lining was a simple tube which could be removed and cut to length periodically. A separate moulded rubber mouthpiece was fitted after assembly of the liner. An additional new feature was the introduction of an air inlet at the mouthpiece 'to assist in conveying milk to the receiver'. It is stated in the patent that this air bleed could be introduced further downstream if desired. Gillies also described a simple tubular 'claw' for connecting the milk and air tubes from the four teatcups with the receiver and pulsator (Fig I 8). This design has survived until the present day. A short length of glass tube was provided in the long milk tube near the outlet of the claw 'so that the flow may be observed and the teatcups removed at the proper time.'

Gillies filed a further patent[12] in June 1903 describing a central pulsator system. He had been using a Lawrence-Kennedy pulsator mounted on the lid of a bucket serving two sets of teatcups. The pulsator, having been designed for pulsating the milking vacuum and therefore able to handle relatively large quantities of air, could serve a larger number of units provided it was used for the extraction of air from the teatcup pulsation chambers. Gillies called this 'intermittent (not pulsating) suction'. Two air pipelines were installed to serve all milking points: one to provide continuous vacuum in the buckets and from these to the interior of the liners, and the other providing intermittent vacuum from

History and Development

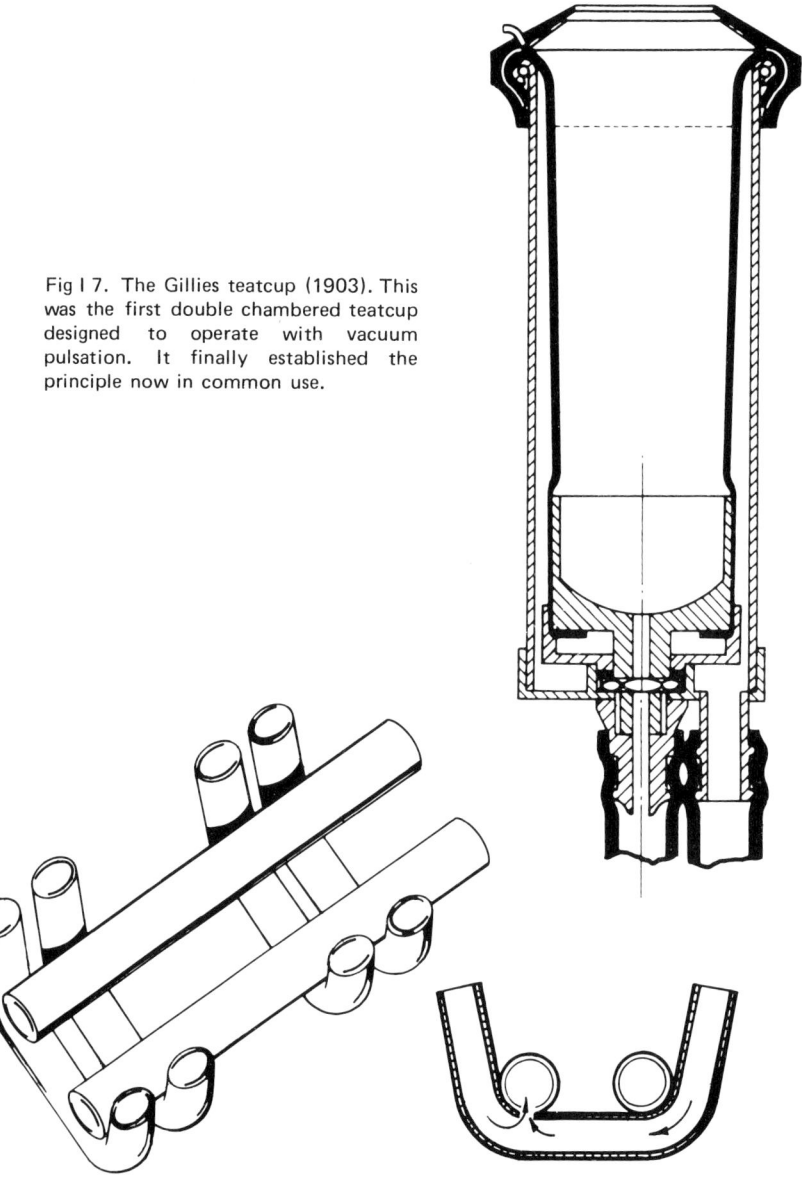

Fig I 7. The Gillies teatcup (1903). This was the first double chambered teatcup designed to operate with vacuum pulsation. It finally established the principle now in common use.

Fig I 8. The Gillies claw (1903). One tube of each pair of branch tubes is connected to the main pulsation tube and the other to the main milk tube, as shown in the section.

the pulsator direct to the pulsation chambers. A spring loaded automatic air inlet valve in the base of each teatcup casing (shell) opened to admit air to the pulsation chamber at the end of the intermittent vacuum phase, and closed again with the occurrence of the next intermittent vacuum phase. The mode of operation of the automatic air inlet valve is not obvious but the purpose was to overcome the difficulty that it is 'practically impossible to carry pulsation any considerable distance.'

It should be mentioned that Gillies' main objective in these patents was to apply external pressure to the teat, thus simulating hand milking. He did not visualize that the liner would, or should, collapse completely under the teat, nor does this seem to have been likely with his particular designs.

Gillies' inventions were accepted by Lawrence-Kennedy and the name was changed to Lawrence-Kennedy-Gillies, or LKG, and this can reasonably be described as the first machine to embody all the basic principles of the bucket milking machine of today. In this form it was marketed in Australasia by J. Bartram & Son (Pty) Ltd and was brought back to Britain to continue the struggle for perfection.

In 1905 it transpired that a New Zealand farmer, Ambrose Ridd of Waipuka, Taranaki, had been working on similar ideas to Gillies. He claimed to have invented the double chambered teatcup (originally made from a ¼ lb cocoa tin with a piece of bicycle inner tube for a liner) some years earlier but as he did not patent this the claim remains unproved. His first patent was in 1905 for air admission and to avoid conflict with Gillies' earlier patent this was purchased by the Bartram Company. However Ridd marketed his own machine in 1907 and formed a group of companies in Australia and New Zealand which competed keenly, even acrimoniously, with other manufacturers for many years. Ridd called his teatcup 'The Rubber Hand' because, he claimed, the milk was extracted from the teat only during the pressure or squeeze phase of the pulsation. He (like Murchland) considered it essential that the teat must remain bathed in milk, simulating wet hand-milking practice.

The Lawrence-Kennedy machine had also been exported to the USA where it was taken up by the Burrell Manufacturing Co. of Little Falls, NY, and became the Burrell-Lawrence-Kennedy or BLK. For many years it retained the original single chambered teatcup but Burrell took out patents in 1905 and subsequently for machines in which double chambered teatcups, somewhat similar to the Gillies, were used.

In 1905 the Wallace machine, made by J. and R. Wallace, Castle Douglas, Scotland, appeared on the market and was exhibited at the Doncaster Royal Show of 1905[13] where it gained a silver medal after a practical trial. This developed further the design by Gillies in which the pulsation air was admitted through a spring-loaded valve on the teatcup. In the Wallace machine each teatcup had its own pneumatic pulsator, a principle which was retained until this make was withdrawn from the market more than 50 years later.

Fig I 9. The Vaccar portable milking machine (1910). This machine was powered by a 2½ h.p. engine and was said to be suitable for milking herds of 200 cows.

In 1906 James Bartram, who had exported the Lawrence-Kennedy machine to Australia, formed the Vacca Company to sell the LKG in Britain. The name soon changed to Vaccar and the Vaccar machine was produced on somewhat similar lines to the LKG. In 1910 this was adapted as a field milking installation (Fig I 9) mainly for export to Denmark and Holland and became the forerunner of the modern field bail. The firm claimed at this time that their sales exceeded the total of all other makes combined throughout the world.

COMPARATIVE TRIALS OF MILKING MACHINES 1913 AND 1916

The use of milking machines slowly increased and more manufacturers in various countries turned their attention to this new field. The Royal Agricultural Society decided to hold a further series of trials at Bishop Auckland in 1913[14]. There were seventeen entries but only eleven actually competed: the Davies & Ransome, Lawrence-Kennedy, Vaccar and Wallace were British; the Omega (Amo) and Manus were Swedish; the Max, Nyeboe and Nielson were Danish; the Bartram came from Australia and the Gane from New Zealand. The Nyeboe, Nielson and Manus were lactactors, or mechanical pressure type machines. All the others were vacuum machines using pulsators and double chambered teatcups: with one exception they were of the bucket type, the Omega being suspended from the cow (Fig I 10). The exception was the Gane which was a type of machine delivering the milk either into containers, or into a metal pipeline from which it was delivered by a "releaser". This type of machine had been developed in New Zealand to avoid the labour of carrying milk. Cannulae and teat siphons were debarred from entry, and the use of antiseptic for cleaning was prohibited. Each machine was allotted four cows, and the trial extended over five days, during which the judges assessed the efficiency of the machines from the operating point of view. Milk samples were taken and tested for bacteriological count and acidity to indicate keeping quality. These tests provided some anomalies: one of the lactators, a type of machine generally discredited for lack of cleanliness, gave the best results. The worst results came from the Lawrence-Kennedy, the make richest in experience.

When all the factors had been taken into account, however, the first prize was awarded to the Omega, and second prize to the Vaccar. Both of these machines milked more cleanly and faster than the other competitors, and were placed 4th and 3rd respectively in the bacteriological tests. The Omega, being a suspended type machine, gained preference because the teatcups could not fall to the floor, and because celluloid tubes were used instead of rubber. At this time, and for some years subsequently, rubber was condemned as hygienically unsatisfactory and mechanically unsound.

Fig I 10. The Omega milking machine (1913).
This machine was awarded first prize in the RASE trials held at Bishop Auckland.

In 1916 the Highland and Agricultural Society of Scotland asked William Burkitt, who had conducted the RASE trials in 1913, to investigate the use of milking machines in Scotland[15]. A questionnaire was sent to 174 farmers known to have milking machines: 135 replies were received and from these installations were chosen for inspection. Ten makes were represented of which the Wallace, LKG, Lister, Amo, Gane and Vaccar had our present day type of double chambered teatcup and a pulsator controlling the squeeze applied to the teat. The Thistle, and its successor the BLK, using a single chambered teatcup and a pulsator to interrupt the vacuum, were also in use and one Murchland, the continuous vacuum machine, still survived. There were also several Amanco plants, imported from the USA. This make incorporated a new idea: it had single chambered teatcups like the BLK, but the intermittent vacuum was obtained by using a simple reciprocating vacuum pump running at 45 strokes per minute. Thus the complication of a pneumatic pulsator was avoided. The Amanco was a little cheaper than other makes (though Burkitt considered the price absurdly high for so simple an installation), and in consequence was being used with a herd as small as 12 cows. In the case of the other makes herd sizes were much larger than average. One unit to about 12 cows was common, this being in accord with modern practice.

The farmers contributing to this enquiry did not provide many suggestions for improvement and Burkitt did not make any serious comparison of the different types of machine. He observed that machines like the Thistle and BLK were now much less used although, when seen at work, they did not appear less efficient than those using the double chambered teatcup. The Amanco, in spite of other criticisms, was not condemned on principle. It is evident that there was still considerable opinion in favour of the single chambered teatcup principle, for Lister ultimately changed over to it.

Further support for this principle came from Thomas Orr in a series of articles published in "The Dairyman" in 1919[16]. He contended that the rubber liner, unless readily removed for cleaning was a needless complication. It is not, he said, a massaging action which is required, but relief to allow the teat to fill with milk and rigid support while the suction is withdrawing the milk. Nevertheless the Thistle and the Lister, which with the imported Amanco were the chief exponents of the single action system in this country, ultimately went off the market.

VACUUM LEVEL AND PULSATION RATE

Operating conditions of vacuum and pulsation rate however have always had a rather vague foundation. Murchland used a vacuum of 11 in of mercury (37 kPa) probably because that was the highest value he could conveniently maintain with his particular form of water column control. It was, however, sufficient to draw milk and probably served as a guide to later inventors. In the 1913 RASE trials a 15 to 16 in vacuum* was provided for all the competitors. Burkitt, in his 1917 report gave values of 12 to 20 in of mercury (41 - 68 kPa) with a 15 to 16 in of mercury vacuum as that most generally used. The lower limit was set doubtless by the need to milk reasonably quickly and completely: the upper limit was probably influenced by pump capacity and power considerations as much as by evidence, accurate or otherwise, of possible harmful effects.

The Nicholson & Gray, Thistle, BLK and later machines using a pulsating vacuum with single chambered teatcups, as well as the LKG and its modern counterparts (excepting only the Wallace) using the double chambered teatcup, adopted a pulsation rate of between 40 and 60 pulsations per minute. The guiding principle was variously stated as the rate at which the calf sucks or the rate at which the cow's heart beats. Neither is quite illogical.

MAJOR DEVELOPMENTS IN THE 1920s

A notable development in machine milking came in 1922 when A.J. Hosier introduced his system of open air dairying[17]. An essential feature of this system was a portable all-weather milking machine installation, still known as the 'Hosier bail'** It combined some of the principles of the 1910 Vaccar portable bucket plant and of the 1913 Gane releaser plant but was designed for milking in the fields all the year round, being moved every day. The Hosier bail was a shed, open on one side and provided with wheels, constructed to form six cowstalls, linked to a portable hut. The latter housed an engine which provided power for the vacuum pump, the pulsator-operating shaft and other auxiliaries such as a dynamo for lighting and a water-pump, together with a boiler for producing hot water or steam for cleaning and sterilizing. The vacuum pipeline and the pulsator-operating shaft extended the full length of the milking shed, three rotary pulsators being positioned between pairs of cowstalls and provided with connections to the vacuum line. A nickel plated milk pipe also ran the length of the milking shed, terminating in a sealed milk can which also was connected to the vacuum line. Between each pair of cows two cocks were placed, one connected with the milk pipe under constant vacuum, the other to

---

* The judges' report refers to a vacuum of 15 to 16 in of water. Similarly Orr (page 7) mentions a vacuum of 15 to 16 lb. Both are obvious misstatements and should be taken to mean 15 to 16 in of mercury (51 - 54 kPa).

** In Australia one meaning of the word 'bail' was a neck yoke to restrain cows during milking. Hosier adopted the word to identify his portable installation, which included stalls to restrain cows.

Fig I 11. The Hosier recorder vessel (about 1926). The bail is shown arranged for milking direct to pipeline: the recorder vessel was connected to the system on recording days. This was probably the first recorder milking machine.

the pulsator. A set of double chambered teatcups, with a claw and rubber tubes was attached and thus either cow of each pair could be milked, the milk being drawn through the pipeline into the can under vacuum. If it were desired to record individual yields, a vacuum-tight milking bucket was interposed between the teatcup cluster and the milk line. Later, Hosier designed a recorder vessel for permanent installation at each milking point (Fig I 11).

Initially the cans of milk were taken back to the farm for cooling but in 1928 Hosier designed an in-line vacuum cooler which acted as a milk receiver and contained a coil supplied with water from a portable tank (Fig I 12). This was followed by the introduction of a brine cooler supplied with brine from a storage tank and refrigeration unit housed with the vacuum pump and dynamo. As a further exercise Hosier designed a portable bulk vacuum tank intended primarily for bulk collection of uncooled milk for cheesemaking (Fig I 13).

One further development in the early history of milking machines should be recorded. This was electrical pulse transmission, introduced by De Laval in USA in 1929 and by Alfa-Laval in Britain the following year. Hitherto, most plants had used unit pulsators operated pneumatically or mechanically, usually with one pulsator to each single or double milking unit. The use of one master pulsator with pneumatic pulse transmission to relay pulsators had been

Fig I 12. The Hosier bail with vacuum cooler (1928).

Fig I 13. The Hosier milk vacuum tank (about 1930). This was the forerunner of bulk collection.

introduced by De Laval in 1917. The new electric pulsation system overcame the problem of pulse attenuation. Electrical pulses were generated by a simple contact breaker incorporated in the vacuum pump in conjunction with a 9V d.c. generator. The pulses were transmitted through a single core insulated conductor attached to the air pipeline to electromagnetic valves connected to vacuum and atmosphere, one at each milking point. The metal air pipeline was normally used to complete the electrical circuit. In the case of bucket plants the vacuum tube (stall tube) had contact ferrules at each end with conductors in the tube wall, the electromagnetic relay then being located on the bucket lid.

THE PATTERN OF DEVELOPMENT

Three periods can be distinguished in the history of machine milking. The first period, up to 1903, covered the exploration of principles, culminating in the pulsated double chambered teatcup that is universal today. During the second period, which lasted for the next 20 years or so, the protagonists of the various alternatives battled for supremacy. Manufacture of the Murchland continuous vacuum machine ceased within a few years followed by the complete abandonment of the mechanical pressure principle sometime during the 1914 - 18 war. Machines using pulsating vacuum with single chambered teatcups disappeared in most countries by about 1920 although a few survived in local markets in the USA for another 30 years.

The third period, from about 1920 until the present day, has been characterized mainly by the development of different types of milking machine installations, as distinct from the principles of mechanical milking. This development had as a primary objective the improvement of the efficiency of labour utilization but a second objective was improvement of hygiene, the need for which had been amply demonstrated by the experience of earlier years. Inevitably these developments have been accompanied by many improvements in detail design and the introduction of new ideas for ancillary components.

The traditional arrangements for milking cows by hand in a cowshed or in the field obviously steered the early inventors towards the bucket type of milking machine installation. Although the possibility of milking directly into a pipeline was visualized by Murchland and by Nicholson and Gray, commercial development of the milking pipeline machine probably started with Gane in New Zealand about 1910. This involved the new concept of bringing each cow to a fixed milking point, a change which was probably more easily accepted in the less rigorous climate of New Zealand. In any case butter was the main dairy product exported from Australasia for many years and cream only was despatched from a high proportion of farms. Therefore continuously conveying the warm milk as it was produced to the centrifugal cream separator had obvious advantages.

Hosier's pioneer work in the 1920s showed that bringing cows to the milking machine was equally applicable to this country and it was a very short step to develop fixed milking parlours from the mobile bail. Initially these were the abreast type and, in this country, incorporated recording vessels. Other developments which followed rapidly in the mid-1930s were the substitution of the transport can for the milking bucket giving the direct-to-can machine, and the tandem arrangement of cows in the fixed milking parlour. However in spite of these innovations the bucket machine constituted the major proportion of sales, probably 80 - 90%.

Soon after the Second World War there was increasing emphasis on labour utilization and efficiency of cleaning. One of the notable developments, in 1953, was the NIRD system of immersion cleaning based on the direct-to-can abreast parlour type of installation. For this the milk-contact components were made from stainless steel to a simplified design so that they could be cleaned simply by immersion in 4% cold caustic soda solution. While this provided an elegant solution to the problems of the smaller herd, increasing herd size led to further development of milking parlour layouts to suit UK requirements, notably the fixed herringbone and later the various types of rotary parlour. However, parlour milking was not acceptable to many farmers particularly in northern England. Portable 4-unit milking trucks with trailing service leads were devised: direct-to-can machines suspended from an overhead runway had a brief spell of popularity. The more common development, which has persisted, involved a permanent milking pipeline serving all the cowstalls with suitable devices to connect the clusters to the milk and vacuum lines. This latter development, the cowshed milking pipeline, received an impetus in the late 1950s when bulk collection began to make substantial progress.

These developments have all contributed significantly to improving output per man but added new problems in hygiene, necessitating corresponding improvements in systems for in-place cleaning.

Perhaps the most interesting point which emerges from this chapter of history is that commercially successful milking machines became established without any clear understanding of the principles involved. Yet even in the early days there was no lack of designs which worked, and the period between the two principal landmarks, Murchland's machine in 1889 and the LKG in 1903, was surprisingly short. It was not until 1941 that a theory on which machine milking could be explained was advanced, by Ely & Petersen[18] and Miller & Petersen[19] in the USA. Since then research into the problems of design and use of milking machines has been developed on an increasing scale, and is now proceeding in England and several countries overseas. In recent years the objective has been to understand more thoroughly the mechanics of teatcup action and the relationship of the milking machine to udder disease. These aspects are discussed in detail in Chapters V and IX.

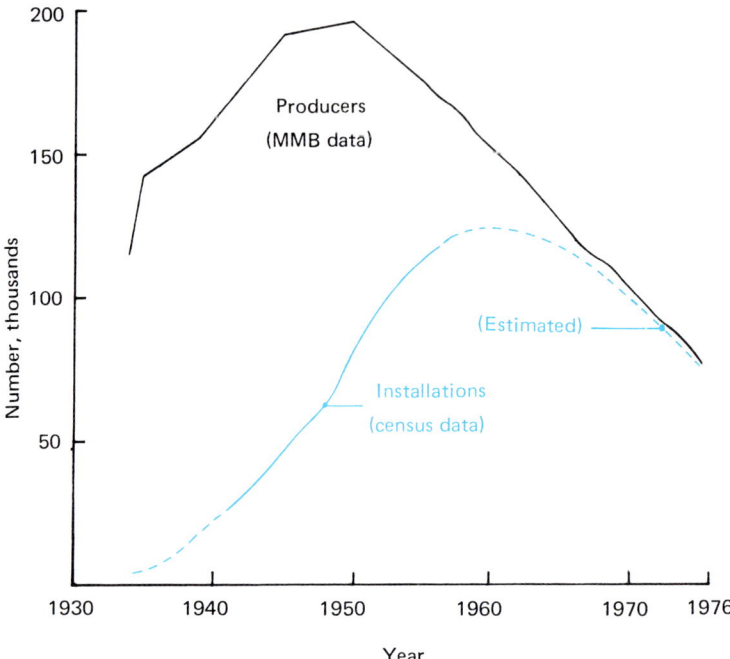

Fig 1 14. Number of registered milk producers and milking machine installations in the UK 1934 - 75.

1. Since 1955 the number of registered milk producers in the UK can be considered to be the same as the number of herds with more than three cows. The number of herds before that date was a little greater than shown by the graph, the difference varying with the implementation of the Agricultural Marketing Acts 1931 and 1933 in England and Wales, Scotland, and Northern Ireland.
2. The numbers of milking machine installations prior to 1940 have been estimated from unpublished data supplied by the leading manufacturers of that time. From 1944 to 1956 the numbers are based on censuses taken by the Agricultural Departments of the countries concerned. The terminal section of the installations graph has been estimated from the known numbers of hand milked herds in England and Wales (ADAS data).
3. The progress of machine milking in the UK is probably best indicated by the percentage of herds so milked at any time, i.e.

$$\frac{\text{installations}}{\text{herds}} \times 100\%$$

In 1975 this figure was nearly 98% for England and Wales and probably 95% for the UK. In the same year the percentage of herds with more than 10 cows was about 85%.

GROWTH OF MILKING MACHINES

There are no accurate statistics to show the extent of use of milking machines in the UK prior to 1942. Circumstantial evidence suggests that by 1920 there were probably about 1000 installations. During the following decade most of the present-day manufacturers in the UK, in addition to Vaccar, became established, viz. Alfa-Laval, Fullwood & Bland, Gascoignes, Hosier and Simplex. This must have increased the sales effort and the number of installations is estimated to have been 7 - 8000 by 1936 and 16 000 by 1939. War conditions rapidly changed the situation for manpower reasons and plants were installed as fast as they could be manufactured — about 5000 per annum. Since 1942 the Ministry of Agriculture, Fisheries and Food has taken a periodic census of installations in England and Wales. A similar census has been taken in Scotland and, later, in Northern Ireland.

Fig I 14 is based on these and other data from the Ministry and the Milk Marketing Boards and shows the rapid increase following the end of the 1939-45 war. It is evident also that since that date the percentage of herds milked by machine has steadily increased to about 96% in 1970 and to over 97% in 1974. No doubt the remainder will continue to disappear but already the proportion of cows not milked by machine is negligible.

REFERENCES

(1) *British Patent* No. 13674 (1851).

(2) *British Patent* No. 2634 (1860).

(3) *British Patent* No. 1688 (1863).

(4) *British Patent* No. 3831 (1881).

(5) Henderson, R. (1890) *Journal of the Royal Agricultural Society of England*, **1**, 645.

(6) Neville Grenville, R. & Rigby, T. (1891) *Journal of the Royal Agricultural Society of England*, **2**, 854.

(7) Marshall, C. (1895) *Journal of the Royal Agricultural Society of England*, **6**, 460.

(8) Drysdale, J. (1898) *Transactions of the Highland and Agricultural Society of Scotland*, 5th Series, **10**, 166.

(9) Courtney, F.S. (1900) *Journal of the Royal Agricultural Society of England*, 3rd Series, **11**, 466.

(10) *British Patent* No. 27894 (1902).

(11) *British Patent* No. 12958 (1903).

(12) *British Patent* No. 13657 (1903).

(13) Hippisley, B. (1905) *Journal of the Royal Agricultural Society of England,* **66**, 80.

(14) Hippisley, B. & Sadler, J. (1913) *Journal of the Royal Agriculture Society of England,* 74, 234.

(15) Burkitt, W. (1917) *Transactions of the Highland and Agricultural Society of Scotland,* **29**, 229.

(16) Orr, T. (1919) *Dairyman,* **42**, 50, 96 and 124.

(17) Hosier, A.J. (1927) *Journal of the Farmers' Club,* **Part 6,** 103.

(18) Ely, F. & Petersen, W.E. (1941) *Journal of Dairy Science,* **24,** 211.

(19) Miller, K. & Petersen, W.E. (1941) *Journal of Dairy Science,* **24,** 225.

Chapter II

BASIC MECHANICS

H S Hall

The milking machine is essentially a system through which two fluids, air and milk, have to be transported. In describing in the following chapters how this is done, and the characteristics of the equipment and techniques involved, the terms 'pressure' and 'vacuum' occur frequently. A clear understanding of their meaning is therefore essential.

Gases and liquids are both called 'fluids' because, unlike solids, they cannot resist forces which tend to cause movement or change of shape. This means that fluids can be made to flow in any desired direction if the necessary flow path is provided, and at any desired rate if the necessary forces are applied.

THE CONCEPT OF PRESSURE AND VACUUM

PRESSURE IN A GAS

The molecules of a gas, for example air, are in constant rapid motion in all directions and if the gas is enclosed in a vessel or system the molecules continually bombard the surrounding walls. The total force of this bombardment on unit area is called the absolute pressure of the gas. It follows that if we have a closed system which contains no gas (a somewhat theoretical concept) the absolute pressure inside will be zero.

It can be shown that for a perfect gas in a closed system:

$$\frac{pv}{t} = \text{a constant}$$

where p = absolute pressure
 v = volume
 t = absolute temperature (°C + 273)

This relationship can be expressed in various ways. For example, if the absolute temperature remains constant the absolute pressure is inversely proportional to volume, or, if the volume remains constant the absolute pressure is directly proportional to absolute temperature. It follows that at any two places in a flow system:

$$\frac{p_1 v_1}{t_1} = \frac{p_2 v_2}{t_2}$$

Atmospheric pressure

The earth is surrounded by a layer of air which we call the atmosphere. Air has mass and is therefore subject to the force of gravity and this, expressed as force per unit area, creates pressure in the atmosphere. As air is gaseous this pressure acts in all directions, on the surface of the earth and on all bodies such as structures, animals or human beings in contact with the atmosphere, although we are not normally conscious of this.

Thus atmospheric pressure is a measure of the weight of a column of air, of unit cross-sectional area, above any point in the atmosphere. It follows that atmospheric pressure decreases with increasing altitude above the earth as there is less weight of air to support. Atmospheric pressure at sea level has a nominal value of 100 kilopascals (kPa) or 1 bar (1000 millibars), equivalent in Imperial units to 14.50 lb/in² *. Using this datum atmospheric pressure at an altitude of 1000 m would be about 89 kPa. In practice the normal range of local weather variations causes atmospheric pressure to vary by up to 2% or so, although extreme values of 925 and 1055 mbar have been recorded in the UK. Low temperature and low humidity both increase atmospheric pressure.

---

* The 'Standard atmosphere', which is the internationally agreed reference for pressure is slightly greater being equal to 101.325 kPa, 1.013 bar or 14.696 lb/in².

Vacuum

The fact that people and animals live, and most machinery operates, in the ambient conditions of atmospheric pressure is a compelling reason to regard atmospheric pressure, rather than zero absolute pressure, as the datum for most practical measurements. The term pressure in this context means pressure above atmospheric pressure, often made more explicit by specifying gauge pressure.

An absolute pressure which is less than atmospheric pressure is called a vacuum and is measured on a scale in which atmospheric pressure at the time and place of measurement is zero vacuum. The maximum vacuum which can be obtained is obviously numerically equal to atmospheric pressure, whatever that may be, but for practical purposes vacuum indicating instruments are usually scaled 0 - 100 kPa or 0 - 1 bar. Scales of 0 - 30 inHg or 0 - 760 mmHg and others are still used. Thus if atmospheric pressure is 100 kPa a vacuum of 10 kPa is equal to 90 kPa absolute pressure and a vacuum of 51 kPa, the value at which most milking machines operate, is equal to 49 kPa absolute pressure.

The general terms free air and expanded air are often used to describe air at atmospheric pressure and under vacuum respectively. In both cases the precise pressure or vacuum may be stated if required.

PRESSURE IN A LIQUID

Liquids, like gases, cannot resist forces tending to cause flow but unlike gases they do not expand indefinitely to fill the enclosing space. Thus a liquid in a partially filled container conforms to the internal shape of the container but assuming it is subject only to the force of gravity it will form a free surface which is always truly horizontal, i.e. in a plane at right angles to the force of gravity. The volume of a liquid is virtually independent of pressure or vacuum and is therefore in effect incompressible. However liquids increase in volume with increasing temperature and *vice versa*.

Hydrostatic pressure

The pressure at any point in a liquid at rest is called hydrostatic pressure and, like atmospheric pressure, is due to the weight of the column of liquid of unit cross-sectional area above the point of measurement. Thus for a liquid of uniform density, hydrostatic pressure is proportional to depth and can therefore be expressed by stating the depth and identifying the nature (and if necessary the temperature) of the liquid. This is called head. The free surface of a liquid is subject to the absolute gas pressure of the atmosphere above it and this must be taken into account together with hydrostatic pressure when using liquid columns as pressure measuring devices and in determining conditions of flow, aspects which are discussed later.

VAPOUR PRESSURE

The molecules of a liquid have a certain freedom of movement, hence the property of fluidity, but unlike a gas they are mostly confined by the surface film so that the volume of a liquid remains relatively constant. However, a proportion of the molecules have sufficient velocity to break through the surface into the atmosphere above the liquid, forming a *vapour*. This process is called evaporation and the vapour molecules behave in some ways like gas molecules creating a pressure which is called vapour pressure. If the liquid is contained in an open vessel evaporation will continue at a rate dependent on temperature until all the liquid eventually disappears.

If the liquid is in a closed vessel some of the vapour molecules return to the liquid, a process called condensation, and ultimately a state of equilibrium is reached and the number of vapour molecules in unit volume then remains constant. The vapour is then said to be saturated and for a given liquid at a given temperature the vapour pressure is constant. Changing the volume of the vapour space will not change the pressure − it results only in further evaporation or condensation to keep the pressure constant. This constitutes the fundamental difference between a vapour and a gas. A vapour can be converted to liquid by increased pressure alone: with a true gas reduction of temperature is also necessary.

Boiling

If the temperature of a liquid is raised so that the consequent saturated vapour pressure is equal to the pressure of the atmosphere above the liquid, the liquid is said to boil and the temperature is called boiling point or saturation temperature. Further addition of heat will not raise the temperature but evaporation will continue until all the liquid has been converted to vapour. However if this takes place in a closed vessel the temperature and the pressure will rise in a precise relationship depending on the nature of the liquid. For example, water at normal atmospheric pressure at sea level boils at 100°C (212°F). At a pressure of 1000 kPa (145 lb/in^2) it boils at 184°C (363.5°F) and at a vacuum of 50 kPa it boils at 81.7°C (179°F). Thus when cleaning milking machine installations with aqueous solutions under vacuum the maximum temperature which can be achieved is limited by the applied vacuum. For the same reasons, at an altitude of 1000 m, where atmospheric pressure is about 89 kPa water is an open vessel will boil at about 96°C (205°F).

MEASUREMENT OF PRESSURE OR VACUUM BY A LIQUID COLUMN

The properties of gases and liquids discussed above provide the basis for an accurate method of pressure or vacuum measurement. If we take a U-tube

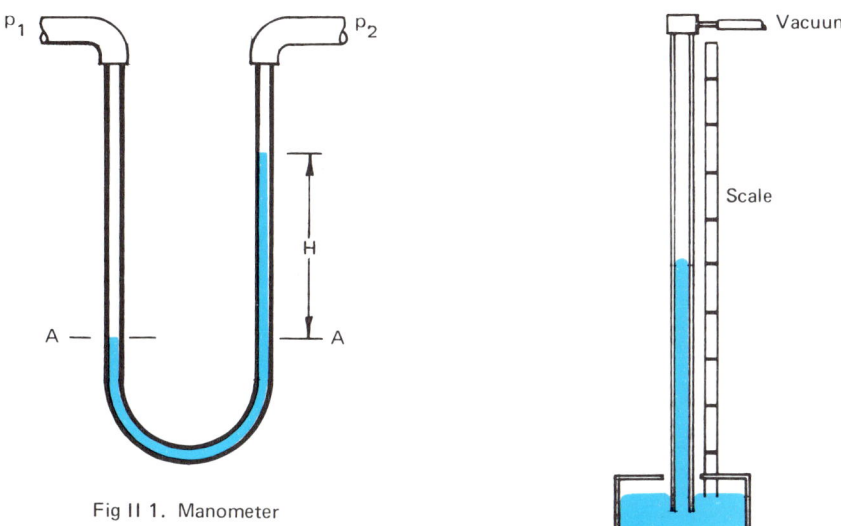

Fig II 1. Manometer

Fig II 2. Mercury manometer for vacuum measurement.

containing a suitable liquid as in Fig II 1 and apply different pressures p_1 and p_2 to the two limbs the liquid will move to an equilibrium position such as that shown in the diagram where the levels differ by a height H. The pressure at line AA must be equal as the liquid is at rest. Therefore the pressure p_1 is balanced by the pressure p_2 plus the hydrostatic pressure of the column H. Thus H is a measure of the pressure difference.

This instrument is called a manometer and is used for the measurement of pressure differences above or below atmospheric pressure. For differences up to about 10 kPa water could be used as the height H at 10 kPa pressure difference would be 1020 mm (40.15 in). For values up to 100 kPa mercury would be used when the height H at 100 kPa pressure difference would be 750 mm (29.53 in).

A practical form of portable manometer for vacuum measurement is shown diagrammatically in Fig II 2. A glass tube is mounted vertically so that the bottom end dips into a pot containing mercury. The pot is vented so that it is at atmospheric pressure. The top of the tube is connected to the vacuum to be measured and a scale is mounted at the side of the tube so that the difference between the mercury levels can be measured directly. The vacuum can then be calculated from the following relationships:

1 mmHg	=	0.133 kPa	=	1.33 mbar
1 inHg	=	3.386 kPa	=	33.86 mbar

If the tube in Fig II 2 is sealed at one end, filled with mercury and inverted with the open end below the surface of the mercury in the pot, the instrument then measures atmospheric pressure and is called a mercury barometer.

FLUID FLOW IN MILKING MACHINE INSTALLATIONS

The transport of air or milk through a milking machine installation follows certain well-defined physical principles but where both fluids are transported together, as in a long milk tube or a milking pipeline, the flow conditions become too complex for theoretical analysis and empirical solutions have to be found.

A milking machine installation consists of a pipework system linking various vessels and other components which together provide the flow paths for air and milk. The forces necessary to move air and milk through the system arise from the fact that the system is maintained at a vacuum. Thus it is atmospheric pressure which forces air, and intra-mammary milk pressure which forces milk, into the system and the combination of these forces causes flow.

To be a continuous operation it is necessary to remove air and milk from the system at the appropriate rates. The air is removed continuously by a vacuum pump. In a bucket or direct-to-can machine milk is removed by disconnecting the milk container: in milking pipeline and recorder machines the milk is removed by a device called a releaser. If this is in the form of a releaser milk pump it will normally be automatically controlled according to the amount of milk to be discharged.

Fig II 3 shows diagrammatically the flow of air and milk through three basic types of machine during normal milking. In the bucket (or direct-to-can) machine milk enters the teatcups and travels through the short milk tubes to the claw where air is admitted, the milk and air travelling along the long milk tube to the bucket (or can). The milk remains in the container and the air separates to pass up the vacuum tube to the vacuum pipeline. The pulsator is usually situated on the bucket lid and admits air which passes along the long pulse tube to the teatcup chambers and then returns to the vacuum tube for extraction through the vacuum pipeline. Air is also admitted to the system through the regulator which is situated on the vacuum pipeline.

In addition to these normal sources of air admission, air is sometimes drawn into the teatcups past the teat; each time a milk container is changed air must be withdrawn to reduce this to the working vacuum; in a poorly maintained machine there may be inward leakage of air at joints or points of damage. All the air admitted passes through the system to the inlet of the vacuum pump which continuously extracts the air by compressing it so that it can be discharged against atmospheric pressure.

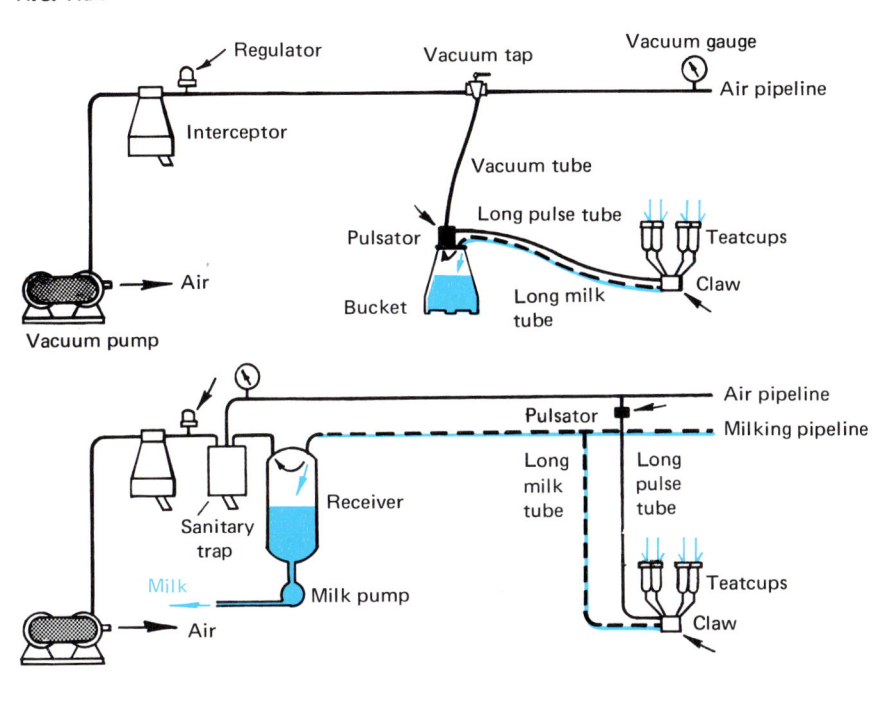

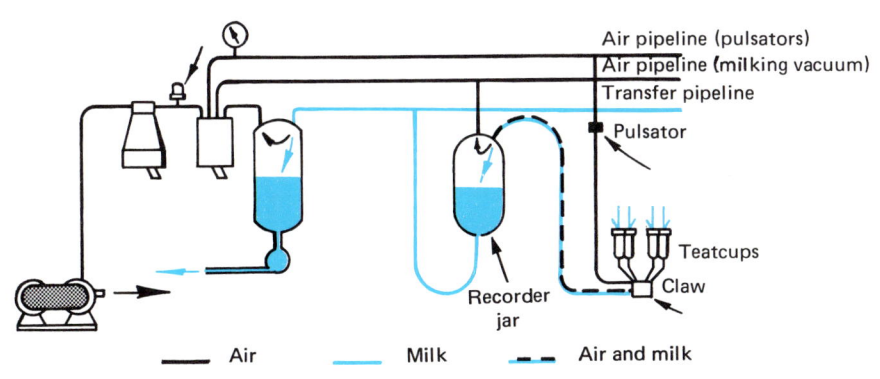

Fig II 3. Flow of air and milk
   top: Bucket machine
   centre: Milking pipeline machine
   bottom: Recorder machine.

In the milking pipeline machine the flow pattern is similar to the bucket machine except that the milk and air from each claw flow through the milk pipeline to a common receiver where the air and milk are separated. There is no further air admission at this point when a motor-driven releaser milk pump is used to empty the receiver but one type of milk pump is pneumatically operated and so admits air direct to the air pipeline. All other types of releaser, for example the pulsator controlled spit chamber releaser (Chapter III) involve the admission of air.

In the recorder type machine the flow pattern is similar to the milking pipeline machine except that the air admitted at the claw is separated from the milk at the recorder jar. Air has to be admitted at this point through a special inlet or through the teatcups at the end of each milking operation to force the milk to the receiver (which is under vacuum). Some air may pass along with the milk as the jar empties especially if the controls are not expertly handled. This air is separated from the milk at the receiver.

THE MECHANICS OF FLUID FLOW

In the absence of friction, the total energy of a fluid at any point in a flow system would remain constant. However, when a fluid flows some of its energy is expended in overcoming friction, the nature of which is discussed later. The total energy can therefore be expressed as:

Potential energy due to position + potential energy due to pressure + kinetic energy due to motion − energy lost as friction.

Excepting the energy lost as friction, the components of the total energy may be transformed one to the other, depending on the flow conditions. Thus for a fluid flowing through a pipe which is horizontal (no change in energy due to position as there is no increase or decrease in head) and of constant cross-sectional area (no change in kinetic energy as velocity is constant) the friction loss must be met by a corresponding loss of pressure energy, that is an increase in vacuum. In a milking machine system therefore the vacuum must be greatest at the inlet to the vacuum pump and it must decrease progressively through the system to the various points of air and milk admission, the intermediate values depending on friction loss up to that point and on any interchange of potential and kinetic energy. This decrease is called vacuum drop.

A fluid at low rates of flow travels in streamline motion where the individual particles flow in lines parallel to the axis of the pipe. The particles in contact with the pipe wall remain stationary: those in the centre of the pipe move at the highest speed. As the mean flowrate increases the coefficient of friction decreases until the flow pattern changes quite rapidly from streamline to turbulent flow. This is characterized by eddies and vortices in the fluid except in the boundary layer which remains streamline. The difference between mean and maximum velocities is reduced and friction increases. The transition from streamline to turbulent flow depends on the pipe diameter and on the velocity, density and viscosity of the fluid. In milking machine practice the conditions are such that air flow where no milk is present is normally turbulent and milk flow where no air is present is normally streamline.

Where air and milk are transported together the flow pattern becomes complex depending on various factors particularly the volume of air relative to milk or air:milk ratio. Air is normally admitted to the claw at a rate of 4 - 8 ℓ/min. The maximum milk flowrate for a very fast milking cow might be about 6 ℓ/min, giving an air:milk ratio of 0.7:1 to 1.2:1. Towards the end of milking when the milk flowrate has decreased to say 0.25 ℓ/min the ratio becomes 16:1 to 32:1.

Flow in long milk tubes

The air : milk ratio becomes important where milk has to be elevated from the claw as in milking pipeline and recorder machines other than those with low level milk pipelines. Elevating a liquid, as distinct from a gas, involves a loss of potential energy due to position and this is compensated for by a change of vacuum. Thus elevating a column of milk in a vacuum system through 1 m height reduces the vacuum by about 10 kPa: if the vacuum at the top of the column is 51 kPa it will be only 41 kPa at the bottom. This vacuum drop is dramatically reduced by the admixture of air. If the air : milk ratio is 1:1 the weight of milk in the column is halved and the vacuum drop becomes only 5 kPa: if it is 10:1 the vacuum drop is only 0.9 kPa.

Liquids cannot flow against gravity (i.e. uphill) except as a column which fills the bore of the tube. This maintains the pressure difference between upstream and downstream which causes flow. Where the tube contains air and milk the liquid forms plugs which are separated by pockets of air in the proportion determined by the air : milk ratio.

Flow in milking pipelines

To ensure drainage milking pipelines are normally installed with a slight fall towards the downstream end. This causes a variety of flow conditions. If the flow of milk is small relative to the pipe diameter the air and milk separate and the air passes downstream above the milk at a higher velocity which is proportional to the air : milk ratio. If the air : milk ratio increases the air velocity increases and this may create waves in the milk surface. This effect may intensify to the point where the waves form plugs of milk as in a rising long milk tube. Note that if the milking pipeline has risers milk plugs must be formed to negotiate the riser. The variation in milk flowrate from individual cows and the interaction of multiple milking units causes the flow conditions in a milking pipeline to change frequently from one form to the other, depending on the absolute and relative flowrates of air and milk and the characteristics of the flow path, particularly pipe diameter.

FLUID FRICTION

We have seen that the particles of a moving fluid, whether in streamline or turbulent flow, move at different velocities. This relative motion sets up internal friction in the fluid, that is forces tending to prevent motion, termed viscosity. The numerical value of viscosity, measured under standard conditions, is a specific physical property of the fluid. For liquids, increasing temperature decreases viscosity: for gases, increasing temperature has the reverse effect. Within the limits of operating conditions in milking machines, viscosity is independent of pressure.

As previously stated the rapid increase of relative motion between particles when flow becomes turbulent produces increased friction. In addition to the turbulence induced by flow velocity, physical changes in the flow path, such as sudden enlargements or contractions of cross-section, bends, tees, cocks and valves, all cause a local disturbance to flow absorbing energy and so further increasing fluid friction.

For homogeneous flow systems the vacuum difference (increasing vacuum in the direction of flow) required to provide the energy to overcome fluid friction can be calculated from first principles. While such calculations serve only a very limited purpose in milking machine systems because important parts of the system carry both air and milk and because the flow conditions are so variable, it is important to understand the factors which influence vacuum drop.

It can be shown that

$$V \text{ varies as } \frac{Q^2}{d^5} \cdot L$$

where V = vacuum
Q = flowrate
d = diameter of pipe
L = effective length of flow path.

Thus halving the fluid flowrate (for example, by using a ring main) reduces the vacuum drop by a factor of 4: doubling the pipe diameter reduces the vacuum drop by a factor of 32. Note that in this expression the effective length of the flow path, L, is the actual length increased by an amount equivalent in frictional effect to the various pipe fittings mentioned above.

The general aim in designing a milking machine flow system and its components is, firstly, to minimize vacuum drop in order to maintain as uniform a vacuum level as possible throughout the system. Secondly, those parts transporting milk must be designed to avoid as far as possible agitation and impact, factors which are conducive to lipolytic taint.

THE MECHANICS OF VACUUM PRODUCTION AND CONTROL

THE VACUUM PUMP

The motive power for creating the flow of air and milk through a milking machine system comes from the vacuum pump which is connected to the system at the downstream end to extract air and so to maintain a vacuum in the system. Typical methods of construction are described in Chapter III.

The vacuum pump is designed to take in a predetermined volume of air (the swept volume) at each revolution. This volume is determined by the dimensions of the pump and is not affected by the pressure or temperature of the incoming air. The air is compressed until its pressure is about equal to the prevailing atmospheric pressure when it can be discharged into the atmosphere. In practice the pump will take in slightly less than its swept volume per revolution because of internal leakage between the inlet and exhaust ports. The actual volume divided by the swept volume is termed the volumetric efficiency of the pump and this decreases with increasing compression ratio.

The vacuum level in a milking machine system is maintained at a nominally constant value by means of a vacuum regulator, the principle of which is discussed below. As the scale of vacuum is measured taking atmospheric pressure as zero, the absolute pressure in the system depends on the absolute pressure of the ambient atmosphere. The expansion of the air entering the system and its subsequent compression for discharge therefore vary.

Effect of pressure

Assuming constant temperature the gas equation given at the beginning of the chapter can be written:

$$p_1 v_1 = p_2 v_2$$

or $$v_2 = \frac{p_1 v_1}{p_2}$$

where $p_1 v_1$ = pressure and volume conditions in the atmosphere

$p_2 v_2$ = pressure and volume conditions in the milking machine

Taking unit volume of air at an atmospheric pressure of 100 kPa and admitting it to a system controlled at a vacuum of 50 kPa:

then $$v_2 = \frac{100 \times 1}{100 - 50} = 2$$

Suppose now that atmospheric pressure is high, say 102 kPa, or low, say 98 kPa:

then $$v_2 = \frac{102 \times 1}{102 - 50} = 1.96$$

or $$v_2 = \frac{98 \times 1}{98 - 50} = 2.04$$

Thus the expansion ratio of the air admitted varies ± 2% over the range quoted. However the swept volume of the vacuum pump is fixed and it will therefore extract more air when the ratio is least, that is when atmospheric pressure is high, and *vice versa*.

In practice the vacuum level throughout the system cannot be uniform, because there would then be no flow. The full expansion of the incoming air indicated by the above equations is therefore not completed until it reaches the position of the vacuum regulator. Similarly there must be some increase of vacuum between this point and the vacuum pump, possibly 2 kPa. This will make the compression ratio for the pump about 2% greater than the expansion ratio, diminishing the pump capacity by this amount.

The effect of short-term changes in atmospheric pressure is too small to be of significance in design but low atmospheric pressure becomes important when we consider milking machines installed at high altitudes. At an altitude of 1000 m the normal atmospheric pressure would be 89 kPa as already mentioned. This also would be subject to the short-term variations considered above. Putting this value for p_1 into the equation we have:

$$v_2 = \frac{89 \times 1}{89 - 50} = 2.28$$

Thus the vacuum pump will have 14% less capacity than at sea level and a correspondingly larger pump must be installed.

Effect of temperature

In considering the expansion of air entering the milking machine it was assumed that temperature remained constant. However there are a number of factors in practice which may increase or decrease the temperature of the air after entering the system before it reaches the vacuum pump. The consequent change in volume is proportional to the absolute temperature ($°C + 273$), that is, about 5% for a 15°C change from 20°C.

In passing through the vacuum pump the air increases substantially in temperature, perhaps 70 - 80°C, due mainly to the work done during compression. This does not affect the volume extracted from the system but it is reflected in the pump performance whatever the ambient temperature may be.

VACUUM CONTROL

The rate at which air is admitted to the milking machine from the various sources already described varies continually during operation. The vacuum pump extracts air at a constant rate, so to keep the vacuum level constant additional air must be admitted at a varying rate automatically controlled to make the total constant. This is the function of the vacuum regulator, various forms of which are described in Chapter III.

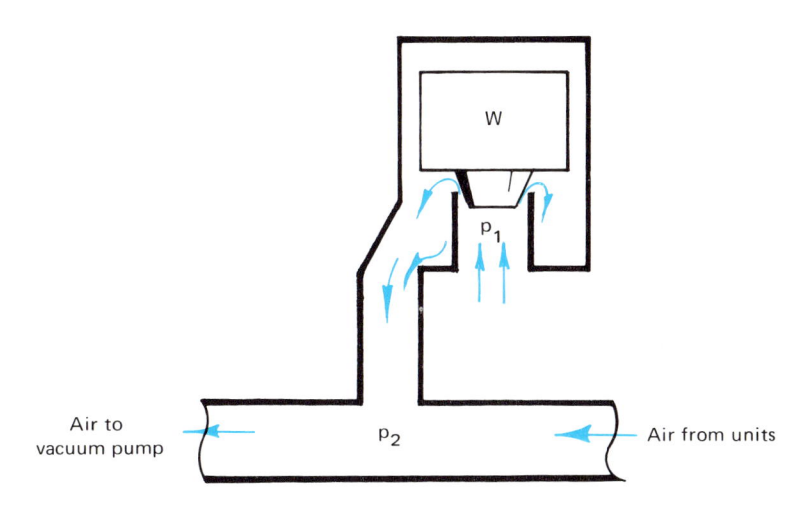

Fig II 4. Principle of vacuum regulator.

The principle of the vacuum regulator is shown in Fig II 4. The regulator has a horizontal orifice open to the atmosphere. The orifice is closed by a valve which is loaded by a weight W or, in some designs, by a spring in compression. The weight is enclosed by an airtight cover.

For equilibrium the upward and downward forces must be equal, that is

$$p_1 a = p_2 a + W$$

where
- p_1 = atmospheric pressure
- p_2 = absolute pressure in the air pipeline
- a = area of orifice
- W = weight of the weight (or force of the spring)

The equation can be written

$$p_1 - p_2 = \frac{W}{a}$$

The expression $(p_1 - p_2)$ is, by definition, the vacuum level: W and a are both constants for a given regulator. The vacuum level therefore will be kept nominally constant. In operation the valve floats above the orifice continuously adjusting the opening to admit sufficient air to keep the vacuum constant. It will be seen that it is necessary for the vacuum to change in order to provide the force needed to adjust the position of the valve. If the air passing into the regulator is plotted against the vacuum level a curve similar to Fig II 5 is obtained.

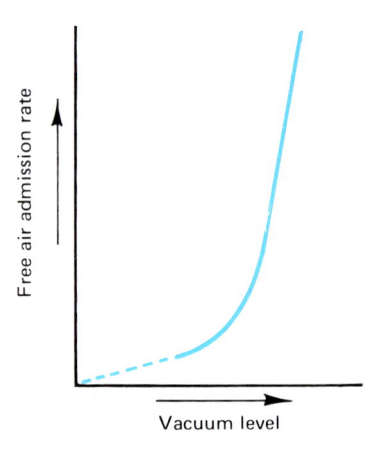

Fig II 5. Vacuum regulator characteristic.

The slope of this curve is a measure of the sensitivity of the regulator, the working range being confined to the steep part of the curve. This is discussed further in Chapter IV.

Chapter III

DESCRIPTION AND PERFORMANCE OF COMPONENTS

D N Akam

Over many years the milking machine has been developed into systems which show considerable diversity. In fact all milking machines have the same basic elements, a vacuum system, the pulsation components, an arrangement for transporting and collecting milk, and the milking cluster which brings the vacuum and pulsation to the cow. The ways in which these elements are arranged for the principal types of milking machines used in the UK are illustrated in Fig III 1. The distribution of types of milking systems in use in England and Wales at the end of 1974 is shown in Table III 1, an important feature being the large proportion of herds milked in cowsheds with bucket units.

In this chapter it is not practical to describe all the many designs of milking machine components that are in use today but examples of current practice are described. These are grouped under vacuum pumps and associated components, the pulsation system, the milking unit, and milk transport. Additionally the auxiliary equipment for use in the event of electricity failure and the equipment for providing hot water for cleaning systems are described.

VACUUM PUMPS AND ASSOCIATED COMPONENTS

PRIME MOVERS

Any form of prime mover can be used to drive the vacuum pump but in practice it is normally an electric motor. An internal combustion engine is

38   *Components*

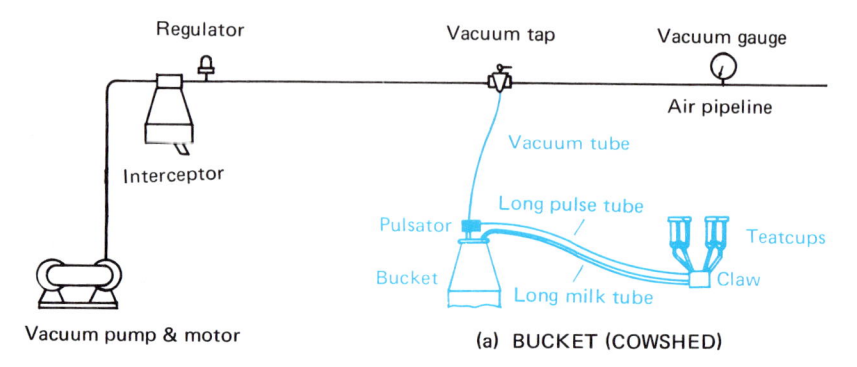

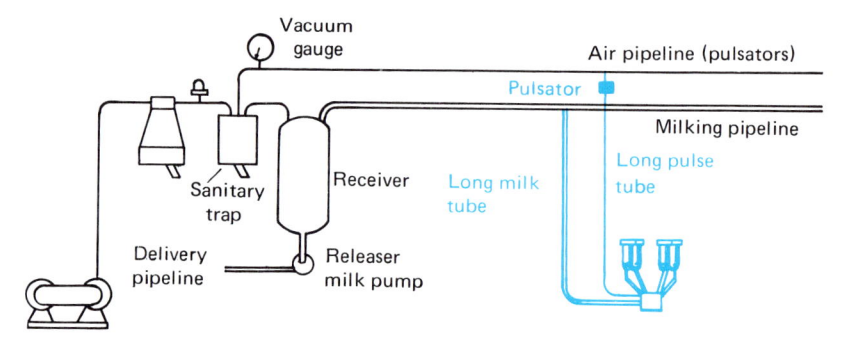

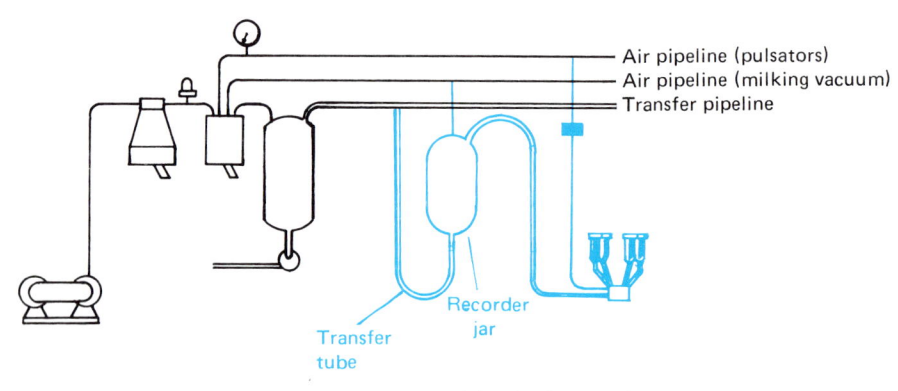

Fig III 1. Principal types of milking machines used in the UK. The assembly of components comprising the unit is shown in blue.

Table III.1 *Distribution of types of milking systems in use in England and Wales December 1974 (from MAAF Report, Dairy Husbandry Dept, 1975)*

	Milking Systems	Number of Herds (and %)	% Cows Milked†
COWSHED	Hand	1 661 (2.6)	
	Bucket	25 788 (41)	
	Pipeline	10 355 (16)	
	TOTAL COWSHED	37 804 (60)	25
PARLOURS*	Rotary	322 (0.5)	
	Fixed		
	Herringbone	9 613 (15)	
	Abreast	13 076 (21)	
	Others	1 031 (1.6)	
	Portable field Bails (mainly abreast)	997 (1.6)	
	TOTAL PARLOURS	25 039 (40)	75
	TOTAL INSTALLATIONS	62 843	

* No figures are available for the proportion of direct-to-can pipeline and recorder jar parlours. Estimated that 5% of the abreast parlours are direct-to-can.

† Estimated from additional data kindly supplied by the MAAF.

necessary for a movable bail and is sometimes used for a fixed installation but in these cases it is mainly for stand-by duty in the event of a failure in the electricity supply (or electric motor failure).

The vacuum pump usually has a V-belt drive. By using a suitable size of motor and pulleys this gives the flexibility of a range of speeds from about 1000 - 1600 rev/min, and a corresponding range of pump capacities to suit several sizes of milking machine. Also this form of drive makes it easier to arrange alternative drive in an emergency.

VACUUM PUMPS

The purpose of the vacuum pump is to exhaust air continuously from the milking machine system, thereby maintaining in it the vacuum necessary for milking, including operation of pulsation. A vacuum pump is an air compressor which in this application takes in air, usually at about 51 kPa (15 inHg) vacuum, and exhausts it at ambient atmospheric pressure. Thus the pump has to compress the air from about half atmospheric pressure to just above atmospheric pressure, a compression ratio of approximately 2 : 1. This is not a particularly

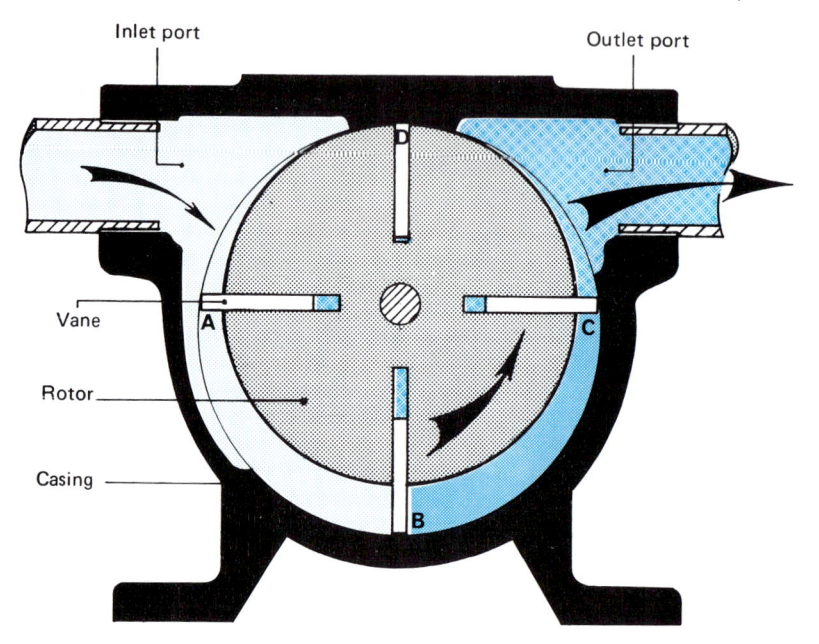

Fig III 2. Diagrammatic cross section of an oil sealed vacuum pump showing the principal components. Light blue represents air at inlet vacuum; dark blue, air being compressed and ejected to atmosphere.

demanding task and several types of pump are suitable. The positive displacement rotary pump is the type most commonly used. It is compact, robust and requires comparatively little maintenance.

Oil lubricated vacuum pumps

Fig III 2 shows a cross-section of a typical rotary vacuum pump. It consists of a horizontal cylindrical casing, with a rotor mounted eccentrically so that it is virtually in contact with the casing at one point of the circumference. The space between the rotor body and the casing is thus crescent shaped, and communicates through the elongated inlet port with the vacuum pipeline, and through the elongated outlet port with the exhaust pipe. The rotor has longitudinal slots, usually four, which house vanes free to slide radially as the rotor turns. The vanes, which are usually made of asbestos fibre composition, are kept in contact with the casing by centrifugal force. In some designs the vanes slide tangentially, the purpose being to reduce frictional losses.

As the rotor turns, pockets of air are enclosed between the vanes and transferred from the inlet to the outlet. In the diagram the space between vanes

A and B is open to the inlet port and is therefore filling with air at inlet vacuum as it increases in volume. Vane B has already passed the edge of the inlet port, so that the pocket of air between vanes B and C is cut off from the inlet. In this position the pocket is decreasing in volume and therefore the pressure of the air enclosed in it is increasing (vacuum decreasing). When the leading vane C reaches the edge of the outlet port the air can escape, as it will have been compressed to a pressure above atmospheric pressure. In the diagram the pocket between vanes C and D is discharging.

Air leakage from the outlet to the inlet is restricted by a film of oil where the body and rotor are closest. Similarly the oil film maintains an air seal at the

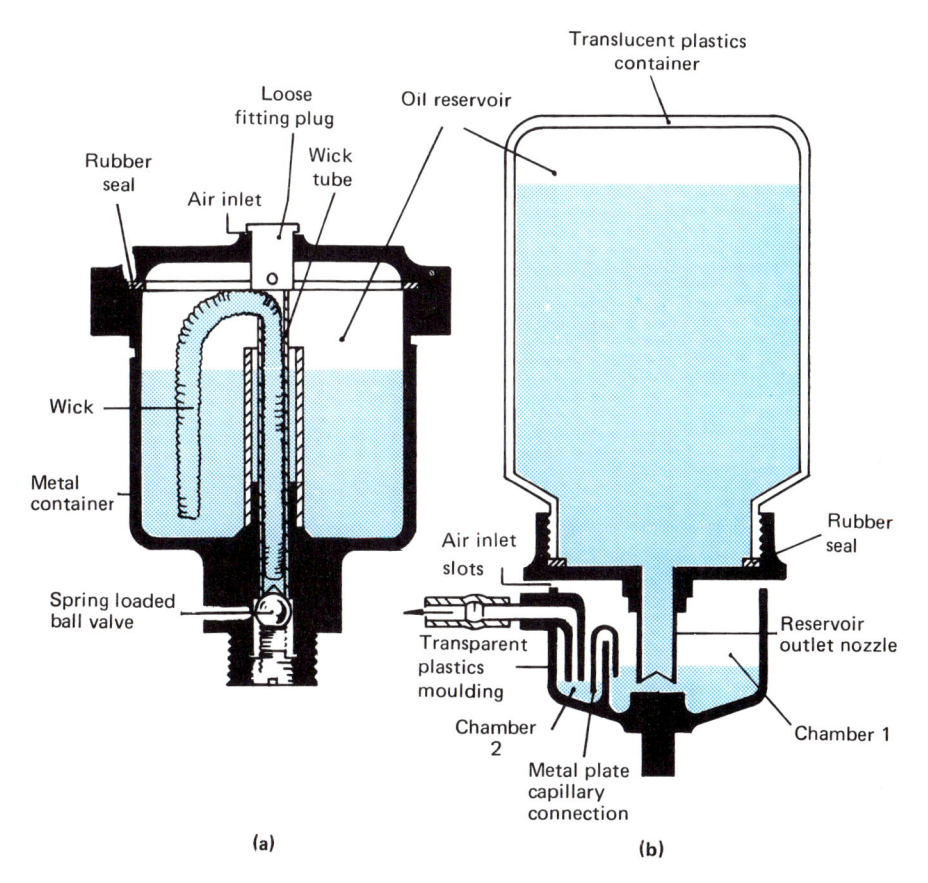

Fig III 3. Vacuum pump lubricators: (a) Gascoigne wick type with ball valve closed (i.e. pump not running); (b) Alfa-Laval plate capillary type, showing oil levels with pump running.

ends of the rotor and between the vanes and slots, in addition to reducing friction and assisting dispersal of heat. As some oil escapes with the outgoing air, a regular supply must be maintained.

Vacuum pump lubrication

Two examples of capillary lubricators used with vane-type rotary pumps are shown in Fig III 3: (a) is an example of the more common wick feed type and (b) is the Alfa-Laval plate capillary type. The reservoirs are sometimes made of transparent or translucent material so that the oil level can readily be observed. The oil flow in Fig III 3a is mainly controlled by the cross sectional area and structure of the wick. In addition the manufacturers of the example shown arrange for the oil reservoir to be under vacuum. As the pump starts the vacuum unseats the spring loaded ball valve allowing the space above the oil in the reservoir to be evacuated. Therefore the loose fitting plug resting on the wick tube moves down under external air pressure and holds the ball off its seat when the pressure above and below the ball equalizes. In 3b the flow is controlled by the difference in oil level in chambers 1 and 2 and the capillary force in the narrow space between the metal plates of the capillary connection. The oil level remains constant in chamber 1 because air can only enter the reservoir when the level falls below its outlet nozzle. Oil flowrates of 1 - 12 ml/h can be set by adjusting the height of the reservoir outlet in chamber 1.

Oil is prevented from flowing into the pump when not in use in (a) by the spring loaded ball valve fitted to the outlet. The central tube above the ball valve fills with oil when the valve is closed. In (b) oil is swept out of chamber 2 which is open to atmosphere, by a current of air flowing into the pump only when it is running. The oil levels in the two chambers equalize when the pump is stopped. Thus with both lubricators there is an immediate flow of oil to the pump when it starts.

In some designs of pump the oil is fed into the end shields or to the pump shaft bearings through a pair of tubes. However, most vacuum pumps have sealed bearings or alternatively separate oiling points are provided for manual lubrication.

The liquid ring vacuum pump

This type of pump has been in use for 50 years in the food and chemical industry and recently has been adapted in design for milking machine use in Australia. The main features of the Australian design are shown diagrammatically in Fig III 4. An impeller with long, almost radial blades is eccentrically mounted in a cylindrical chamber. As the impeller rotates, water fed continuously to the chamber takes up a ring shape under considerable centrifugal force and rotates concentrically with the casing. Thus the air pockets between adjacent blades and the water ring enlarge and contract creating pressure changes in a manner similar to that in the vane type rotary pump.

Crescent shaped inlet and outlet ports are cast in a plate fitted between the pump impeller chamber and the inlet and outlet cavities. The ports are located to hold the water ring at the appropriate depth to give optimum performance in relation to the radial offset of the impeller. As the water leaves a pocket the space fills with air at inlet vacuum and is subsequently compressed and ejected at the outlet port.

Power requirements appear to be similar to those of equivalent vane pumps. The main advantages are little wear, no regular consumption of oil as only the spindle bearings are lubricated, unaffected by carry-over of liquid from the milking system, and a wide range of capacities for one size of pump because it can be rotated over a wide range of speeds. A disadvantage is that in cold climates provision must be made to prevent frost damage.

Cooling of vacuum pumps

The heat produced in compressing air and overcoming mechanical friction in oil lubricated pumps must be dissipated. This is usually assisted by radiation fins on the outside of the casing, and in some cases cooling is improved by the addition of a fan mounted on the pump drive shaft to increase air flow across the pump casing. Overheating may occur, particularly when the casing is not finned, if air

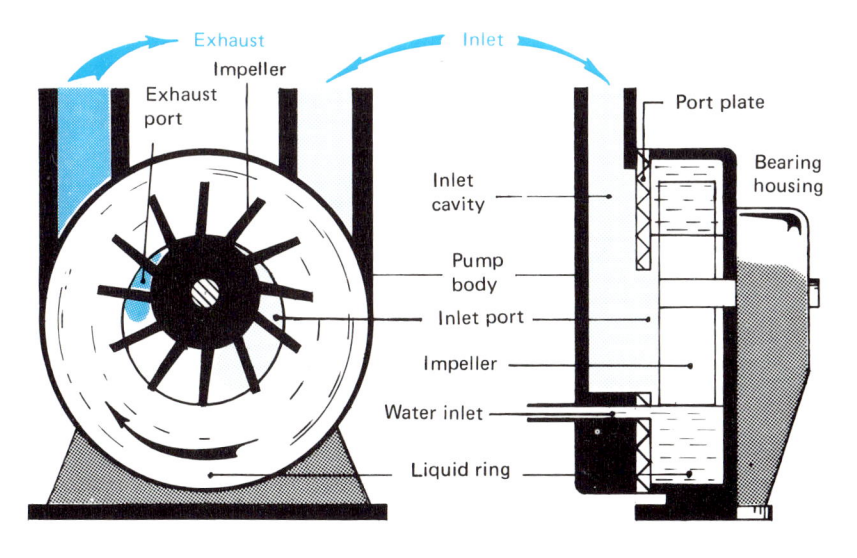

Transverse section — Longitudinal section

Fig III 4. Diagrammatic sections of a liquid ring vacuum pump showing the principal components (Hydrovac Products).

flow through the pump is restricted, for example by operating it at an abnormally high vacuum level.

Exhaust systems

The exhaust from a lubricated pump should discharge outside the building and a silencer should be fitted. The ideal exhaust system is short and slopes continuously downwards from the pump. Where downward slope is not possible a trap for oil and condensed water vapour is necessary to prevent fouling of the surroundings. A satisfactory noise level can almost always be achieved by directing the outlet in a favourable direction, for example away from reflecting walls, and fitting a silencer of adequate performance. Some loss of pump performance will accompany an increase in silencing because of increased back pressure.

The vacuum pump will run in reverse when switched off, with a risk of sucking in foreign particles from the exhaust system unless there is some arrangement to prevent reverse rotation. This can be achieved by providing a large air inlet cock to be opened before switching off or fitting a non-return flap to the exhaust pipe outlet.

Performance characteristics

The pumping capacity of a vacuum pump depends first on its swept volume and speed of rotation. It also depends on the pressure conditions at the inlet and outlet. In addition, internal air leakage in the pump between the exhaust and the inlet side will increase as the pressure difference across the pump increases at higher inlet vacuum levels and this will reduce pump capacity.

The volume of air extracted by a vacuum pump is stated as volume of free air, that is, volume at ambient atmospheric pressure. This corresponds roughly to the volume of air the pump discharges because the pump is exhausting to atmosphere.

The higher the inlet vacuum level the lower will be the volume of air discharged by the pump. This is mainly due to reduction in volume as air at different vacuum levels is compressed to ambient atmospheric pressure. For example, at sea level it will be 40% of its volume when originally at 61 kPa (18 inHg) vacuum, 50% at 51 kPa (15 inHg), and 60% at 41 kPa (12 inHg).

A typical performance characteristic of a vacuum pump is shown in Fig III 5, as the curve of volume of free air plotted against inlet vacuum. At 0 kPa vacuum level (atmospheric pressure) the volume of 495 ℓ (17.5 ft^3)/min is substantially a measurement of swept volume. This declines to less than half at 50 kPa inlet vacuum, 200 ℓ (7 ft^3)/min, and air flow becomes zero at a vacuum level of 90 kPa (26 inHg) due to internal air leakage from the high pressure to the low pressure side of the pump.

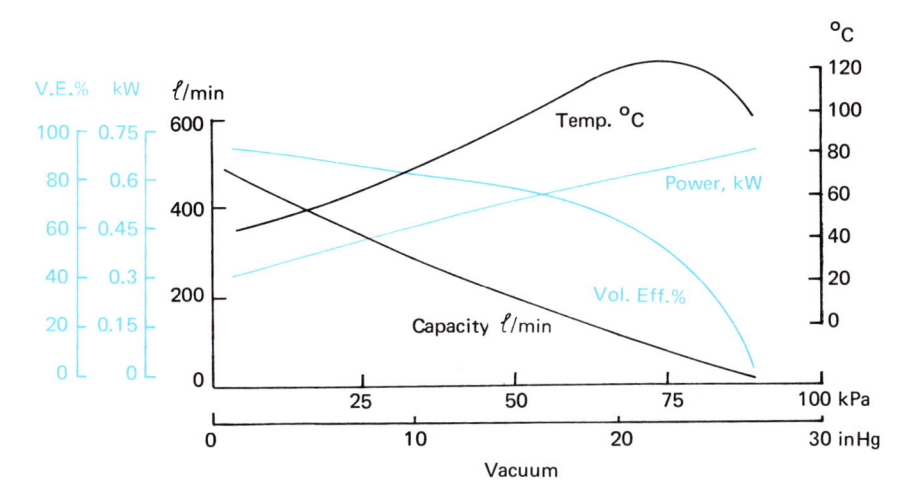

Fig III 5. Vacuum pump performance curves, showing capacity (ℓ/min free air) at a fixed speed, motor output power (kW), volumetric efficiency (%), and exhaust air temperature (°C), plotted against vacuum level (kPa and inHg).

Also plotted in Fig III 5 are typical curves of power consumption and temperature, corresponding to the air flow curve shown. Different sizes and designs of pump will, of course, give different values, but the general trend of the curves is the same.

Capacities of pumps available for milking machines vary from about 150 - 1300 ℓ/min (5 - 45 ft³ min) free air. These require motor shaft output powers of about 0.4 - 4 kW (0.5 - 5 hp) to drive them, taking into account starting load and losses in the belt transmission.

Within limits, a different speed gives almost a proportional air flow and a corresponding increase or decrease in the power requirement. This must be allowed for if the speed of the pump is increased to obtain greater capacity. For milking machine installations the air capacity required depends on several factors, such as the number of units and operators. Requirements for most plants fall in the range 200 - 700 ℓ/min (8 - 25 ft³ min) free air. As a rough guide to arrive at vacuum pump sizes, one milking unit requires approximately 60 ℓ/min (2 ft³ min) of free air.

INTERCEPTORS

The vacuum pump is always protected by an interceptor (Fig III 1), a device formerly called a sanitary trap and still sometimes referred to as a vacuum tank. The primary purpose is to intercept liquid, normally water or cleaning solutions but occasionally milk, drawn through the vacuum pipeline with the air, and to stop it from entering the pump. The interceptor vessel should have a minimum effective capacity of about 15 ℓ (3½ gal) and may be similar to a milking machine bucket or may be a plastics moulding clipped to a special fitting in the pipeline. It should have a float valve incorporated as an additional precaution against flooding of the pump, and a self-sealing flap valve on the outlet which automatically opens when the pump is stopped, so discharging any liquid which may have accumulated. The interceptor should be installed close to the vacuum pump, in a position where the liquid discharged cannot do any harm, for example to the electric motor, and should be easy to inspect and clean.

SANITARY TRAPS

In addition to the interceptor, in milking pipeline and recorder milking machines a sanitary trap is fitted. It separates the milk contact part of the system from the air system to prevent contamination by movement of liquid from one to the other. As an aid in detecting leakage of liquid into the air system under fault conditions (for example, a split liner), all the main pipelines should be connected separately to the sanitary trap as in Fig III 1. For this reason the vessel should be transparent, have provision for drainage, have provision for automatic closing of the connection to the vacuum pump on filling with liquid, it should be installed within sight of the milker, and be not less than 3 ℓ (0.75 gal) effective capacity. The sanitary trap is the central part of the vacuum system and is the obvious site for fitting the vacuum gauge and regulator. However, its capacity would then need to be very considerably increased above the minimum figure quoted.

REGULATORS

A vacuum regulator is an automatic valve designed to maintain a steady vacuum in the milking machine in spite of varying air usage throughout a milking. It is arranged so that a pre-set force holds the valve closed until the reduced pressure of the vacuum system overcomes this force allowing atmospheric pressure to lift the valve off its seat. The regulator is fitted in the air pipe line between the interceptor and the milking installation (Fig III 1) in a position giving reasonable accessibility for cleaning and testing.

An acceptable vacuum regulator should have the following performance characteristics. When tested under milking conditions with one unit only working the vacuum level should not increase by more than 2.0 kPa above that when all units are working. The air leakage through the regulator should not

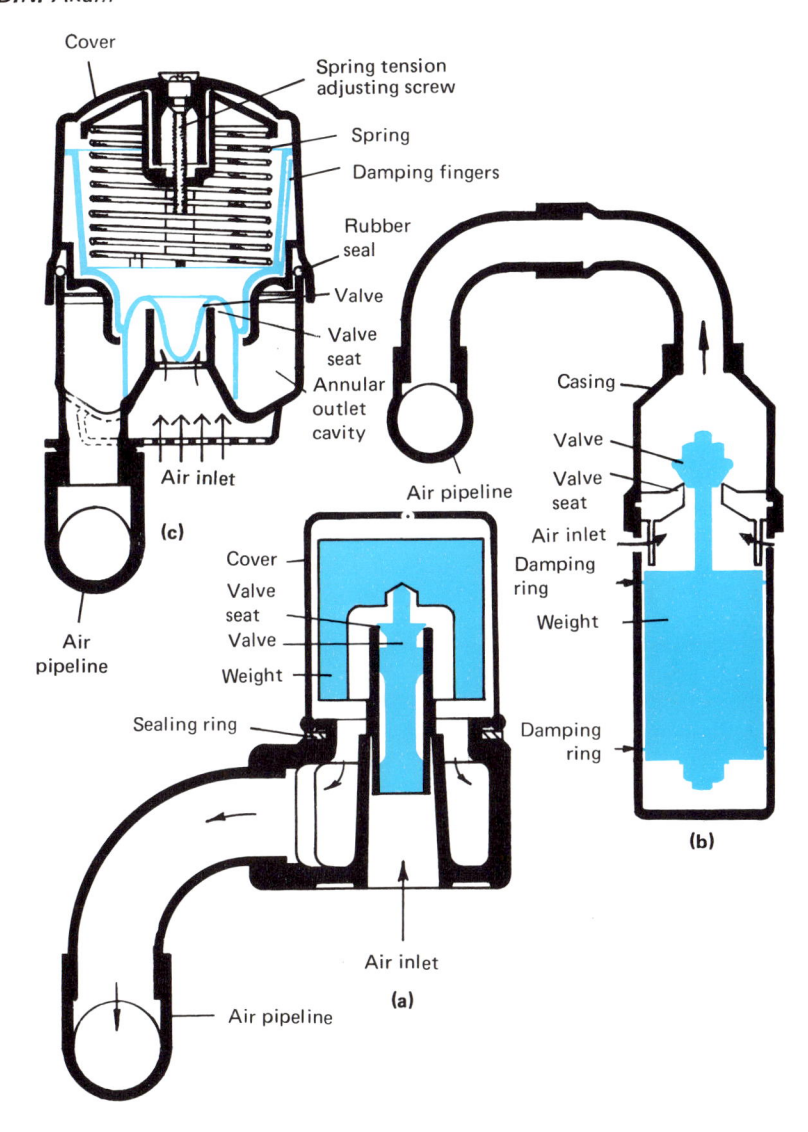

Fig III 6. Diagrams of dead weight and spring controlled vacuum regulators, moving parts shown in blue: (a) Gascoigne dead weight, 560 ℓ (20 ft^3)/min; (b) Fullwood suspended weight, 700 ℓ (25 ft^3)/min; (c) Ruakura spring controlled, 1400 ℓ (50 ft^3)/min free air.

exceed 35 ℓ /min (1.25 ft^3) of free air or 8% of pump capacity whichever is the greater at a vacuum level of 2.0 kPa (0.5 inHg) below the normal working level in the pipeline.

There are two basic types of regulators, weight operated and spring operated, the latter being essential for portable plants. The weight operated regulator must be fitted in a tee piece in a vertical position, in such a manner that moisture from the vacuum system cannot enter the regulator, as in Fig III 6. It requires to be rigidly mounted and as free from vibration as possible.

Weight operated regulators

The construction of a typical weight operated regulator is shown in Fig III 6a. The valve rests on the seating and supports a cast iron weight, which is so shaped that it remains stable in the position shown. The under side of the valve is open to atmosphere. The upper side of the valve and the weight are enclosed by a cover, with a sealing ring, the space inside the cover being in communication with the vacuum pipeline. The valve and seating are usually made of brass, and are renewable. In an alternative form of construction the weight is suspended from the valve instead of being supported by it (Fig III 6b). As weight operated regulators depend on gravitational force they must be rigidly mounted, and are therefore not suitable for portable machines.

Spring operated regulators

To make the regulator independent of gravity, and therefore suitable for use on a portable plant, a spring is used instead of a weight. This incurs a fundamental disadvantage for, whereas the force exerted by a weight remains constant, the force exerted by a spring increases proportionally with deflection. This disadvantage can be minimized by suitable design to reduce the spring deflection to a minimum.

In the design of spring operated regulator illustrated Fig III 6c a large diameter spring is used to improve sensitivity. The valve movement is friction damped by means of two or three plastics fingers in contact with the cover and this together with careful design of the entry and exit airways of the valve allows stable operation over a narrow throttling range to be achieved. The cover, valve and valve seat of this regulator are made of plastics.

Power operated regulators

The regulators so far considered have direct acting control in that the valve opens and closes to admit more or less air in response to changes in milking machine vacuum operating on one of the valve faces. Thus the sensing of system vacuum changes occurs in the same location as changes in air flow to correct deviations from the required vacuum level. An alternative design is to sense the system vacuum level upstream from the valve mechanism (further away from the

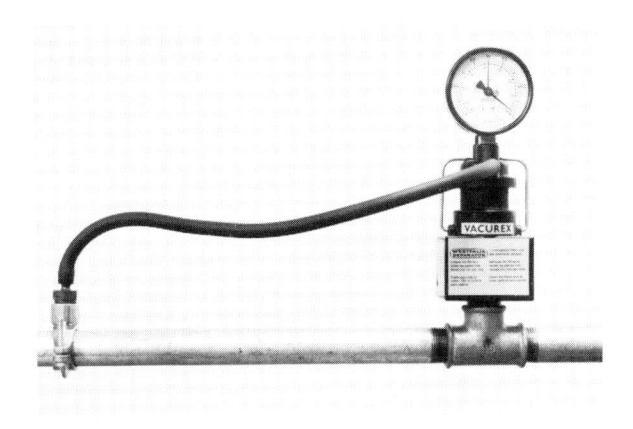

Fig III 7. Westfalia Vacurex remote sensing vacuum regulator.

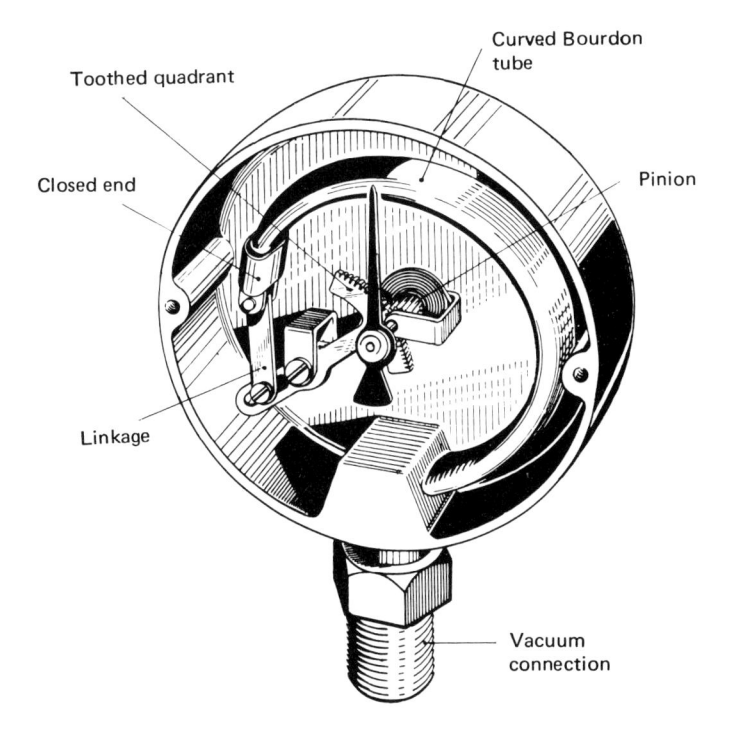

Fig III 8. Bourdon vacuum gauge.

vacuum pump) where more stable vacuum conditions might be expected. Fig III 7 shows a regulator with remote sensing. Raising and lowering the main air inlet valve in response to changes in system vacuum is done by means of a vacuum actuator in the head of the regulator.

VACUUM GAUGES

The vacuum gauge indicates the vacuum in the pipeline system, and is fitted to all milking machine installations in a position so that the operator can at any time check that the vacuum pump and regulator are operating correctly.

The Bourdon type of gauge is almost universally used. The operating mechanism is shown in Fig III 8. The sensing component is a curved tube, one end of which is closed, the other end being in communication with the vacuum pipeline. The tube is elliptical in cross section, and reduction of internal pressure causes it to assume a smaller radius of curvature. Since the open end is fixed, the closed end therefore moves, taking up a position which is a measure of the level of internal vacuum. This movement is transmitted through a linkage to a toothed quadrant. The quadrant meshes with a pinion, the spindle of which projects through the dial and carries a pointer. The gauge should not be less than 75 mm (3 in) diameter. Hitherto, vacuum gauges have normally been calibrated in barometric units with a scale 0 - 30 inHg or 0 - 760 mmHg. These units are likely to be replaced by SI units in which scales will be 0 - 1 bar or 0 - 100 kPa. In each case the normal operating vacuum will be at about half-scale and should be indicated by suitable marking. Gauges used in the UK read from left to right i.e. clockwise, whereas Continental practice is anti-clockwise. Some gauges are provided with a locking screw for transport and some have zero adjustment. Gauges fitted to portable installations require damping to protect them against damage by vibration when in operation.

AIR PIPELINES AND FITTINGS

Air pipelines in milking machines are constructed from a variety of materials. To a large extent this is decided by the type of milking machine installation. Glass or stainless steel is used where the air pipeline forms part of the cleaning circuit. Generally the pipeline connecting the vacuum pump to the milking machine is constructed from 25 mm (1 in) or 32 mm (1¼ in) bore screwed galvanized steel tube. This type of tube is also used to construct the air pipeline for bucket milking in cowsheds and the pulsator air pipeline in milking parlours, where rigidity and mechanical support of other pipelines and components is required. In cases where there is a large number of vacuum taps or pulsator points the pipeline circuit is looped to form a ring, so reducing vacuum drop. Bends, rather than elbows, are used where it is necessary to change direction of the pipeline and it is important to deburr pipe ends before assembly. Other accepted practices are to fit tees at the bottom of rising pipes to facilitate draining and to fit plugs or caps to dead ends for ease of cleaning. In some instances where lightness is an advantage or structural strength is not important plastics tube and rubber slip joints are used. Care must be taken, however, to confirm that the plastics tube chosen is capable of withstanding any internal pressure between 68 kPa (20 inHg) vacuum and 2 bar (28 lbf/in^2) gauge at all temperatures that may exist in the pipeline. Plastics tubing has the advantage of a better surface finish than galvanized steel, giving less resistance to airflow. It is easier to keep clean as it does not corrode but must be supported at more frequent intervals than galvanized steel.

Pressure difference must exist between any two points in the vacuum pipeline, otherwise air would not flow from the milking equipment to the pump. Thus the vacuum at the milking units is always lower than at the vacuum pump. The difference depends on the amount of air flowing, and the frictional resistance of the pipe, which in turn depends on effective length and diameter. It is usual to specify that the air pipeline between the vacuum pump and the regulator is of such a bore that the vacuum drop does not exceed 2.5 kPa (0.75 inHg). The pulsator air pipeline is not normally so closely specified. All air pipelines are usually required to be self draining to automatic drain valves (Fig III 9). These are designed to close when the pipeline is under vacuum, they are constructed of a corrosion resistant material and usually consist of a plunger fitted with a fat resistant plastics or rubber sealing washer and a removable cap for cleaning purposes.

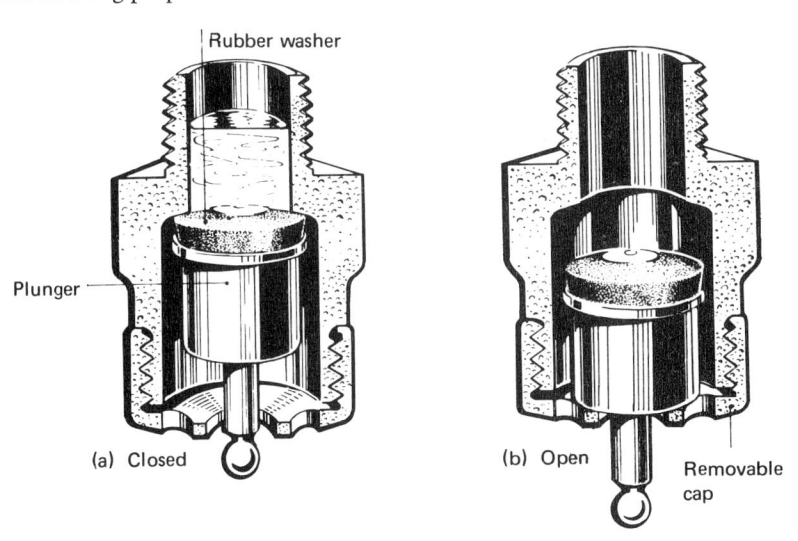

Fig III 9. Typical self draining valve.

Vacuum taps

Vacuum taps (formerly known as stall cocks) are used in bucket and direct-to-can milking machines for connecting the milking unit to the air pipeline. They are rigidly connected to the pipeline, usually screwed into a tee-piece. It is important that the connection is made in the upper part of the air pipeline to minimize the possibility of moisture draining down to the milking unit (see Fig III 10) and the taps are designed with stops to indicate the fully open and fully closed position.

Another design of vacuum tap (Fig III 11) eliminates the operation of turning a valve, as the action of pushing a rubber tube onto the nipple pushes back a soft rubber plunger uncovering the vacuum port (Fig III 11a). Pulling off the rubber tube pulls back the plunger, sealing off the vacuum (Fig III 11b).

PULSATION EQUIPMENT

Pulsation is the opening and closing of the liners in the teatcups which occurs when a valve mechanism, the pulsator, alternately connects the pulsation chambers to the vacuum system of the machine and to the atmosphere. The liner will open when there is more or less equal vacuum within the liner and in the pulsation chamber external to it. It will close when pressure is increased in the pulsation chamber as air is admitted. Most pulsators belong to one of two broad classes: relay pulsators which are caused to operate at a set frequency and pulsator ratio by a central pulsator controller; and self contained pulsators in which each pulsator has a separate built-in controller. Pulsators more usually pulsate the four liners of a cluster together (simultaneous pulsation), although self contained pulsators are better adapted to pulsating liners in pairs (alternate pulsation).

TYPICAL PULSATORS

The simplest pulsator to describe is the electromagnetic relay pulsator. A typical self contained pneumatic pulsator and a pneumatic relay pulsator are also described as examples of the commonest types.

Electromagnetic relay pulsator

The longitudinal section of a Fullwood relay pulsator in Fig III 12 shows the valve in the downward position (solenoid not energized), shutting off the vacuum connection to the pulsator air pipeline. The long pulse tube nipple is consequently connected by the annular space around the valve to the air inlet passage which is central in the electromagnet. When the electromagnet is

D.N. Akam 53

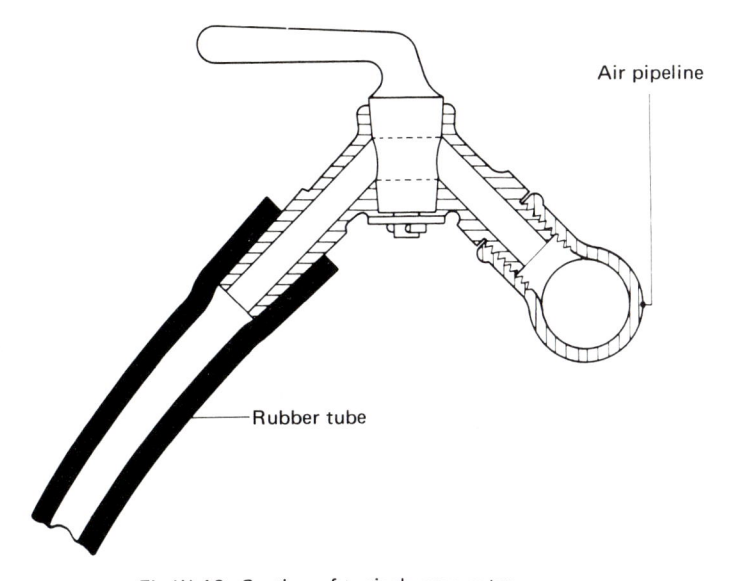

Fig III 10. Section of typical vacuum tap.

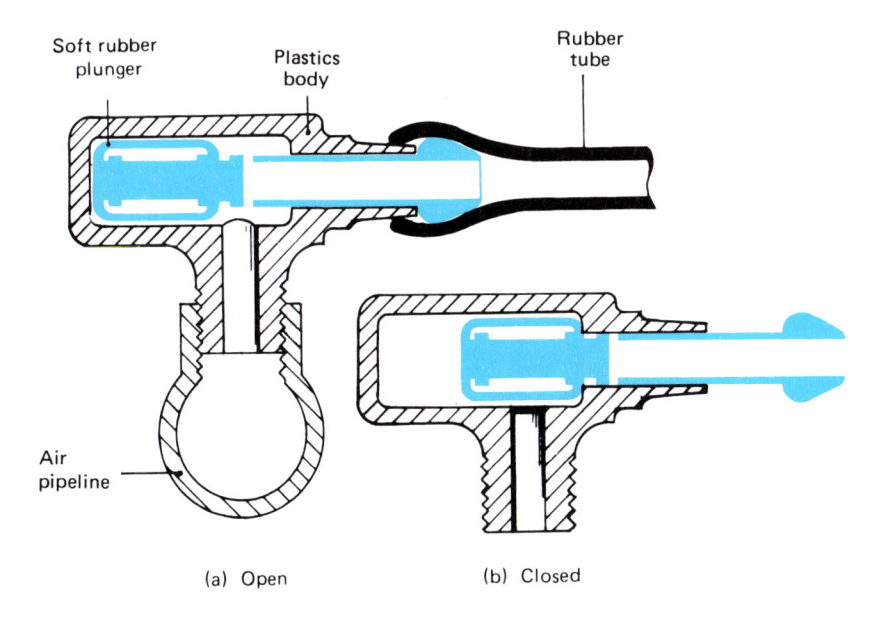

Fig III 11. Section of the Alfa-Laval automatic vacuum tap.

energized the valve, being made of magnetic material, moves onto the upper valve seat, shutting off the connection to atmosphere and connecting the long pulse tube nipple to the pulsator air pipeline. Fullwood also make a large capacity electromagnetic relay pulsator, capable of providing pulsation in as many as four clusters.

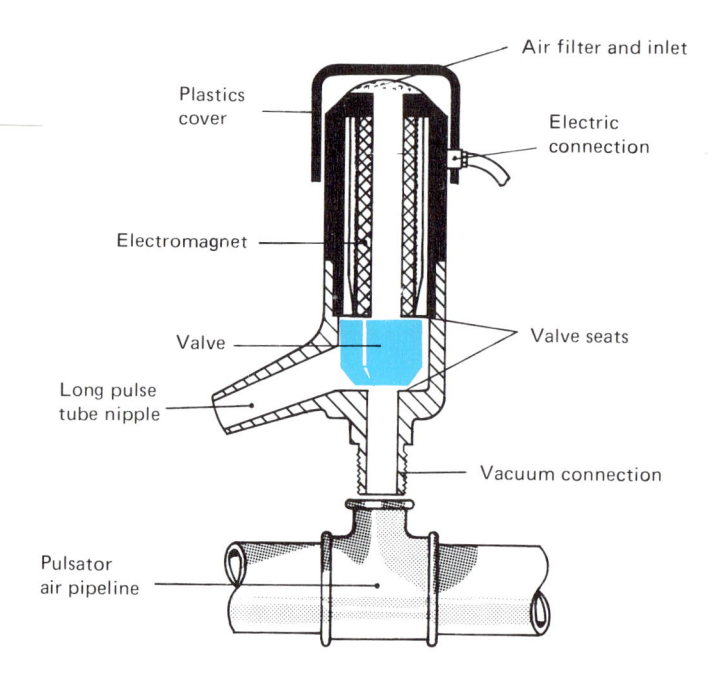

Fig III 12. Section through the Fullwood electromagnetic relay pulsator.

Electromagnetic relay pulsators are connected to the 12 V d.c. output of the pulsator controller, and each pulsator requires a current of 0.5 to 0.75 A. The electric circuit is usually completed through the metal air pipeline.

Self contained pneumatic pulsator

The Christensen pulsator shown diagrammatically in Fig III 13 is divided into three chambers. The middle one carries the long pulse tube spigot and is alternately connected to the vacuum system and atmosphere by simultaneous movement of the air and vacuum valve assemblies. The long pulse tube spigot is shown connected to atmosphere in Fig III 13a and to the vacuum system in 13b.

It will be noticed that the operation of the valves also causes the lower chamber of the pulsator to be under vacuum when the middle chamber is connected to the atmosphere as shown in 13a, and the reverse as shown in 13b. At the moment when such a reversal has just taken place and the middle chamber has attained atmospheric pressure as in 13a, the upper chamber will also be nearly at atmospheric pressure. It is separated from the middle chamber by a rubber diaphragm to which the valve assembles are attached. The upper chamber is connected to the lower chamber by a passage way restricted by a regulating screw, and air therefore is withdrawn from the upper chamber at a rate determined by the setting of the screw. Upward movement of the diaphragm is, however, resisted by the pressure difference across the valves holding them shut as shown in 13a. Eventually sufficient force will build up across the diaphragm to unseat the valves and cause them to flick over to the positions shown in 13b. Air will then start moving into the upper chamber as the second part of the pulsation cycle begins. It will terminate with a similar reversal of the positions of the valves and the beginning of a new pulsation cycle.

The Christensen pulsator described has, essentially, a 50% pulsator ratio. Other self contained pneumatic pulsators, for example those using slide valves, are more readily adapted to providing wider pulsator ratios.

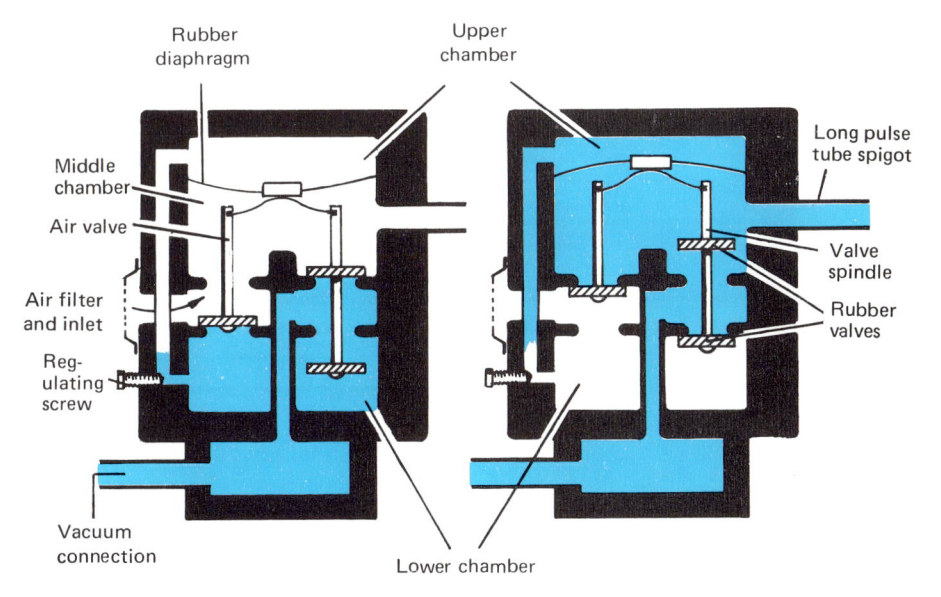

Fig III 13. Diagram of the Christensen self contained pneumatic pulsator (blue = vacuum).

Pulsation rate can vary from the set value because the restriction formed by the regulating screw becomes partly blocked by dust. This is overcome in some designs of pulsator by transferring a fluid (liquid or gas) to and fro through a restriction in a sealed system. The four chambered, double diaphragm pulsator controller built into the Alfa-Laval Hydropulse pulsator is shown in Fig III 14. Liquid is transferred from one end chamber to the other through the central tube and restrictor. The connecting tube is driven to and fro as the pressure conditions in the inner pair of chambers are alternated between atmospheric pressure and machine vacuum. A flick over rotary valve driven by the connecting tube effects this reversal of pressure. The main pulsator slide valve is also operated by the movement of the connecting tube.

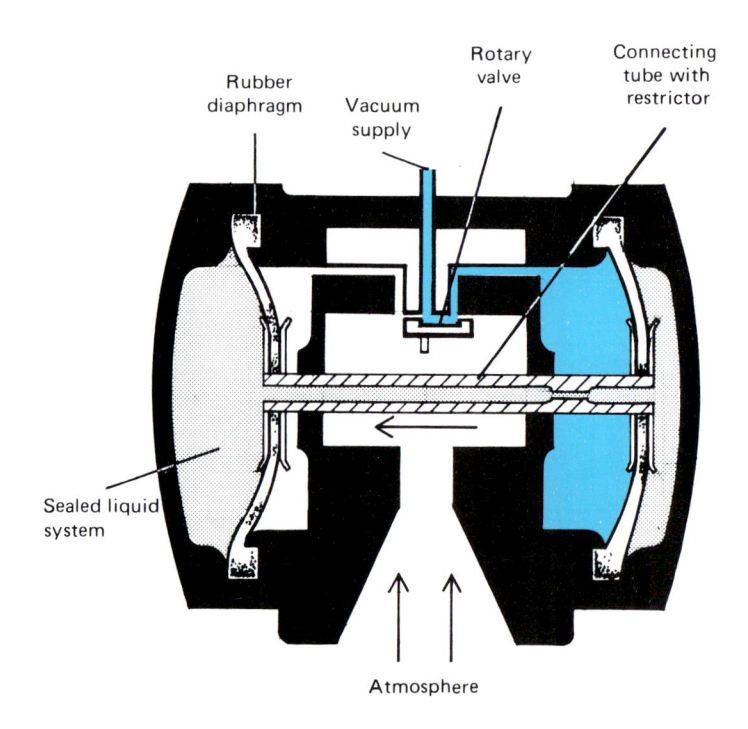

Fig III 14. Diagram of an Alfa-Laval Hydropulse pulsator showing the principle of the sealed liquid controller, with the diaphragms and connecting tube moving to the left.

Pneumatic relay pulsator

A pneumatically actuated relay pulsator, analogous to the electromagnetic relay pulsator already described, is shown in section in Fig III 15. The chamber above the diaphragm is alternately connected to atmosphere and evacuated, thus moving the pulsator valve to the positions shown in 15a and 15b. Several types of pulsator controller with pneumatic transmission, which may themselves be pneumatic self contained pulsators, are available.

Fig III 15 also shows a restrictor on the air inlet to the pulsator. These may be fitted to pulsators of all types to regulate the rate of collapse of liners and are usually referred to as chokes.

(a) Pulse tube spigot connected to vacuum

(b) Pulse tube spigot connected to atmosphere

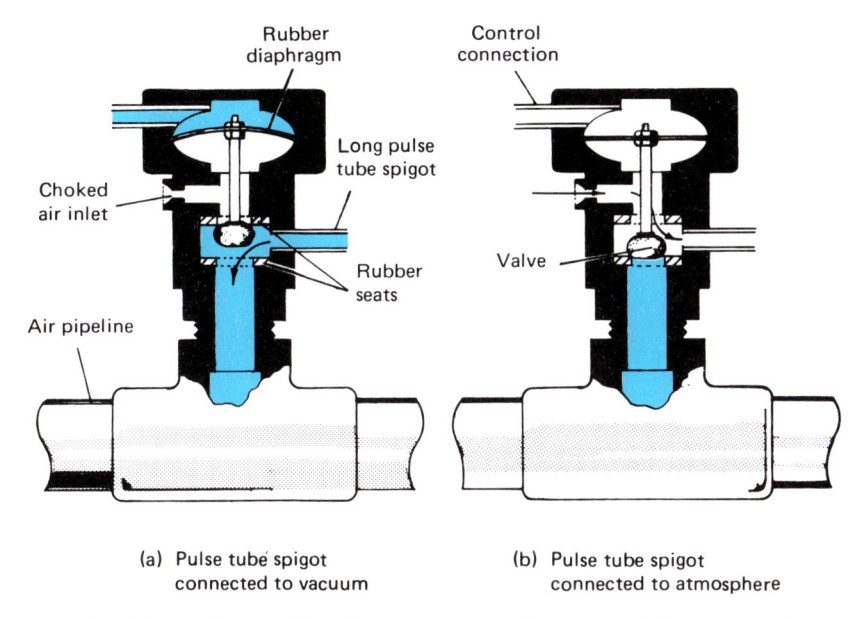

Fig III 15. Diagram of the Vaccar pneumatic relay pulsator (blue = vacuum).

PULSATOR CONTROLLERS

Electric controllers

Electric pulsator controllers are now electronic devices. They are powered from the mains electricity supply which is transformed and rectified to drive a timing circuit and a power amplifying stage to operate electromagnetic relay pulsators. For stand-by purposes they are designed to operate also from a 12 V lead acid storage battery.

Pneumatic controllers

Some devices use rotary valves driven through gearing from the vacuum pump shaft. Self contained pulsators, including those in which the timing component is a pendulum, are also in current use, as are electromagnetic relay pulsators.

A mechanical pulsation system

Another pulsation system driven through gearing from the vacuum pump shaft or electric motor, is purely mechanical. A pulsator slide valve at each milking unit is linked to a common push rod running the length of the parlour. The rod is moved to and fro, thus operating the pulsator valves, by means of a crank and connecting rod driven from the reduction unit. These have not been widely used in the UK.

PROPERTIES OF PULSATORS AND PULSATOR CONTROLLERS

The desirable properties of pulsators and their control devices whether built-in or remote are reliability, adequate port size, low maintenance demand, and not wasteful of vacuum pump capacity. The pulsator is a simple valve mechanism and, provided these properties are adequate, the way in which it is driven is not important. Self contained pneumatic pulsators apparently have natural advantages for use with portable equipment such as bucket units and milking pipeline units in cowsheds. Even here, however, electric pulsation has penetrated from time to time.

The valve mechanism may waste vacuum pump capacity in two ways. First, a considerable amount of air may enter the vacuum system if the connection from the long pulse tube nipple to air or vacuum is reversed slowly, for example if the valve in Fig III 15 moved slowly. Both the air and vacuum ports would then be open simultaneously for an appreciable time. Second, leakage at the vacuum port may be severe. It is rarely tested for. Many relay pulsators now have rubber seats to minimize air leakage.

Port sizes of the pulsator, together with length and bore of the long and short pulse tubes, and bore of nipples determine the pulsation chamber wave form (see illustration of pulsation chamber vacuum record, definition in 6.7, Terminology and Definitions). It has already been mentioned that phase c, decreasing vacuum phase, is sometimes lengthened by choking the air inlet hole of pulsators. More often, however, there is some difficulty in getting the increasing and decreasing phases short enough to allow a desired wide pulsator ratio at a fairly high pulsation rate of 50 or 60 c/min. Length of the long pulse tube seems to be the limiting factor more often than port size or other restrictions to air flow. A length in excess of 2 m (7 ft) imposes a substantial restriction which cannot usually be overcome by increasing the bore. Generally there is no improvement if the bore is greater than 9 mm (0.375 in). Usual bores of long pulse tubes are 6 - 8 mm (0.25 - 0.31 in) for clusters with alternate pulsation, and 8 - 9 mm (0.31 - 0.375 in) for clusters with simultaneous pulsation.

MILKING UNITS

The assembly of milking machine components which is replicated to enable more than one cow to be milked at a time is termed a milking unit and consists essentially of a cluster and pulsation equipment. Units adapted for use in different types of installation require additional components, mainly because of the different ways in which milk from individual cows is brought together for cooling and despatch from the farm. Thus units may be considered as falling into three main classes, adapted (a) to bucket or direct-to-can milking, (b) to milking into recorder jars which are subsequently discharged to a milk transfer pipeline, and (c) to milking direct to pipeline in cowsheds and parlours (Fig III 1).

In this section the main components of the cluster are first described and then the different types of unit and their associated special components. The pulsation equipment of units has already been described above. Milk metering and automatic cluster removal are included here.

THE CLUSTER

The cluster consists of four teatcups (each having a shell, a rubber liner and short milk and short pulse tubes), the claw, the long milk tube and the long pulse tube (Fig III 16).

Shells

Teatcup shells are normally made from stainless steel tube. Plastics have been used occasionally for many years, but even modern plastics offer no great advantages and their light weight is in fact a disadvantage. The New Zealand disposable sealed milk unit (a rubber liner sealed into a plastics teatcup) has a detachable rubber covered weight provided for each teatcup.

The shell is cylindrical in form, the ends being shaped to suit the particular design of liner. A side tube provides the necessary connection to the annular space (pulsation chamber) formed when the liner is fitted into the shell. This tube is usually attached about midway between the ends. The actual position is not important, the main consideration being ease of manipulating the teatcups during milking.

Liners

A liner is a flexible sleeve having a mouthpiece, a barrel and an integral or separate short milk tube, with the physical properties necessary for efficient milking. Natural rubber deteriorates when in contact with fat, and liners made entirely of this material therefore have a limited life due to absorption of fat from the skin of the teat and from the milk. A surface treatment is sometimes applied to reduce fat absorption. Synthetic rubbers or mixtures of synthetic and natural rubber, resistant to fat are now widely used. Further details are given in Chapter XI.

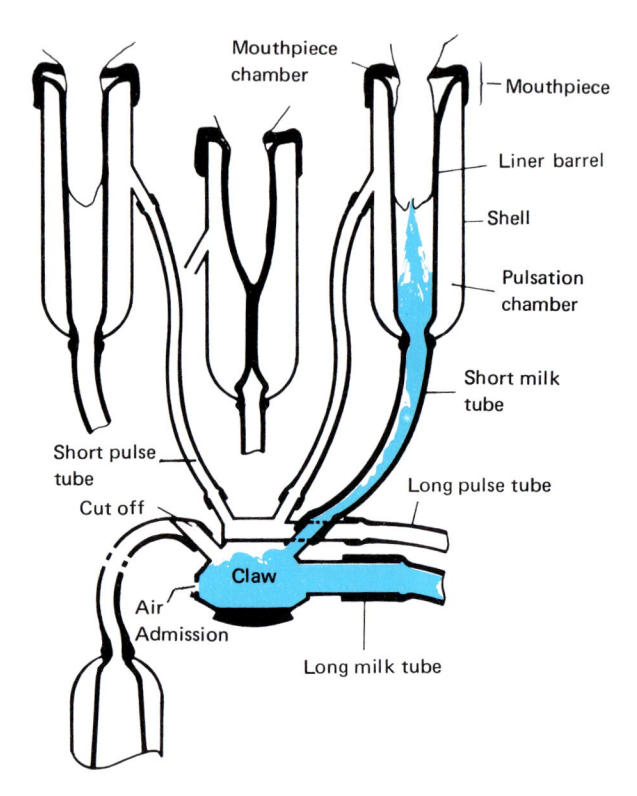

Fig III 16. Diagrammatic cross section of a teatcup cluster showing the main components.

Whatever the design, the liner must be shaped: (a) to provide an airtight joint at both ends of the shell — in some cases the seal at the lower end is made by, or with the assistance of, the short milk tube; (b) to provide a mouthpiece which will fit on the teat so as to prevent as far as possible undue amounts of air being drawn into the system; (c) to be reasonably easy to clean; (d) to milk out as completely as possible.

Three general types of liners are now used with a wide variety of mouthpiece sizes and shapes (Fig III 17).

(a) Moulded with integral short milk tube. The mouthpiece, barrel and short milk tube are moulded in one piece, as illustrated in Fig III 17a. Longitudinal

tension can be varied to some extent, but this is generally small. Liners of this type have been made with the barrel wall graduated in thickness, being thin at the top, on the incorrect assumption that this would cause the liner to collapse progressively from the top. Ribs are sometimes incorporated.

(b) Moulded with separate short milk tube. Liners are made with the mouthpiece and barrel moulded in one piece. A separate short milk tube is attached as with extruded liners, the seal at the lower end being made in a variety of ways. Two examples are shown in Fig III 17b. In some cases a special tool is required to fit the liner into the shell, so as to get a longitudinal tension of a few tens of newtons.

(c) Extruded. The liner is manufactured as a plain tube, the mouthpiece being formed by inserting a ring larger in diameter than the tube (Fig III 17c). A short milk tube is attached to complete the liner. The chief advantages of this type are that the mouthpiece can be varied within limits to suit individual preference and, by using natural rubber and cutting the liner to length periodically, considerable tension can be maintained in the liner during its useful life. Tensioned liners milk faster but require regular routine maintenance.

Gascoigne	Alfa-Laval	Westfalia	Fullwood	Alfa-Laval
Moulded; integral short milk tube		Moulded; separate short milk tube		Extruded
(a)		(b)		(c)

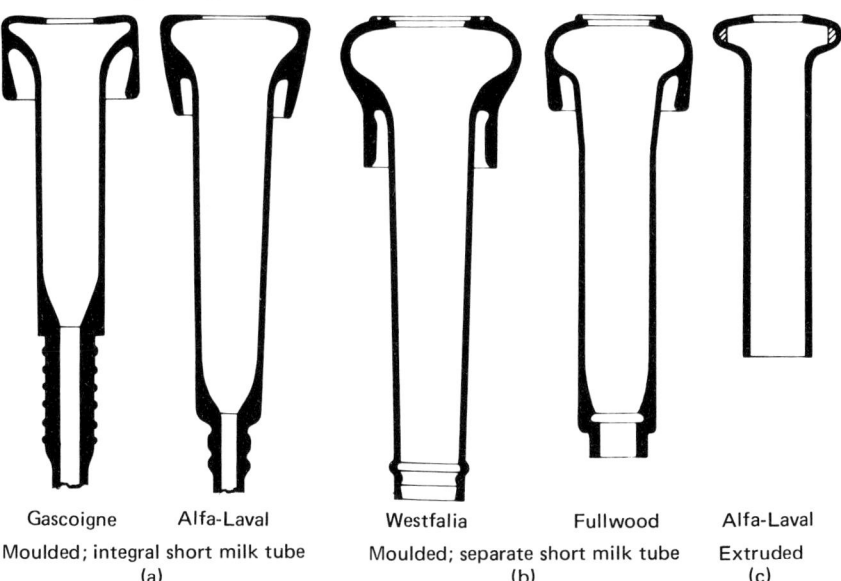

Fig III 17. Cross sections of 3 common types of liner.

Claw

On all types of machine except the suspended bucket type a component is required to connect the short pulse tubes and the short milk tubes from the four teatcups to the long pulse tube(s) and long milk tube respectively. This component is called the claw. A wide variety of designs and methods of manufacture are used, three being illustrated in Fig III 18. Where the four liners are pulsated together (simultaneous pulsation), only one long pulse tube nipple is required. Claws are generally made from stainless steel, the nipples being force fitted, welded, or hard soldered into the milk and air chambers.

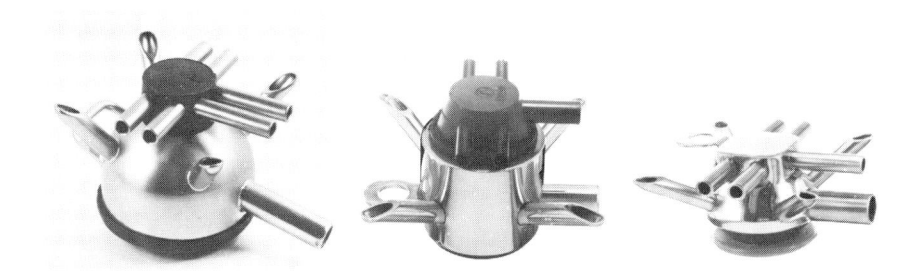

Fig III 18. Three examples of stainless steel claws. *Left to right* Fullwood, Simplex, Vaccar.

The milk chamber and milk tube connections should be fully accessible for cleaning and free from crevices or sharp edges. A rubber bung or cap is normally fitted to close the milk chamber, as it is preferable to avoid screw threads. The four short milk tube nipples are usually set at an angle, scarfed and rounded at the end, so that when the assembly is held ready for attachment to the udder the teatcups will hang downwards so sealing the short milk tubes by their own weight — termed cut off as illustrated in Fig III 16. A disadvantage of the scarfed rounded ends is that there is some restriction to milk flow at this point. The ends of these tubes should be free from sharp edges which may cause piercing due to accidental impacts. The milk chamber volume of the different claws used in the UK vary from 50 ml to 120 ml, except the one shown in Fig III 20 which has no chamber.

Air admission. A slot or hole that is dimensionally stable is normally provided so that air can enter the milk chamber when the cluster is in operation. This is the air admission hole. It helps to remove milk from the cluster and, when milk is elevated above the claw outlet nipple, reduces hydrostatic head by breaking up the milk column. The size of the air admission hole is usually equivalent to a circular hole of about 0.8 mm (0.031 in) diameter, sufficient to pass about 7 ℓ (0.25 ft^3)/min of free air at the normal vacuum level of the milking machine.

Automatic shut off valve. Two designs of claw are available incorporating automatic vacuum shut off valves which are designed to close should the cluster fall off during milking. Besides preventing contamination, loss of vacuum due to excessive air intake is avoided. The shut off valves are also available for manual operation during cluster changing, and provision is made for locking them in the open position for in-place cleaning.

The Alfa-Laval claw has a rounded solid plunger which is held in the open position for milking by the pressure difference across the external retaining disc, as shown in Fig III 19a. When the cluster falls off the cow, the air entering the teatcups reduces this pressure difference and the plunger drops to the vacuum shut off position shown in 19b.

The Gascoigne claw has a plug valve which is shown in the down position for milking in Fig III 20a. If the cluster falls off the plug strikes the floor and moves upwards, shutting off the vacuum as shown in 20b.

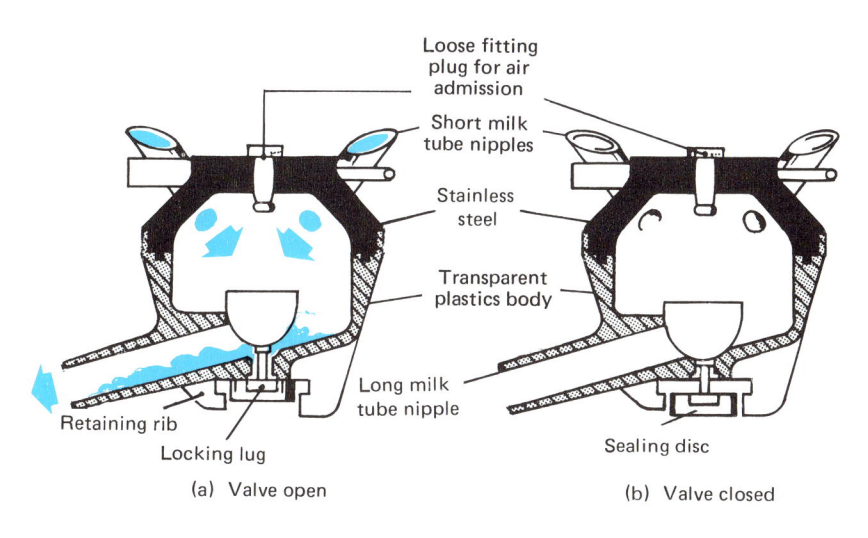

Fig III 19. Diagram of the Alfa-Laval claw with automatic vacuum shut off. The valve can be locked in the open position for washing.

BUCKET AND DIRECT-TO-CAN UNITS

The cluster is attached to a lid, which has a gasket to seal it to the milking bucket or milk can. The assembly is connected to the air pipeline of the vacuum system by the vacuum tube. Bucket units, being portable, have a pulsator fitted to the lid, but in direct-to-can installations the pulsator is normally mounted on the stall pipework. The number of direct-to-can milking machines in use is very small compared to the number of bucket milking machines (Table III 1) and with the steady move to bulk milk collection from all farms in the UK it is likely that direct-to-can milking will diminish even further.

Components

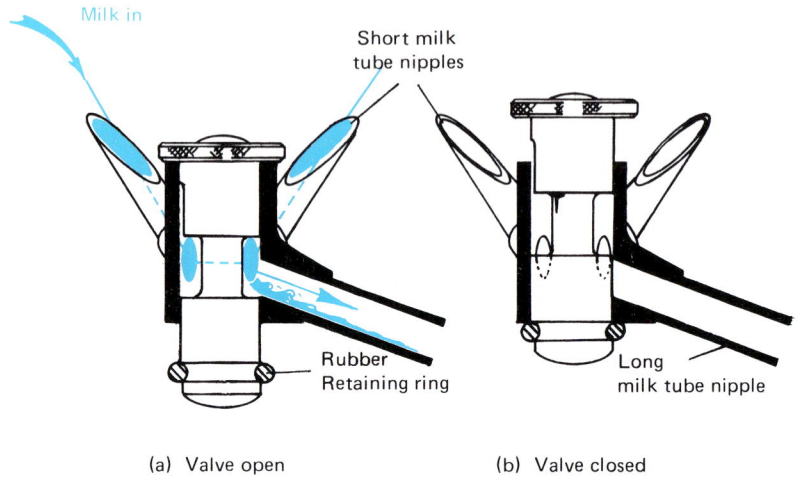

(a) Valve open (b) Valve closed

Fig III 20. Diagram of the Gascoigne automatic claw. The shut off valve can be locked in the open position for washing.

Lid assemblies

The lid must be robust enough to withstand the vacuum applied to the bucket or can and may be made of rubber, nylon or stainless steel. Most lids are made of stainless steel sheet with a thick rubber gasket attached to the underside. The gasket is usually kept in place by a turned down edge which also serves to stiffen the lid. Typical lid assemblies are illustrated in Fig III 21.

The lid carries connections to admit milk and to exhaust air. The milk inlet is usually a simple tube but may incorporate a cock for isolating the teatcups from vacuum and may incorporate a sight glass. Both these items increase the difficulty of cleaning and it is more usual to cut off the vacuum by kinking the long milk tube, operating a valve in the claw (Figs III 19 and 20) or operating a pinch clip fitted onto the long milk tube (Fig III 22). For milking direct-to-can, the milk inlet nipple should be extended below the lid and fitted so that the milk is directed onto the can wall. This reduces the danger of buttering.

The connection on the lid for exhausting air receives the vacuum tube and usually incorporates a form of non-return valve. The main purpose of the non-return valve is to retain an adequate level of vacuum in the bucket should a serious drop of vacuum occur for a short time in the air pipeline. The non-return valve and housing is usually referred to as a moisture trap, but it can probably do little to restrict movement of contaminated moisture from the vacuum tube into the bucket. The best safeguard against contamination of milk in this way is to

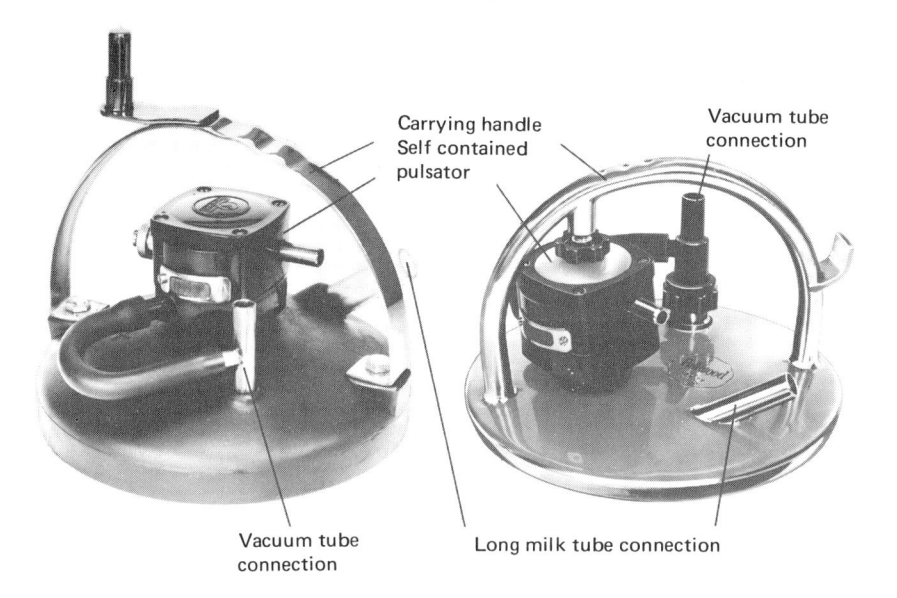

Fig III 21. Typical bucket lid assemblies with self contained pneumatic pulsators. Left, Simplex moulded rubber; right, Fullwood stainless steel.

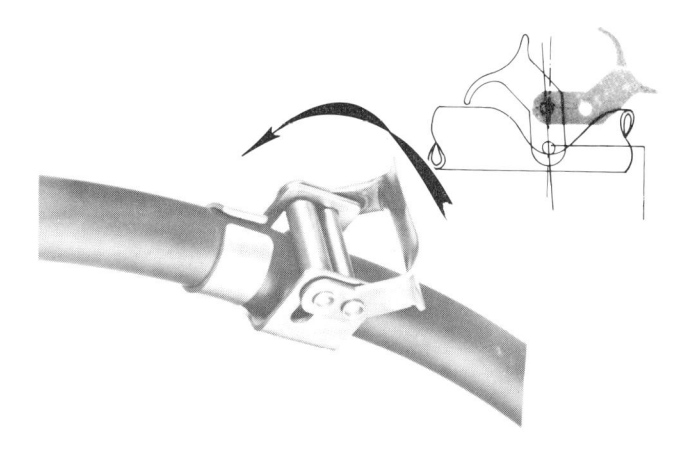

Fig III 22. An example of a pinch clip for shutting off the rubber long milk tube. The photograph shows the clip in the open position and the line drawing shows the clip locked closed in the over centre position.

install and maintain the air pipeline correctly (as already described) and to clean the vacuum tube daily.

In most designs of bucket unit the lid provides a mounting for a self contained pulsator or an electric relay pulsator. The vacuum connection for this may be taken from the milk container but is normally from a branch on the stall tube or moisture trap so that contaminated air from the pulsator does not pass directly into the bucket. The pulsator is usually easily removed to simplify the cleaning of the other components.

UNITS WITH RECORDER JARS

In most parlour installations in the UK, milk is passed from the teatcups into a rigidly mounted all glass recorder jar, which enables the milk produced by each cow to be seen, if necessary rejected, and measured. The current international thinking on recommendations for recorder jars is that the minimum capacity should be 23 ℓ (5 gal), they should withstand an internal vacuum of 102 kPa (30 inHg), and should have a permanent graduated scale on the glass at intervals of 0.2 kg, with a heavier scale mark and a numerical value at each kg. It is also recommended that the jar should be mounted so that the bottom of the scale will not be more than 1.6 m (63 inches) above the operator's floor. The accuracy of calibration, and the fixing and method of use, need to be such that 95% of records of individual cow yields do not have an overall error from all causes exceeding 0.2 kg (0.5 lb) or 5%, whichever is the larger, to satisfy National Milk Records requirements.

During milking the recorder jar is connected to vacuum and subsequently milk in the jar is discharged to the transfer pipeline (under vacuum from the receiver), by closing off the connection to vacuum, admitting air to the jar, and opening the connection to the transfer pipeline. The general arrangement of a recorder jar in the pipeline system is shown in the diagram in Fig III 1, and an example of a unit in Fig III 23. The operations of closing off the vacuum and opening the transfer valve are accomplished by the use of pinch valves operated by a single lever. During washing, the air pipeline (milking vacuum) becomes the supply line for cleaning fluids. Both valves involved in milk transfer then need to be open, with the operating lever in the centre position. Also the teatcups are usually connected into the cleaning circuit by means of jetter cups, and a spreader device to encourage liquid to flow evenly down the inside surface of the jar is fitted into the vacuum-washing fluid inlet spigot at the top centre. The normal arrangement of pipelines and components of recorder milking machines during milking and washing are shown in Chapter X.

Milk sampling

For taking milk samples from recorder jars most manufacturers make provision for fitting a sampling tap made of either rubber, plastics or metal into the rubber elbow or other connection at the jar outlet, and for closing the vacuum connection to the jar independent of milk transfer.

Recorder jar outlet valve

An ideal fitting for the outlet of the jar would be a multi-purpose device which incorporated the following features.

1. A fixed datum from which the jar is calibrated.

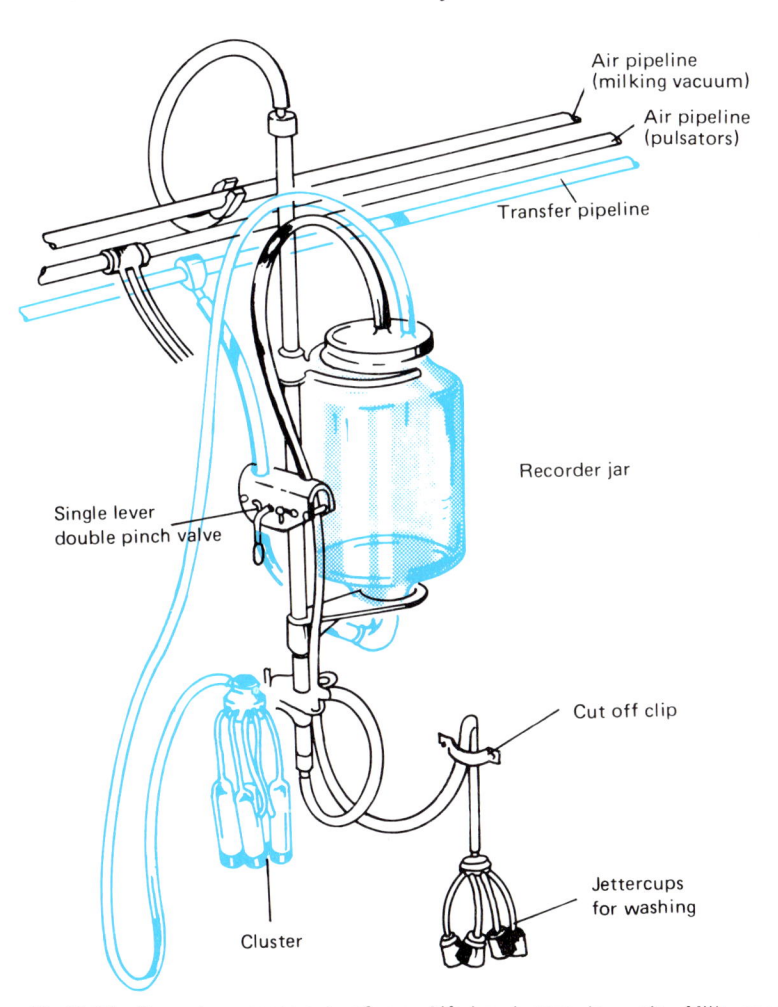

Fig III 23. General arrangement of an Alfa-Laval recorder unit. Milk contact components are shown in blue.

2. A combined change over transfer and vacuum cut off valve.

3. Provision for cutting off vacuum to the top of the jar, without operating the transfer valve, to enable a milk sample to be taken.

4. For sampling: (a) an air-bleed hole 0.8 mm (0.031 in) diameter for continuous mixing of milk in the jar throughout milking; (b) a sampling orifice 5 mm ($^3/_{16}$ in) diameter.

5. A port for rejecting milk not suitable for transfer to the milk room of at least 19 mm (¾ in) bore.

6. Automatic draining after washing.

7. Easily dismantled, suitable for heat disinfection, self cleaning as far as possible.

All these features were incorporated in an experimental valve designed at the NIRD. Fullwood and Bland Ltd. use a milk transfer control valve based on this design (Fig III 24). It incorporates a sliding plate used for milk sampling, air mixing and milk rejection, and has a single lever operated double pinch valve, mounted in a plastics box, to cut off the vacuum and milk transfer tubes.

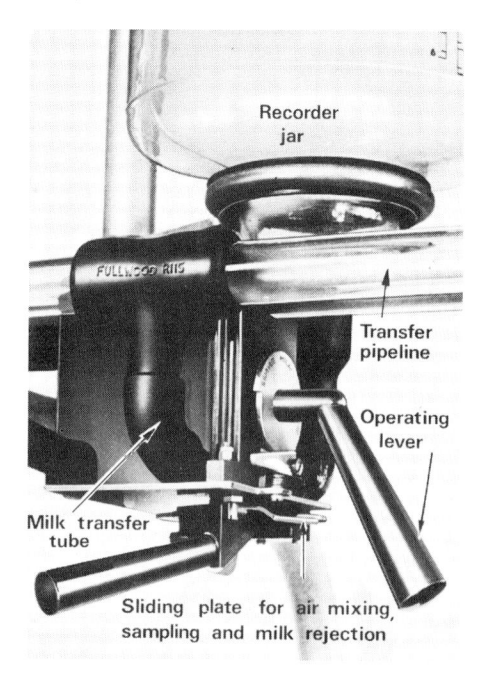

Fig III 24. Fullwood milk transfer control valve in a recorder milking machine with high level jars.

D.N. Akam

UNITS FOR MILKING PIPELINE MACHINES

The main additional items associated with milking direct to pipeline are milk flow indicators and milk meters.

Milk flow indicators

With all units, except those using transparent recorder jars, there is usually some provision for seeing when milk flow from an udder is coming to an end or has stopped. Often this is in the form of transparent components or inserts, such as glass transition pieces joining liners to short milk tubes, transparent covers on claw bowls and domes on milk inlet connections to bucket units, and short lengths of transparent tube inserted in the long milk tube. These do not usually give an instantly clear indication of the state of flow because of difficulties of lighting and obscuring of the transparent surface by a film of milk. Larger transparent areas are much more useful, for example transparent long milk tubes provided they stay transparent, and small interceptor vessels in the long milk tube (never widely used in the UK), while recorder jars are entirely satisfactory. Two sophisticated milk flow indicators successfully used in the UK, particularly with milking pipeline installations, are the Ruakura milk flow indicator from NZ and the Alfa-Laval Alfa-Flag indicator.

Ruakura milk flow indicator. As shown diagrammatically in Fig III 25, the device has three chambers, an indicating chamber at the front, a metering chamber, and an outlet chamber. Incoming milk falls into the central metering

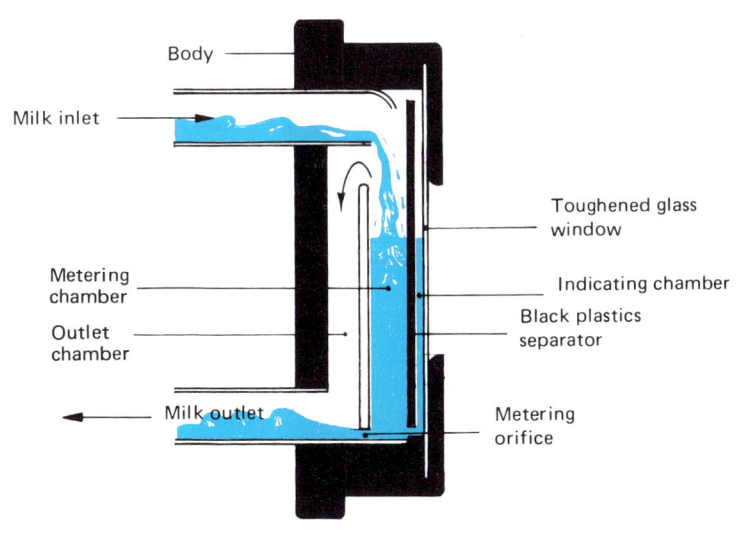

Fig III 25. Diagram of the Ruakura milk flow indicator, when the flowrate has declined enough to allow partial clearing of the window.

chamber and for the greater part of milking most of it overflows into the outlet chamber. The indicating chamber is connected top and bottom to the metering chamber so for the greater part of milking the window shows white. There is a metering orifice connecting the metering and outlet chambers at a low level, which discharges all the milk from the metering chamber once flowrate has fallen enough. At this stage the window begins to clear. When the head of milk in the metering chamber has fallen to a level at which the window is half cleared, the flowrate will be about 0.25 kg (0.5 lb)/min. Another design, the Milk-Rate indicator, is available which does not require to be fixed and is suitable for pipeline milking in cowsheds.

The Alfa-Flag milk flow indicator. The diagram in Fig III 26 shows that the indicator consists of a float and float chamber, the guided float lifting an arm which rotates a pointer on the outside of the chamber. The milk enters the float chamber near the top through the left hand tube and flows out through the top of the right hand tube as well as through three metering holes at different levels in that tube. The height of the milk in the chamber is proportional to the flow, once it falls below a designed rate, and is then indicated by the pointer on the outside of the case. The outside of the case has three calibration marks at flowrates of 0.6 kg (1.3 lb)/min or more, 0.4 kg (0.9 lb)/min and 0.2 kg (0.45 lb)/min or less. The manufacturers recommend that stripping be started when the pointer falls below the middle mark.

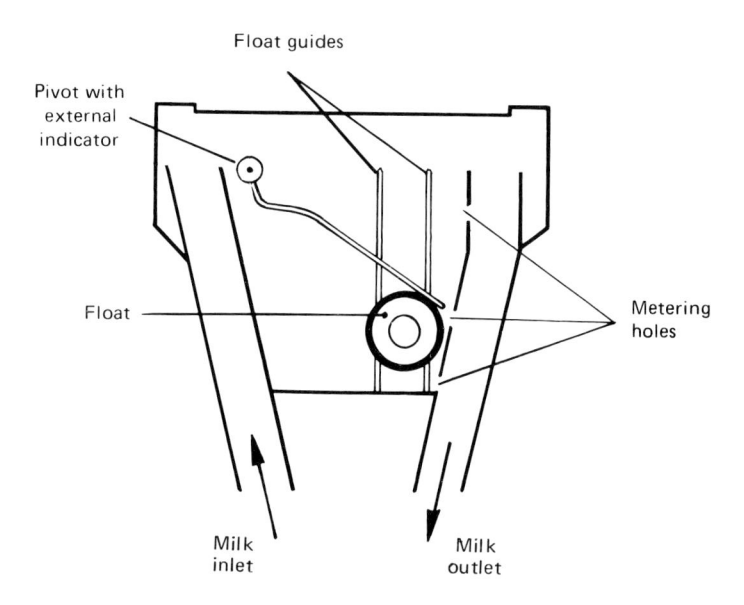

Fig III 26. Alfa-Flag milk flow indicator.

D.N. Akam

Milk metering

Milk meters are available to measure individual cow yields in the absence of recorder jars. A constant proportion (about 2.5%) of the throughput is divided from the main flow and collected in a measuring flask which is calibrated to indicate yield. Since the milk in the flask is a proportional sample it is suitable for chemical analysis. These meters are manufactured in rigid clear plastics and being light in weight are suitable for use with milking pipeline machines in cowsheds and by field technicians of milk recording schemes. Milk meters can also serve as end-of-flow indicators.

Four meters are approved by the UK Milk Marketing Boards for the National Milk Records Scheme. This means they satisfy the test requirements that nineteen out of twenty consecutive samples are within the range of ± 0.25 kg (0.5 lb) or 5% of the true value whichever is the greater, and for fat samples within 0.1 butterfat percentage units of the true value. The four meters are the Hartstone, Waikato, Milkoscope and Meterite.

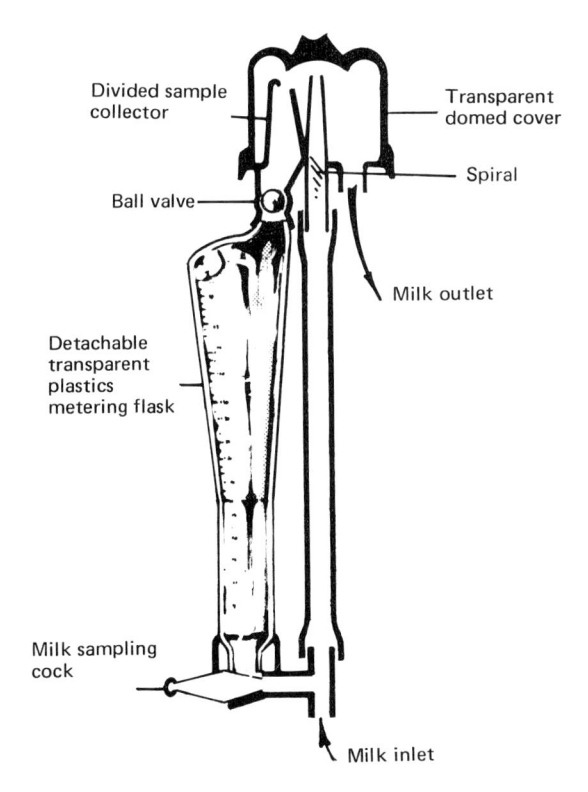

Fig III 27. Diagram of the Supamilk Waikato milk meter.

The Tru-Test meter marketed by Fullwood & Bland (Hartstone) and the Supamilk marketed by Alfa-Laval (Waikato) are similar in operation in that the milk enters the meter through the inlet tube and impinges on the underside of the domed cover, causing the milk to be deaerated and spread into an umbrella of liquid. In the Waikato meter a swirling motion is imparted to the milk by four spiral ribs on the inlet nozzle wall before striking the domed cover. A fixed angular proportion of this curtain of milk is intercepted by an orifice and so enters the measuring flask. The Waikato meter (Fig III 27) has a ball non-return valve above the measuring flask. This allows removal of the measuring flask without loss of vacuum. The remainder of the milk passes into a chamber and then out to the main milk line.

The Gascoigne Meterite meter is similar in action but requires a separate filter fitted to the inlet side. Milk enters the meter through the filter element and inlet tube ending in a shallow strongly divergent inlet cone and strikes a glass spreader plate placed 1 mm (0.040 in) above the rim of the cone. Milk is evenly distributed through the circular gap formed and a proportion of it enters the inlet aperture of the sample collecting chamber and then drains into the measuring tube. The remainder of the milk flows down to the outlet tube and then to the main milk line.

The Foss Milkoscope is different in operation from the other meters (Fig III 28). The incoming milk collects in the air separating chamber, air escaping through an air tube into the outlet chamber. A plastics ball, restrained to move in a vertical direction, floats up on the surface of the milk until it closes the mouth of the air tube. The milk is then drawn up the transfer tube through a restricting cone into the outlet chamber, a small proportion of it flowing through a metering orifice to the measuring tube. The air separating chamber is thus emptied and the pressure in the two chambers then equalize, releasing the ball from the mouth of the air tube so that the process is repeated. Thus, fixed quantities of comparatively air free milk flow under reproducible conditions through the metering orifice. The manufacturer sets the proportion of milk transferred to the measuring cylinder by means of a screw, the tapered point of which extends into the restricting cone. This meter and the Meterite require air admission holes, preferably in the claw.

Under field conditions, the meters differ in the care and attention required to maintain accuracy. For this reason meters are said to be more satisfactory when used by technicians of recording schemes than for general farm use. They are generally not successfully cleaned by circulation methods as practised on farms, and require careful handling to produce accurate results. Their effect on milking vacuum should be minimal and the requirements at present being considered by international committees are that the vacuum drop across a milk meter should not be more than 3.0 kPa (0.9 inHg) at a flowrate of 3 kg (6.6 lb)/min.

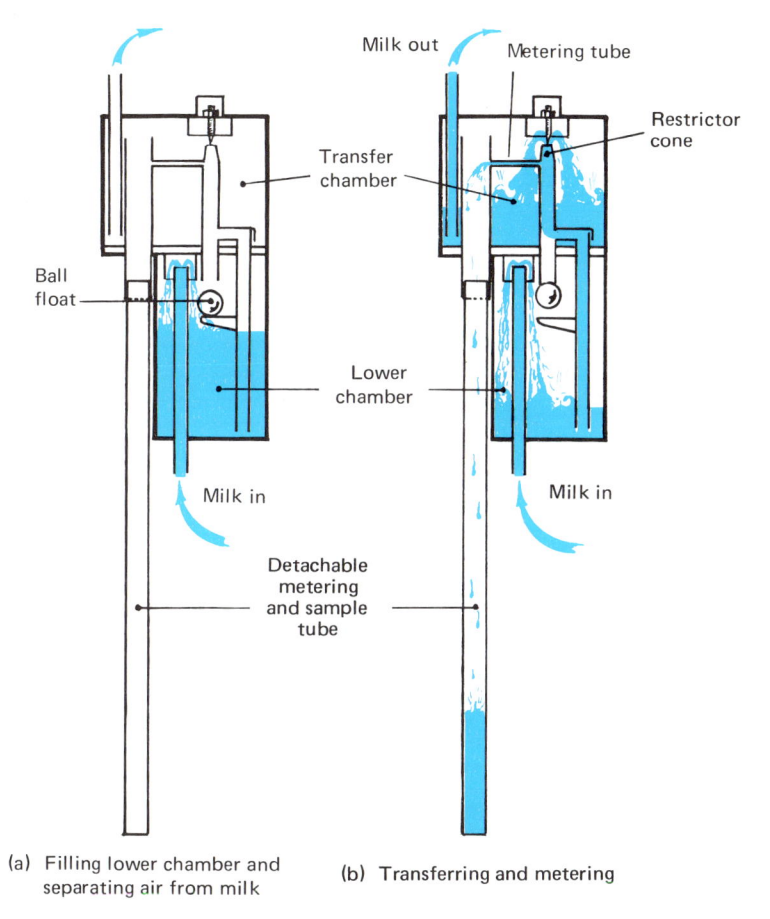

(a) Filling lower chamber and separating air from milk

(b) Transferring and metering

Fig III 28. Diagram of the Foss Milkoscope milk meter.

AUTOMATIC CLUSTER REMOVAL

Fully and semi-automatic cluster removing equipment is designed to remove the cluster from the cow at the end of milking. Most use a piston and cylinder for the power unit. The cylinder is connected to the vacuum supply of the milking machine by a valving arrangement, and the piston to the milking cluster by a nylon cord. The cluster removal sequence is initiated either by a lever mechanism or by an end of milk flow detector. Lever actuated cluster removers in fixed milking installations are usually operated manually and in rotary milking installations from a fixed position by the rotation of the platform.

Detection of end point of milking

Detectors of end point of milking are the same in operating principle as the milk flow indicators already described (Figs III 25 and 26). Milk from the long milk tube enters a small metering chamber which has a by-pass arrangement to accommodate the main flow of milking. A metering orifice will detect a predetermined end of milking flow rate either as the level or weight of milk in the metering chamber. The level or weight of milk is used to actuate the cluster removal equipment either by a magnetic float and reed switch operating an electric valve, a float operating a diaphragm valve, or the movement of a counter-balanced container operating an over-centre pinch valve. The last two systems have the advantage of being entirely mechanical. Milk flow detectors are arranged to activate the cluster removal equipment when the milk flow from the cow has been less than 0.25 kg (0.5 lb)/min for a period varying from 15 to 45 s. The delay is to avoid premature cluster removal in cases where flowrate is variable near the end of milking. To avoid premature cluster removal at the start of milking before milk flow is established, the detectors are usually made inoperative for the first 2 min of milking by built-in delays.

Detectors of end point of milking are vulnerable to blocking of the metering orifice. A self cleaning orifice has been devised by Alfa-Laval. A central tube shaped to fit a conical seat in the chamber outlet, has a notch in the bottom edge to form a metering orifice. The tube is flanged at the top and surrounded for most of its height by a sliding float (Fig III 29a). During the main flow of milking the central tube is lifted off its seat and the metering orifice is thus self cleaning (Fig III 29b). At the end of milking, when flowrate falls to a pre-determined level, a magnet in the float operates an external reed switch.

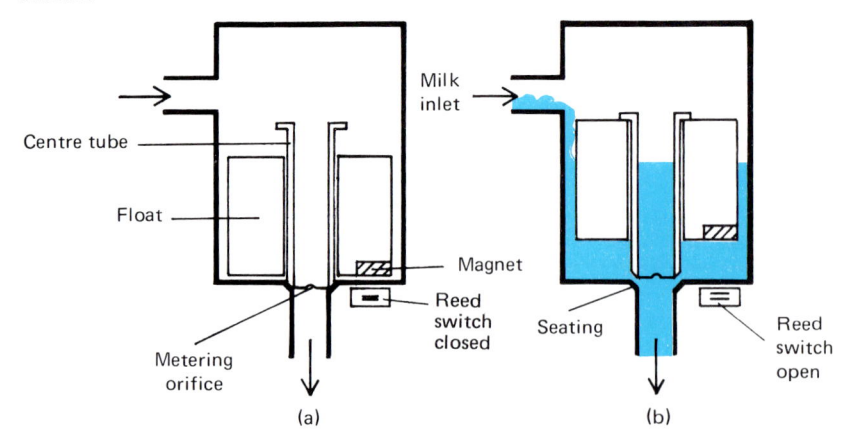

Fig III 29. Sectional drawing of Alfa-Laval end of milk flow detector: (a) no milk flowing; (b) milk flowing, central tube lifted off its seating.

D.N. Akam

Operating sequence

The operating sequence of cluster removal is as follows. Vacuum is applied to the cluster removal cylinder, causing the piston to move; the initial movement tensions the nylon cord, and closes off the milking vacuum to the teatcups by operating a valve either in the claw or long milk tube. When vacuum in the teatcups has declined sufficiently due to normal air admission at the claw, continued tension on the nylon cord pulls the cluster off the cow and swings it clear. A typical example of an automatic cluster remover is shown in Fig III 30.

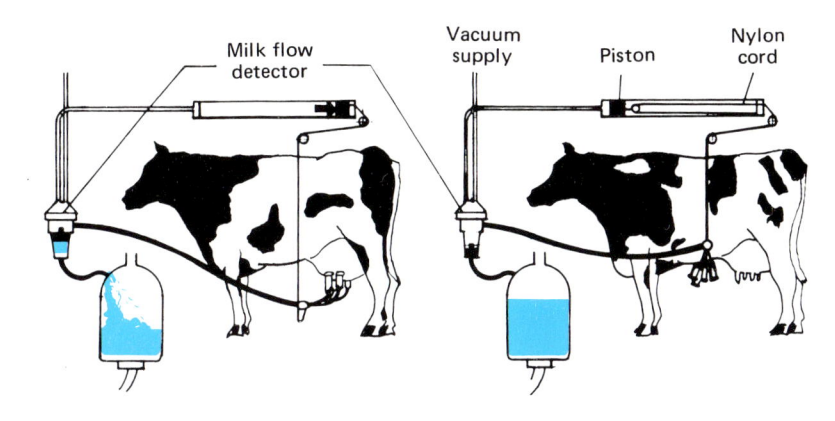

Milking. Flow detector float in raised position holds the vacuum valve to the cylinder closed.

Cluster removed. Vacuum from recorder jar shut off by the float and cluster returned to atmosphere by air entering admission hole. Vacuum valve to cylinder open and ram has withdrawn the cluster.

Fig III 30. Waikato automatic cluster remover.

Another cluster remover utilizes a hinged arm supporting the long milk tube, the initial movement of which folds or pinches the long milk tube cutting off the vacuum supply to the cluster. The continued movement of the arm pulls the cluster off and away from the cow. Another remover has an arm supporting the cluster under the cow during milking. When the vacuum to the cluster is closed off, the teatcups fall away but are held out of contact with the floor until a lever mechanism swings the cluster from under the cow.

All types of automatic cluster removers are suitable for milking pipeline machines where the milk is passed directly into the pipeline, but for recorder milking machines additional valving arrangements are necessary for milk transfer to admit air to the jar and open the connection to the transfer pipeline. To prevent large quantities of air being drawn into the machine after the recorder jar has emptied, arrangements such as a ball floating on the surface of the milk are used to seal off the milk outlet.

MILK TRANSPORT SYSTEMS

Transport of milk in the UK from the cowshed to the farm dairy for cooling and storage was for many years mainly by the milker carrying a bucket; this is still so today on many farms milking by hand or using bucket milking machines. The number of trips to the dairy carrying milk can be greatly reduced by tipping the milk from the milking bucket into 10 gal transport cans, mounted on a trolley. A more recent system used for transporting milk from a cowshed is to have a milking pipeline around the shed which has the dual function of supplying vacuum to the teat for milking and of transporting milk to the receiver under vacuum in the dairy. The cows are milked direct into this pipeline thus eliminating the use of buckets.

In milking parlours cows milked in small batches are sometimes milked direct into 10 gal transport cans but it is more common to milk direct into a milking pipeline or into a glass recorder vessel which is then discharged into a milk transfer pipeline under vacuum from the receiver. The steady growth of bulk collection has favoured the installation of pipelines for milk transport. This change will be intensified when bulk collection of refrigerated milk from farms becomes universal in the next few years.

MILKING PIPELINES

Cluster connections to milking pipelines in cowsheds

Several designs of quick acting connections to the milking pipelines in cowsheds are available. As well as due attention to the requirements of milk hygiene, one hand operation has been a desirable feature. In the most elaborate devices the connections of the long milk and long pulse tubes are combined into one fitting, which may also carry a pulsator.

Alfa-Laval RTS milk cock. This connects the long milk tube only, and is used in conjunction with the vacuum tap shown in Fig III 11. The cock is made of nylon and the pipeline fitting consists of a rotatable crescent shaped valve, shown closed in Fig III 31a. By inserting the mating milk nipple carrying the long milk tube into the square socket and rotating it through 90° the valve is opened making the connection to the milking cluster as shown in 31b.

Gascoigne Milkveyor valve. This also connects the long milk tube only, as shown in Fig III 32. It consists of a plate clamped to the side of the stainless

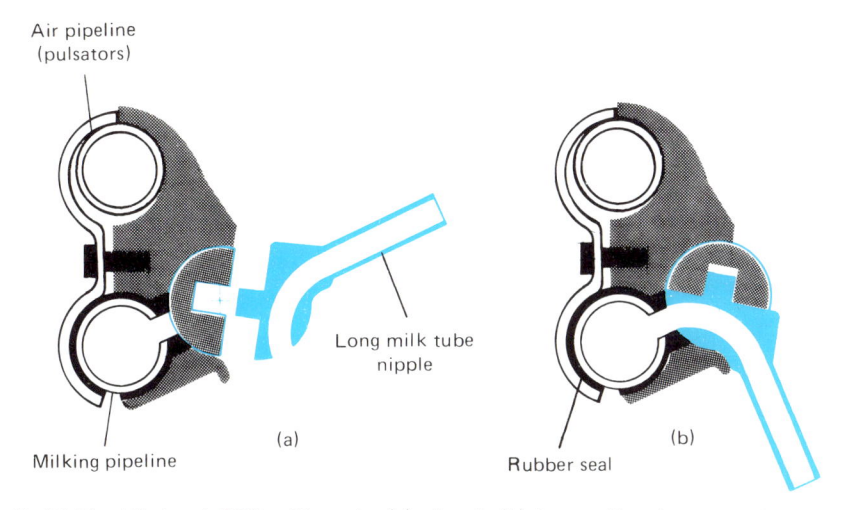

Fig III 31. Alfa-Laval RTS milk cock: (a) closed; (b) long milk tube connecting nipple attached.

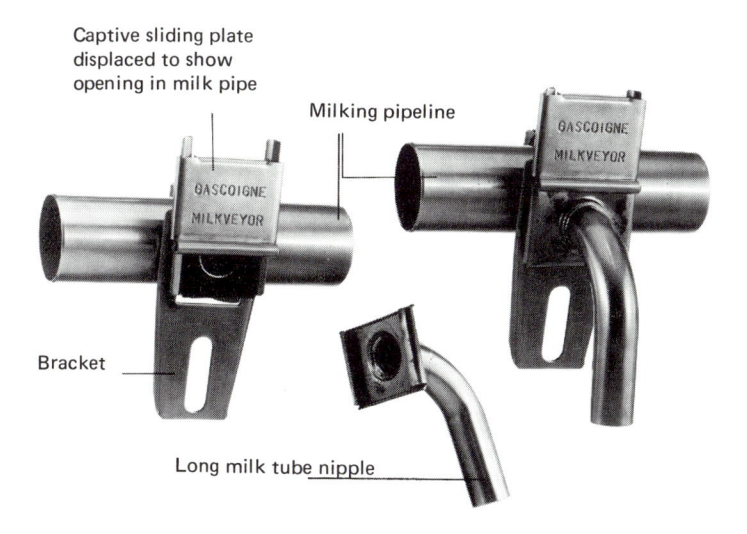

Fig III 32. Gascoigne Milkveyor valve.

steel milking pipeline with a rubber moulding to form a seal. A hole through all three items makes a permanent opening into the milking pipeline. The rubber also forms a seal for a captive sliding plate which closes the opening into the pipeline. Attached to the long milk tube is a similar plate with a hole in the centre connecting to the long milk tube. This detachable plate is hooked under the bottom edge of the captive plate and both slid upwards until the orifices in the milking pipeline and detachable plate are in register. Pulling down on the detachable part at the end of milking pulls the captive plate back over the hole, and at the end of the movement the long milk tube and detachable plate come free.

Fullwood Dari-Line milking point. The long milk tube and long pulse tube are connected simultaneously by a single plug and socket type fitting (Fig III 33). The socket consists of a moulded rubber connection. It has two ports, one giving access to the milking pipeline, the entry of which is covered by a rubber flap, and the other connecting to the output of an electromagnetic pulsator fixed on the pulsator air pipeline. In addition the rubber moulding houses a microswitch for energizing the relay pulsator. The plug has two nipples for the long milk and long pulse tubes, and a central locating prong. These are moulded in an H-shape with the prong extending beyond the nipples. When the prong is inserted into a hole in the centre of the pipeline socket, the milk nipple pushes up the rubber

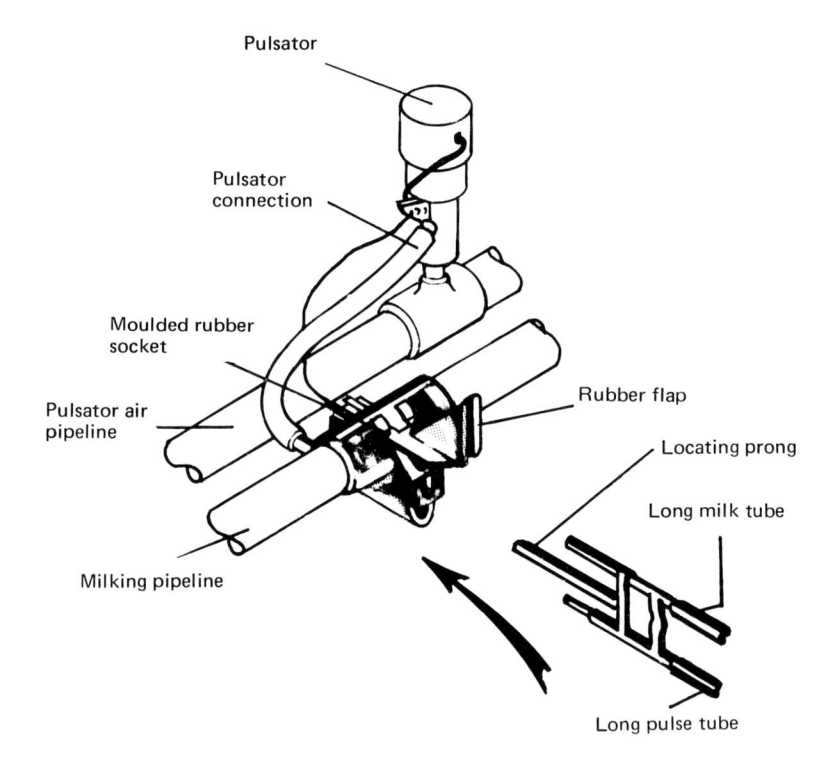

Fig III 33. Fullwood Dari-line milking point.

flap, and the pulsation and milk line connections are made at the same time as the centre prong actuates the microswitch energizing the pulsator. Self contained pulsators mounted on each of the plugs in use may replace the larger number of electric pulsators, one of which is required for each pair of stalls. This eliminates the microswitch and the pulsator air supply socket is then connected directly to the pulsator air pipeline.

Alfa-Laval Alfa-Matic S system

The Alfa-Matic S system is designed to provide a more stable vacuum at the claw when milking direct to a high level milking pipeline, for example in a cowshed. The milking pipeline is connected to a higher than normal vacuum source (74 kPa, 22 inHg), from which a stabilized vacuum for the cluster is derived by a stabilizing device. The pulsator air pipeline is operated at a vacuum level of 51 kPa (15 inHg), which also plays a part in stabilizing cluster vacuum. The connections to the stabilizer, the pulsator, and the cluster are shown in Fig III 34.

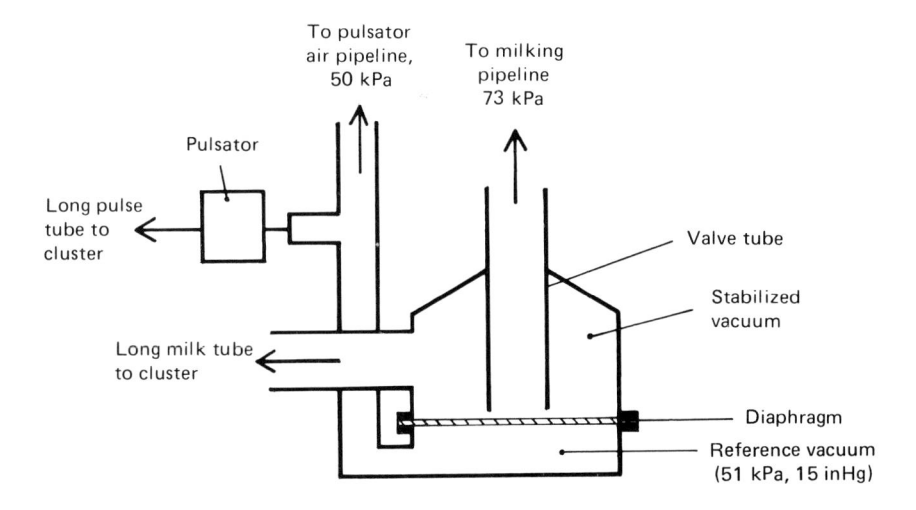

Fig III 34. Diagram of the stabilizer of the Alfa-Matic S system showing the pulsator air pipeline connections to the underside of the diaphragm and to the pulsator, the milking pipeline connection to the valve tube terminating above the diaphragm, and the long milk tube connection to the stablized vacuum chamber.

When vacuum in the upper chamber of the stabilizer falls, the diaphragm will be deflected towards the higher reference vacuum (lower pressure) in the lower chamber. This will uncover the end of the valve tube, allowing milk and air to pass to the milking pipeline operating at high vacuum. When vacuum is restored in the upper chamber, the diaphragm will move back to close the high vacuum valve port.

RELEASER MILK PUMPS AND RELEASERS

The receiver

In most British designs of milking machine, pipelines carrying milk terminate at a glass receiver. It acts as a liquid-air separator, and feeds the releaser. To extract milk from the receiver under vacuum in the parlour or the milk room a releaser milk pump is normally used. The inlet of the pump is connected to the bottom outlet of the receiver and either the level of milk in the receiver or the weight of the receiver and contents is utilized to operate the control circuit of the pump. The receiver should have a minimum capacity of 23 ℓ (5 gal). Most manufacturers in the UK supply 27 ℓ (6 gal) glass jars.

Centrifugal releaser milk pumps

The most common type of pump used is a high capacity electrically driven stainless steel centrifugal pump, with a pumping capacity of 6300 - 4500 ℓ (1400 - 1000 gal)/h at 0 and 3 m (0 and 10 ft) head respectively, extracting milk from a milking machine operating at 51 kPa (15 inHg). The pump is fitted with a simple blade impeller and a rubber non-return flap valve on the outlet (Fig III 35). The hygienic shaft seal consists of either a spring loaded carbon or ceramic faced seal and a rubber lip seal on the shaft, or a spring assisted double lip seal pressed into the back plate of the pump. All these pumps are easily dismantled for inspection and servicing, and are provided with a drain tap either in the pump body or in the outlet pipe.

Diaphragm releaser milk pumps

Mechanical. These pumps are more common in North America, Australia and New Zealand than in the UK. A large diameter (approx. 225 mm; 9 inches) rubber diaphragm is reciprocated by a connecting rod and crank shaft, usually driven by an electric motor by V-belt and pulley at speeds from 400 to 800 rev/min. The milk chamber, equipped with inlet and outlet valves, is on one side of the diaphragm (Fig III 36). The pumps are available with one or two chambers. The second chamber is horizontally opposed to the first and driven by its own connecting rod attached to the crank. Thus for one revolution of the crank shaft the pumping capacity of the pump is doubled.

When extracting milk from a receiver at a vacuum level of 51 kPa (15 inHg) the maximum pumping capacity, depending on strokes/min, is approximately 2000 ℓ (450 gal)/h for a single ended pump and twice as much

D.N. Akam

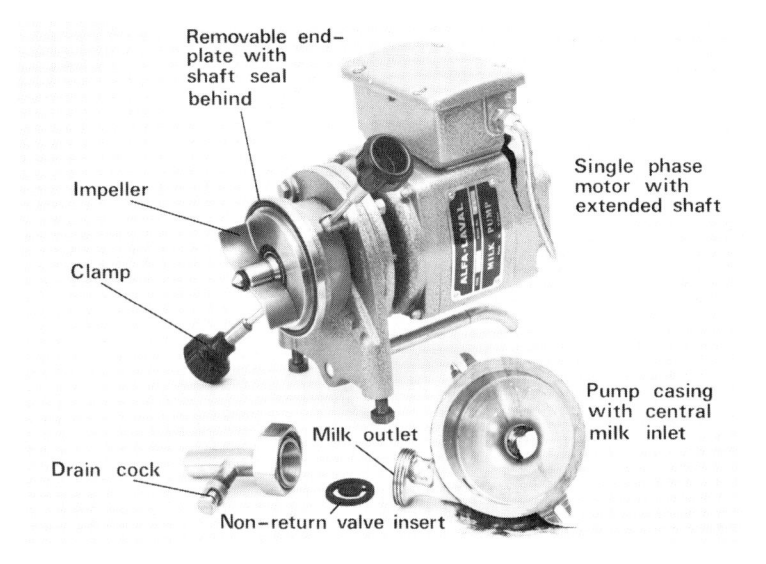

Fig III 35. Alfa-Laval milk pump.

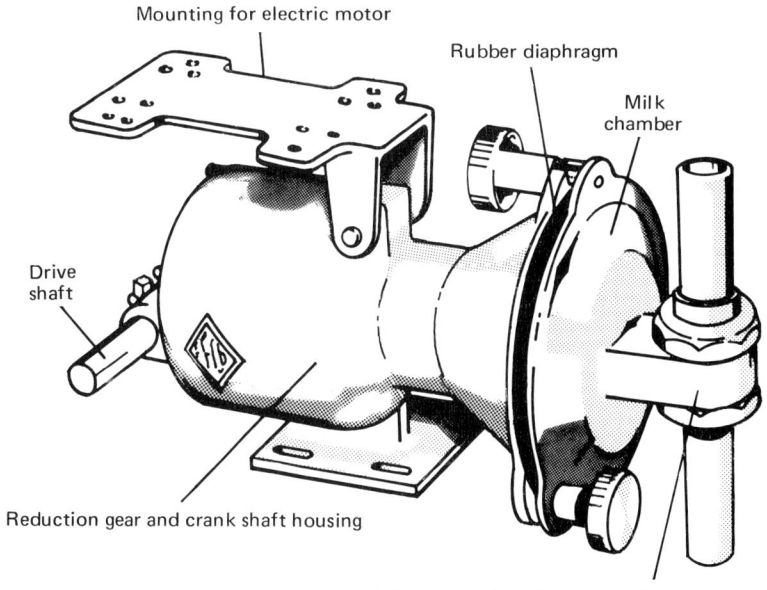

Fig III 36. Diagram of a Kingston single ended mechanical diaphragm milk pump.

for a double ended pump. Diaphragm pumps are run continuously throughout milking and therefore do not require liquid level controls.

Vacuum operated. A design of diaphragm releaser milk pump that is available in the UK is vacuum driven using a pulsator operating at 50 pulsations/min. It has an air demand of 140 ℓ (5 ft^3) free air/min, thus requiring a higher capacity vacuum pump compared with an equivalent installation using an electrically driven milk pump.

Fig III 37 shows a diagram of a vacuum driven diaphragm releaser milk pump which has three air chambers, A, B and C and a milk chamber. The milk chamber is separated from chamber C by a diaphragm and is connected to the milk system pipe work through inlet and outlet flap valves. Chambers A and B are separated by one large and one small diameter diaphragm with an air space in between. All three diaphragms are connected together by a spindle supported by two nylon bushes. Chambers A and C are connected to a pulsator.

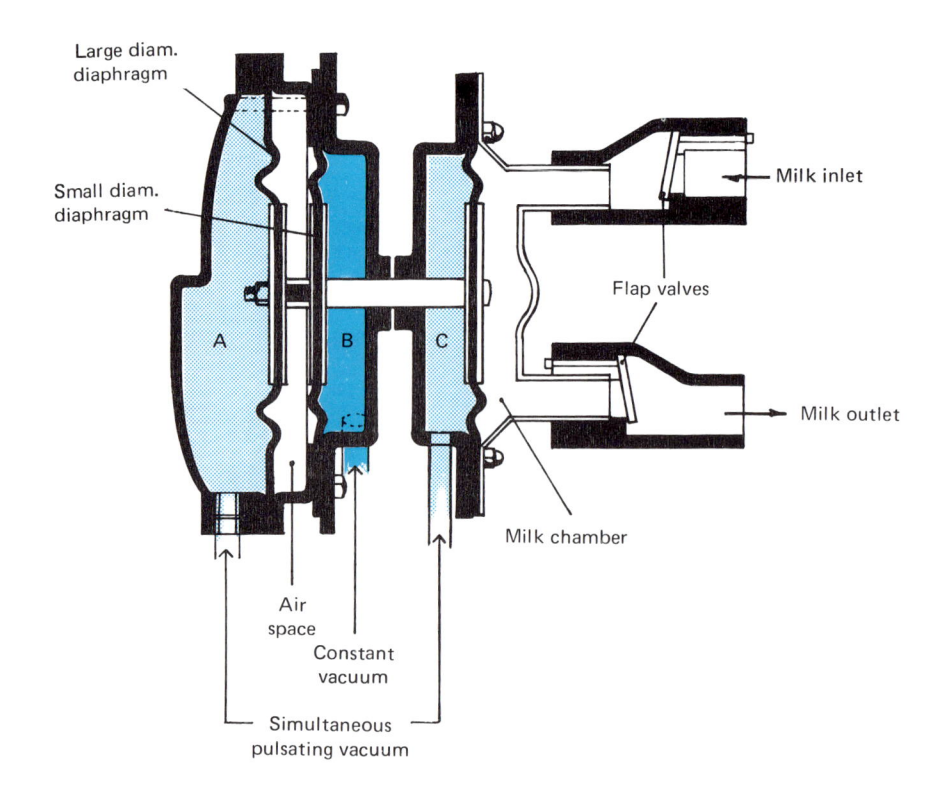

Fig III 37. Diagram of a Fullwood vacuum operated diaphragm milk pump.

When operating, chamber B is held at constant vacum. Simultaneously, chambers A and C are alternately at atmospheric pressure and vacuum. With chambers A and C at the same vacuum level as B, because of the greater effective area of the larger diaphragm the spindle will move towards A, thus reducing the pressure in the milk chamber. The outlet flap valve closes and milk pushes past the inlet valve. With chambers A and C at atmospheric pressure the spindle will move towards the milk chamber, increasing the pressure in this chamber, closing the inlet valve and pushing milk past the outlet valve.

The pumping capacity of this pump when extracting milk from the system at a vacuum level of 51 kPa (15 inHg) is 2000 and 1150 ℓ (450 and 250 gal)/h at a discharge head of 0 and 2 m (0 and 7 ft) respectively. With the limited capacity of these pumps, they are generally unsuitable for large milking installations although a special model is available which incorporates a double milk chamber.

A virtue of the pneumatic diaphragm pump is that in the event of electrical failure, no stand-by arrangements are needed. Also the pump suction can be 600 mm (2 ft) higher than the outlet of the receiver, which is sometimes useful.

Spit chamber releaser

This type of releaser is little used in the UK. A diagram of one of the two models available is given in Fig III 38, and its operating principle is described in the legend.

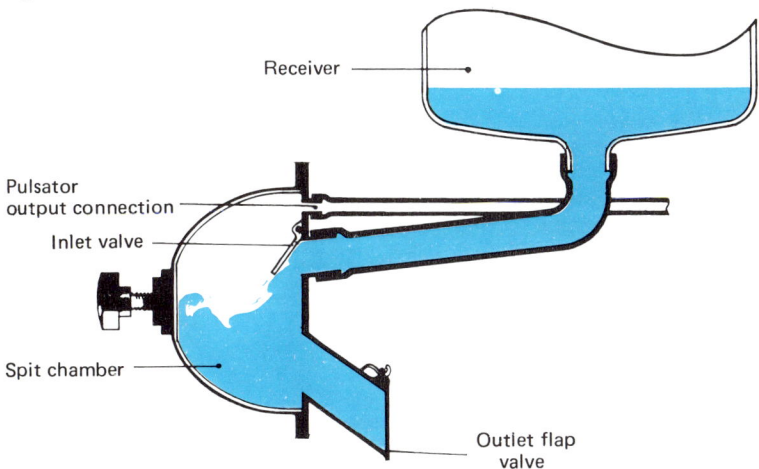

Fig III 38. Diagram of the Fullwood pulsator operated spit chamber releaser. With the receiver jar and spit chamber under vacuum as shown, the outlet flap is held closed and milk flows by gravity past the inlet valve into the spit chamber. With the spit chamber at atmospheric pressure the inlet valve closes and the milk flows by gravity past the outlet valve.

Depending on the size of the chambers and the frequency of operation these releasers have capacities of 900 - 2700 ℓ/h (200 - 600 gal/h). The requirements of a spit chamber releaser to achieve a fast throughput are good air separation in the upper chamber and rapid opening and closing of the inlet flap valve. A virtue of this type of releaser is that no additional stand-by arrangements are required other than for the vacuum supply.

LIQUID LEVEL CONTROLS

Centrifugal pumps are switched by liquid level control devices designed to start the milk pump when the milk in the receiver rises to a predetermined level and to switch it off when the receiver is nearly empty. In most cases the liquid level controls are fitted with over-riding facilities so that the pump can be run continuously for emptying the system of milk and for circulation cleaning. Receiver jar lids are fitted with spreader plates that can be swivelled into position under the milk inlet for cleaning purposes.

The reasons for not completely emptying the receiver are to keep the pump primed and to avoid excessive aeration of the milk. Three types of switching device are available: (1) electrical probes; (2) magnetic float switches; (3) weight operated switches. These are normally powered from the a.c. mains supply and, apart from two designs, the control circuits operates at 12 V d.c. This makes them suitable for operating from a 12 V battery to switch a stand-by 12 V d.c. series motor, powered by the same battery and belt coupled to the milk pump in the event of an interruption of the mains electricity supply.

Probe liquid level controls

One long and one short stainless steel rod pushed through holes in the rubber lid of a rigidly mounted glass receiver jar are connected to a 12 V d.c. electrical circuit. When the milk level in the jar reaches the short probe, a low current circuit is completed through the milk and pump casing. The signal is amplified to provide sufficient power to operate a relay to start the pump motor and complete a maintaining circuit which is not broken until the milk level is clear of the end of the long probe. A third probe is used by some manufacturers to provide a more reliable return path than the pump casing.

The Alfa-Laval two probe system is very similar in operation except that the control circuit is low voltage a.c. operating the milk pump through a titling mercury relay. To operate from a battery in an emergency, the leads from the probes are plugged into a separate additional d.c. control circuit.

Magnetic float switch

Two reed switches are housed in a stainless steel guide tube, which is fastened into the lid of the receiver and extends almost to the bottom of the vessel. The switches are operated by a ring magnet concentric with the guide tube and contained in a stainless steel ball float. One reed switch is actuated by the magnet in the float when the jar is approximately two-thirds full and the other when the jar is nearly empty. The milk pump is switched by a 12 V d.c. circuit and relay.

Weight operated switch

The receiver is connected to the milk and vacuum pipelines by flexible rubber tubes and is suspended from a spring loaded weight operated tilting mercury switch. The increasing quantity of milk in the jar compresses the spring and operates the pump switch. As the jar empties it moves upwards and switches off the pump before the jar is completely empty.

VACUUM-PRESSURE SYSTEM

A method of milk transport in a milking parlour with recorder jars, which does not require either a milk pump or a releaser, is used by Weycroft Macford in the Weymac V-P system. Air pressure at 60 kPa (0.6 bar or 10 lb/in^2) is used to move the milk.

As in a conventional installation, the recorder jar is under vacuum for milking. To transfer the milk to the dairy, the long milk tube from the cluster is closed off and the recorder jar is pressurized from a compressed air supply by the operation of a three-way pinch valve. The air pressure above the milk forces it out past a non-return valve at the bottom of the recorder jar into the milk transfer pipeline from which it is discharged direct to a bulk milk tank. It is claimed that the V-P system eliminates vacuum fluctuations when transferring milk and increases the rate of milk transfer. Also, much higher temperatures can be used for cleaning (Chapter X). Disadvantages are that non-return valves are not easily cleaned, pipe joints must be screwed or clamped, and it is necessary to carry clusters through to the milk room so that they may be attached to the outlet end of the cleaning circuit where the pressure is low.

STAND-BY EQUIPMENT

There is very little information on the duration and frequency of interruptions in the mains electricity supply to farms but in the UK in recent years industrial disputes have made them more frequent and of longer duration. It has been said that interruptions occur four times more often with farms than with the average consumer because of rural isolation, although all interruptions would not necessarily be at milking time.

Stand-by equipment can vary from a substitute supply at mains voltage from a diesel alternator costing several hundred pounds to a simple mechanical drive to the vacuum pump from a tractor costing one-fifth as much. It is difficult to justify a large capital expenditure on emergency equipment. However, past records of frequency and duration of mains failure, vulnerability of overhead supply lines to weather conditions, and the size of the herd will be the deciding factors for the individual farmer.

Ideal stand-by equipment is reliable and inexpensive, and requires little or no maintenance. It is usually required to provide an alternative drive to the vacuum pump, and sufficient electricity for some lighting, electric pulsation and the milk pump.

VACUUM SUPPLY

There are three recognized ways of ensuring adequate vacuum pumping capacity when there is an electric power failure.

Stationary engines

A petrol or diesel engine is positioned so that a belt drive to the vacuum pump pulley can easily be arranged. The standy-by engine requires adequate maintenance and periodic running to ensure reliability in an emergency.

Tractor PTO

The vacuum pump is driven from a tractor PTO by means of a special drive unit to obtain the correct vacuum pump speed as in Fig III 39. This is relatively cheap and requires very little maintenance, but good access for the tractor is necessary.

Tractor air intake

A diesel tractor is quite capable of pumping air from the milking machine. The volume of air taken into a tractor engine depends on the engine size and speed. A 4 ℓ engine running at 1200 rev/min will take in about 800 ℓ (28 ft^3) of free air/min with the air intake manifold vacuum at 51 kPa (15 inHg). The actual figure will depend on engine condition. NIRD experience suggests that the following method is practical and effective. Its advantages are that the tractor does not need to stand as near as with PTO drive to the vacuum pump, and the tractor is a reliable source of 12 V electricity for powering essential electrical equipment. It is also a cheap stand-by arrangement.

Connecting the tractor to the milking machine. Provision is made for connecting a strong convoluted hose (rubber or plastics) to the pipe joining the interceptor to the vacuum pump. A sensible arrangement is to have a flexible

D.N. Akam

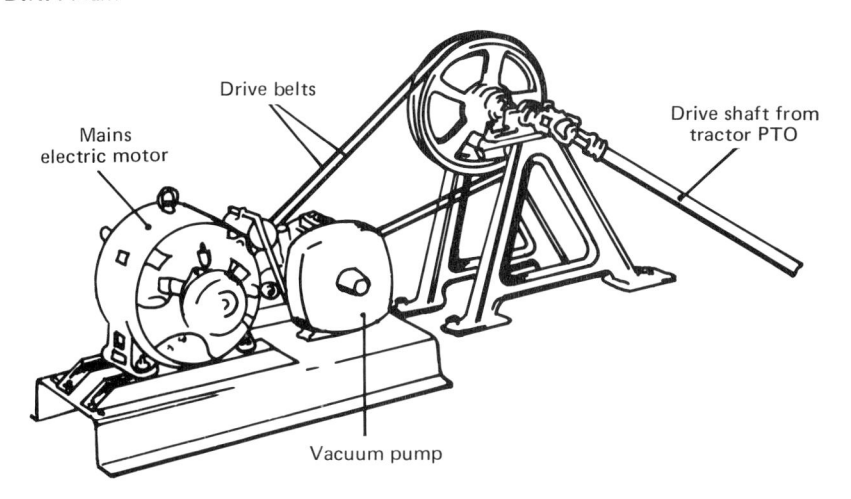

Fig III 39. Fullwood vacuum pump drive unit from PTO of tractor.

connection in this pipe which can be removed and the hose substituted. The other end of the hose is connected to the air intake pipe of the tractor by a snug fitting tee. The remaining branch of the tee is equipped with an adjustable throttle valve (a rubber milking machine butterfly valve is satisfactory) to supply the air demand of the engines in excess of the requirements of the milking machine (Fig III 40).

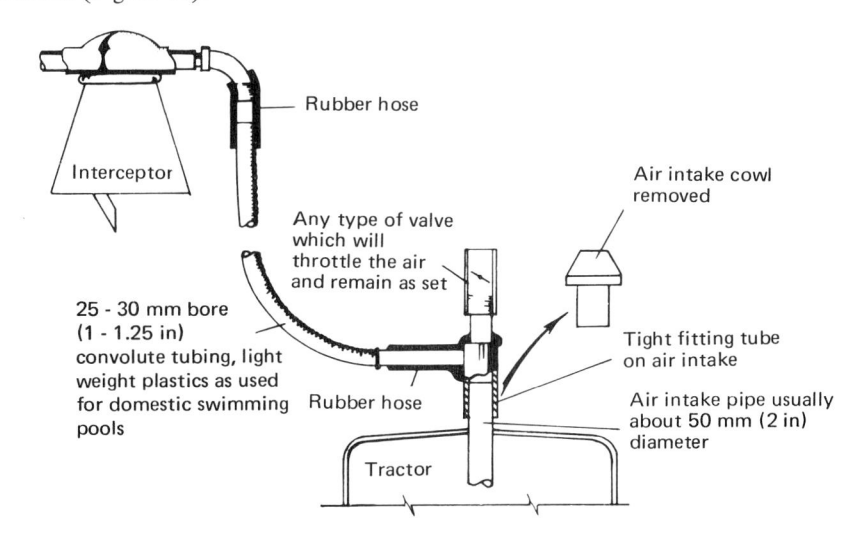

Fig III 40. Sketch showing flexible connection from air intake of a tractor to the vacuum system of a milking machine.

Engine settings. When first setting up the emergency system, warm the engine to operating temperature with the butterfly valve full open. Set engine speed to say 1200 rev/min. Close the butterfly valve progressively until, with all parts of the milking machine working, about the normal amount of reserve air appears to be entering through the regulator. If the exhaust shows black smoke there is insufficient combustion air. Reduce reserve air or increase engine speed and open the butterfly valve until the exhaust clears. Note these settings for the future.

Other points. It is very important that if an air filter on the tractor is removed it is replaced at the end of the operation. Modern tractor air filters are often difficult to remove and it may be more satisfactory to retain the filter in the sytem. Oil bath type air filters should be drained before operating – if not the oil should be renewed afterwards. Where any soft rubber elbows or connections are included in the air intake system these may be prevented from collapsing by inserting a thick wire spiral to give mechanical support to the rubber.

ELECTRICITY SUPPLY

An electricity supply is not essential for farms without electrically operated pulsation or milk pumps, for example bucket milking machines in cowsheds, lighting then being provided by portable lanterns. Many farms, however, will need electricity and requirements will vary from a small low voltage supply to a complete mains voltage stand-by supply.

Full a.c. electricity supply

A mains voltage 25 kVA alternator can be used to supply all the milking parlour electricity circuits through a change over switch which automatically isolates the mains supply. The alternator unit is driven by coupling it direct to a tractor PTO. With some designs the alternator can be mounted on the tractor by a three point linkage making it mobile for other uses on the farm. This system is the most expensive but for the large herd is probably the most satisfactory method (Fig III 41).

Selective a.c. electricity supply

A less expensive system is to use a mains voltage 5 kVA alternator to provide electricity to selectively wired circuits, for example lighting, pulsation, milk pump and feeders. The vacuum pump and alternator are driven by an extension shaft from the tractor PTO by belts and pulleys or a gear unit to obtain the correct speeds (Figs III 42 and 43).

Low voltage d.c. supply

This can be from a battery kept specifically for stand-by purposes or from the battery on the tractor that is being used to provide a source of vacuum, or from

D.N. Akam

Fig III 41. Froment mains voltage alternator driven by tractor providing full a.c. electrical supply.

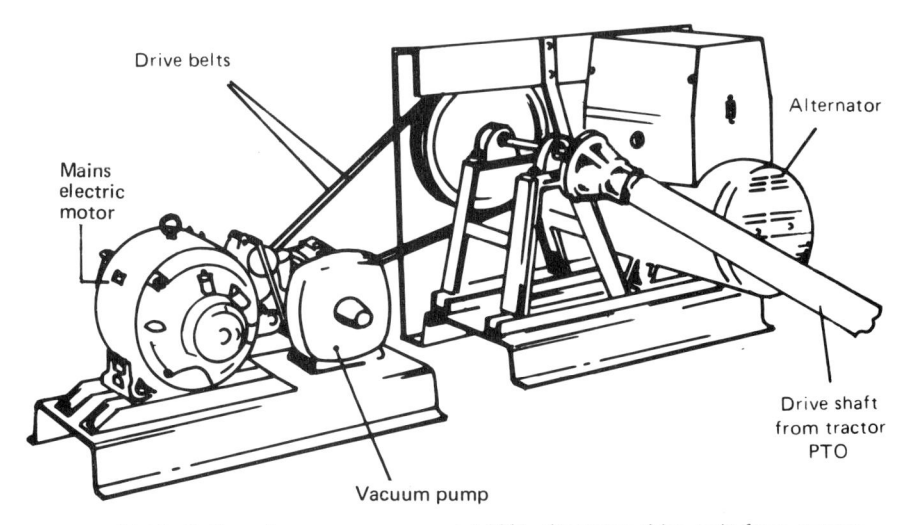

Fig III 42. Fullwood vacuum pump and 5 kVA alternator drive unit from tractor PTO.

a d.c. generator driven by the stand-by diesel engine. This low voltage supply is then used for lighting, operating pulsators and feeders, and if necessary to drive the milk pump. For the purpose of driving the milk pump the shaft of the a.c. mains motor on the milk pump is extended and fitted with a pulley. In the event of an interruption in the mains electricity supply, the milk pump motor is isolated from the mains supply and a belt fitted to drive the milk pump plus its a.c. motor from a 12 V d.c. motor mounted adjacent to the milk pump (Fig III 44). As the 12 V d.c. motor will have an electrical load of about 70 A when pumping milk, it is necessary to use heavy cable for the connections to the battery. The further away the battery is situated, the larger must be the cable size (Table III 2).

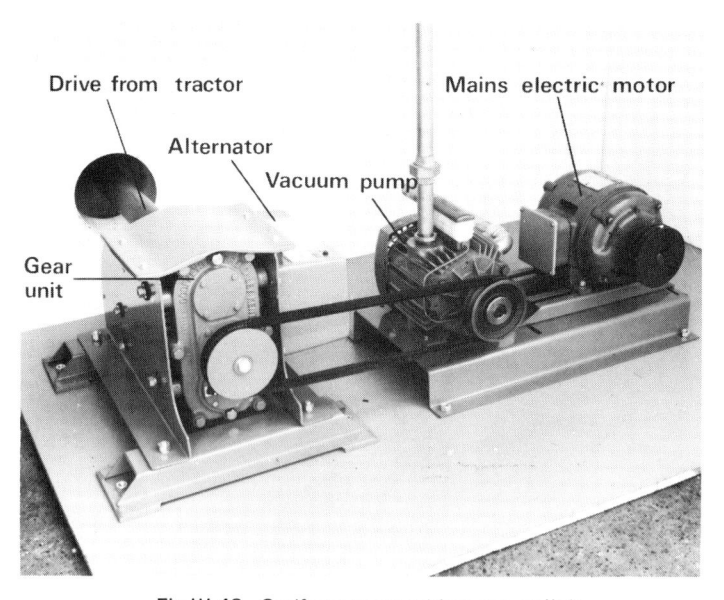

Fig III 43. Godfreys tractor driven power link.

COMPARATIVE EQUIPMENT COSTS OF STAND-BY SYSTEMS

Approximate costs of four different examples of stand-by systems are given in two forms in Table III 3. First, the costs are given as a percentage of the equipment costs of three types of milking parlour. Secondly, figures are given for the relative costs of the four systems.

D.N. Akam 91

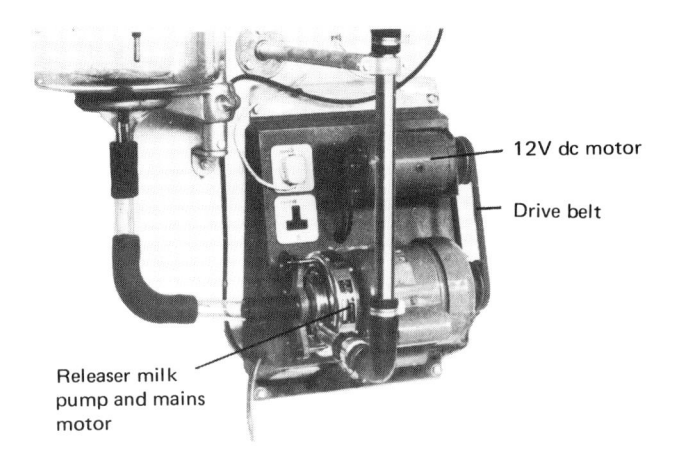

Fig III 44. Fullwood battery powered stand-by motor mounted adjacent to the releaser milk pump.

Table III.2 Electric cable sizes for connecting a 12 V d.c. milk pump stand-by motor to a battery on a tractor.
Cable specification: aluminium single core 7 strand, P.V.C. covered.

Distance to tractor*	Cross section area	Rating	Total length reqd.	Cost ratio
m	mm^2	A	m	
8	25	94	16	1.0
15	50	140	30	2.6
22	70	150	44	6.0

* These are maximum distances allowing for a voltage drop of 1.5 V for a 70 A load. Larger voltage drops than this will give a severe reduction in pumping speed.

Table III.3 Costs of stand-by equipment as a percentage of the total capital outlay for milking equipment including a bulk milk tank but excluding buildings.

MILKING INSTALLATIONS

Stand-by system	5/10 Herringbone %	20/20 Herringbone %	20 point Rotary abreast %
A	1.7	0.7	
B	2.5	1.0	
C	11.0	4.4	4.0
D		5.6	5.0

	Description of stand-by systems	Cost ratio
(A)	Using air intake of a tractor engine for vacuum. Fittings plus flexible connecting tube; 13 m run of 50 mm^2 aluminium cable i.e. 26 m total; auxiliary milk pump drive, panel mounted with starter relay and 12 V 70 A d.c. motor.	1
(B)	Vacuum pump drive unit, complete with tractor PTO drive shaft. Auxiliary milk pump drive, panel mounted with starter relay and 12 V 70 A d.c. motor.	1.5
(C)	Vacuum pump and alternator drive unit, complete with tractor PTO drive shaft, 5 kVA mains voltage alternator, and change-over switch for selectively wired circuits.	6
(D)	25 kVA mains voltage alternator, complete with tractor PTO drive shaft and mains change-over switch.	8

MISCELLANEOUS COMPONENTS

MILK FILTERS

The purpose of milk filters is to remove extraneous matter which is not a normal constituent of milk as drawn from the cow and which consists of hair, superficial tissues, earth, dung, debris from bedding materials and feedstuffs. Such material can enter the milk either because it has not been removed from the teats prior to applying the teatcups or because of contamination when teatcup clusters fall to the floor or the teatcups brush the floor when being applied to the udder. It is not possible to remove dirt particles smaller than about 70 µm diameter since finer filters will begin to remove the larger fat particles and become choked. Filtering of milk in relation to hygienic milk production is discussed in Chapter X.

Pad filters

With can collection and bucket milking, filtration is usually by means of filter pads placed on perforated discs in a hopper on the milk can or in a D-pan mounted above a surface cooler.

With bulk collection and bucket milking, the hopper filter is retained for use with the bulk tank which is usually provided with a hole in the cover to take a filter. Farmers, when changing over to pipeline milking, may continue to use the same filter with the delivery pipe discharging into it.

Sock filters

Flow rates when some form of pipeline machine is used have generally proved too great for pad filters. Farmers have either dispensed with filtration or used sock filters at the end of the delivery pipe discharging into the bulk tank (Fig III 45). These can be provided with a device to prevent the sock falling into the milk if it bursts or becomes detached from the milk delivery pipe. Several

Fig III 45. Blow sock type milk filter on the end of the delivery pipe discharging into a bulk tank.

manufacturers produce an in-line filter, which is a sock filter fitted in a holder placed in the delivery pipe from the milk pump (Fig III 46). The flow through these filters is generally outward, with the result that if the sock breaks the dirt already filtered out, and the sock itself, can enter the bulk tank.

A design in which the sock is clamped between a rubber tapered plug and the filter body appears to be more positive. Some manufacturers have developed inward flow filters, supported on wire frames. These have the advantage that if the sock splits it will remain in place in the filter body. Obviously the stitching, the weave and the material the socks are made of are important, or they will break when the back pressure increases as the sock gets choked with soil. Materials used are plain woven cotton, brushed cotton, and rayon of about 68 g/m^2 (2.0 oz/yd^2) weight, brushed cotton and thick rayon being the most successful. Most socks will filter 450 - 1350 ℓ(100 - 300 gal) of milk without excessive back pressure if the content of soil is moderate.

Fig III 46. Alfa-Laval sock filter fitted in a holder placed in the delivery pipe from the milk pump.

Multi-service in-line filters

A multi-service as opposed to a filter with a single service disposable element has been developed by the NIRD[1]. The overall costs should be lower because of the absence of a disposable element. It is effective as a filter, is free from any danger of pieces breaking away, and has elements which can be quickly changed during milking, particularly if the filter is sited in a milking parlour near to the inlet of the receiver. A diagrammatic vertical section of the filter is shown in Fig III 47. The milk enters through a tangential port in the casing, passes through two pierced nylon or stainless steel filter elements having aperture sizes of 120 and 70 μm respectively, and is discharged through the base. Two sets of filter elements are used alternately, one being soaked in 5% caustic soda solution between milkings and exchanged for the other at the end of milking before circulation cleaning of the milking machine.

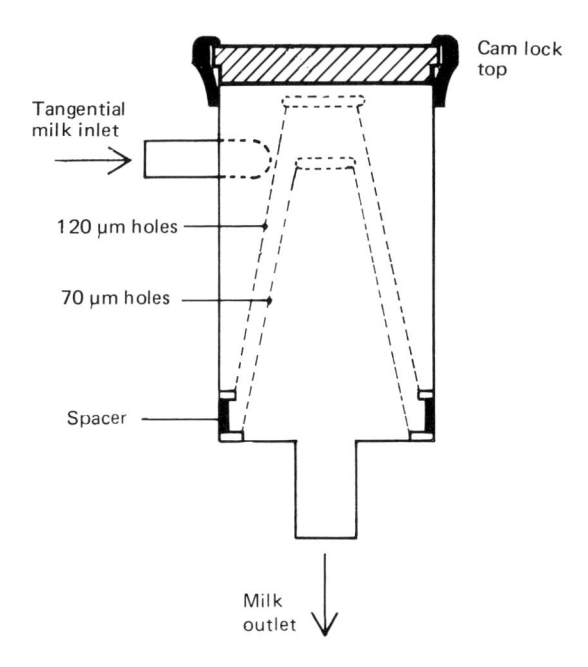

Fig III 47. Diagrammatic vertical section of a multi-service in-line milk filter showing the two layered (coarse and fine) inward flow filter element (nylon or stainless steel).

Filter for detecting clinical mastitis

An in-line filter developed by NIRD[2], that is fitted into the long milk tube of each cluster, retains milk clots enabling clinical mastitis to be readily detected and also acts as a coarse filter for removing extraneous material (Fig III 48).

Filter to protect milk meters

When milk is metered from refrigerated farm bulk milk tanks into road tankers a coarse filter is fitted in the suction line to protect the meter. This filter will remove a proportion of the dirt in milk, depending on its fineness. The universal use of suitable filters in suction hoses could be used to monitor the hygienic quality of production methods − but only if the use of filters on farms were disallowed.

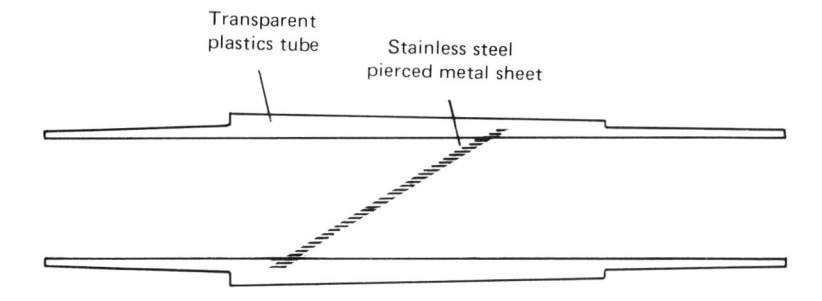

Fig III 48. Diagrammatic section of an in-line filter designed by NIRD for fitting in the long milk tube to enable clinical mastitis to be detected by the retained clots (see Chapter IX).

ROTARY MILKING INSTALLATIONS

Rotary milking parlours have the same basic equipment as any other type of parlour milking installation. Items that are peculiar to them are directly attributed to the fact that the milking units and stalls are mounted on a circular raised rotating platform. Rotary joints are therefore necessary in the vacuum and milk pipelines to the dairy, as are electrical slip rings if electricity is required on the moving platform for items such as pulsators, automatic cluster removers and milk pumps.

In addition, installations with recorder jars require milk transfer valves that can be operated automatically. A pinch valve mechanism is commonly used (see section on the milking unit). They are either mechanically or vacuum operated at a fixed position relative to the platform movement or linked with automatic cluster removing equipment and end of milk flow detectors.

WATER HEATERS

Equipment used for providing hot water varies from a simple free-standing insulated or uninsulated tank which is hose filled and has a manually operated electric immersion heater, to an insulated tank automatically filled and heated, permanently connected to the mains water supply and the milking machine[3]. Water heaters are expected to operate with the minimum of servicing and to be economic. The quantity of water and temperature required will be decided by the cleaning system chosen and the type of milking installation, for example, hand washing a few bucket units or the in-place cleaning of a pipeline machine having many units. Details of the method of cleaning and disinfection are given in Chapter X.

The choice of method of heating water depends on cost and convenience. Electric heating has been the most popular as the equipment is cheap to install and convenience of operation has not been outweighed by running cost for the comparatively small amounts of hot water required. The main alternatives of oil and liquefied gas heating involve higher capital, are less convenient and require more maintenance. However, comparative costs are now unfavourable to electricity and alternative methods of heating may be expected to become more important in the future. Moreover, all methods of heating are now more expensive and thoughts are also turning again to waste heat recovery from the condensing units of refrigerated farm bulk milk tanks (as described in Chapter XII) and from the vacuum pump exhaust air.

The temperature to which water needs to be heated for various purposes in milk production may be summarized as follows: for udder washing, 40°C (105°F); for hand washing of milk equipment, 45°C (115°F); for recirculation cleaning combined with chemical disinfection, 85°C (185°F); and for acidified boiling water cleaning and disinfection, over 96°C (over 205°F). At the two higher temperatures the equipment used is specially adapted in respect to deposition of scale.

Types of water heater

Free standing. A simple insulated or uninsulated free standing open top tank of 55 - 90 ℓ (12 - 20 gal) capacity (a domestic wash boiler) will fill the needs of small herds using 3 or 4 bucket milking machine units. The boiler is usually positioned under a cold water tap or filled by hose, and electrically heated either with manual or thermostatic control after initial switching on. The equipment is cheap to purchase and instal.

Permanently connected water heaters. For the larger farm using bucket or direct-to-can milking equipment which is hand washed, an insulated storage heater of 100 - 350 ℓ (23 - 80 gal) capacity is suitable. A commonly used type has its own built in cold water supply cistern fitted with a ball valve. The hot water tank is, however, refilled by a hand operated valve (Fig III 49). This is turned off at the beginning of washing to prevent entry of cold water so that all the hot water can be drawn at the set temperature. The storage vessel has a vent pipe at the top which terminates above the water level in the cistern to enable air movement in and out on emptying and filling. To reduce heat losses and, possibly, deposition from hard water these water heaters are fitted with two thermostats, one set at 60°C (140°F) and the other at 82°C (180°F). The lower setting is for controlling the water temperature between milkings, and the higher setting thermostat is switched in at the beginning of milking to raise the water to the required temperature in time for washing. After washing and refilling the heater is then returned to the lower temperature setting.

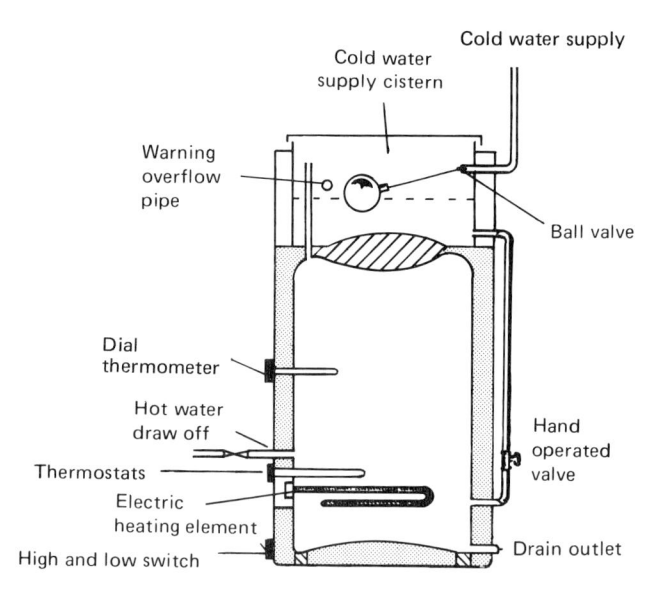

Fig III 49. Diagram of an electric water heater fitted with its own built in cold water supply cistern. The hot water tank is refilled by opening the hand operated valve.

For milking pipeline and recorder milking machines which are cleaned in place at high water temperatures, the water heater is permanently connected to a cold water supply but in addition is permanently connected to the milking machine, a precaution which is necessary for the safe handling of very hot water. It is usual to specify that these heaters should be capable of supplying hot water up to 100°C (212°F), and are provided with a 38 mm (1.5 in) bore drainage outlet, a draw off pipe of not less than 25 mm (1 in) bore with a valve to control hot water discharge, and an overflow pipe greater in diameter than the cold water inlet. When used for near boiling water temperatures the water heater should ideally be of the open top type provided with a loose fitting lid making easy access for descaling. A vent pipe of at least 32 mm (1.25 in) bore, fitted above maximum filling level, is satisfactory for venting vapour to the outside of the building. An example of an open top electric water heater for in place cleaning is shown in Fig III 50. Chapter X on cleaning and disinfection shows how these water heaters are connected into the cleaning systems.

Heaters providing water for udder washing are usually small in volume, enclosed, and thermostatically controlled. They have high heating capacity, for example 4 kW which is sufficient for about 90 cows/h at 1 ℓ (0.25 gal)/cow. These heaters are best sited centrally in the milking area and permanently piped to washing hose outlets.

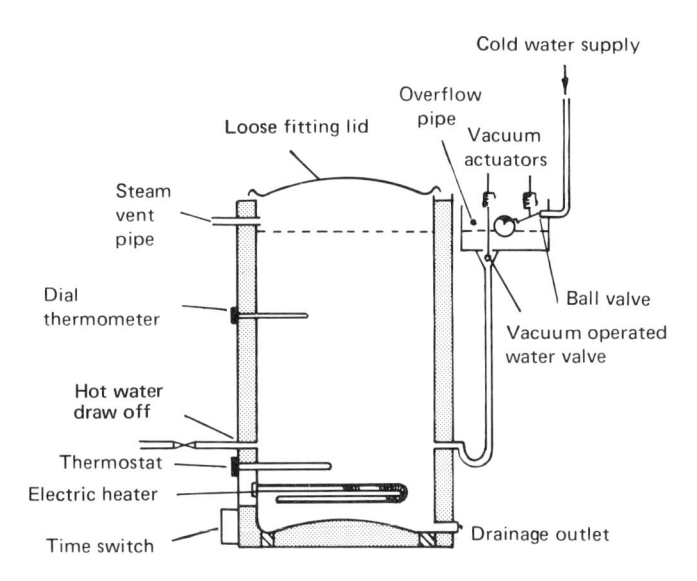

Fig III 50. Diagram of an open top electric water heater suitable for in-place cleaning at near boiling water temperatures. Two positions of a vacuum actuator are shown to close a water valve or, alternatively, to hold the ball valve closed until the vacuum pump is switched off.

Electric heating

The heating elements are usually 3 or 6 kW depending on the quantity of water to be heated. They are sheathed in an alloy of copper and nickel and have a larger surface area than the normal domestic type of element, thus giving longer life. The elements are installed some distance above the bottom of the heater to allow sufficient space for accumulation of scale. The hot water draw off pipe is positioned above the heating elements so that the elements are always covered with water.

Oil heating

A pressure jet burner complying with BS 799[4] may be used to heat water in an insulated storage tank, either indirectly by pumping the boiler water through a heat exchanger in the tank or directly by playing the oil flame onto the hot water vessel. Direct heating is more common. Oil burners have a higher heat output than can be conveniently got from electric elements so that rates of heating are higher.

An example of a suitable oil fired water heater is shown in Fig III 51. Oil fired water heaters may now be sited in a room in which milk is handled or stored provided that the oil supply, air for combustion and products of combustion are isolated from that room.

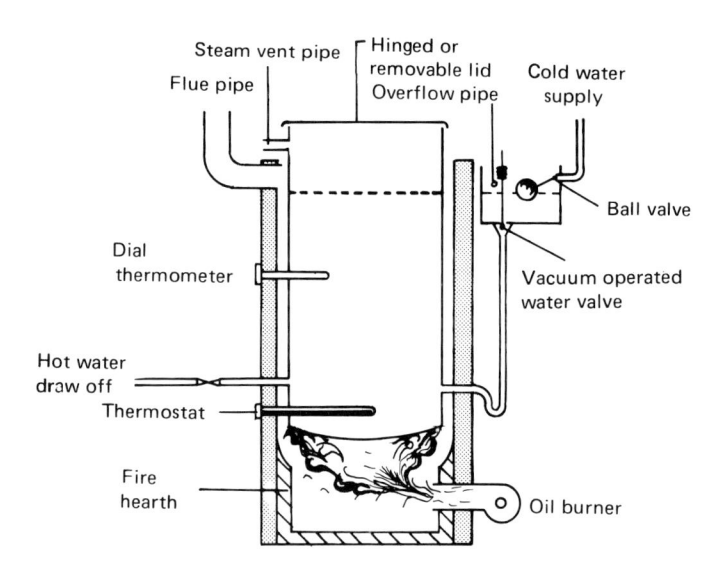

Fig III 51. A diagram of an open top oil fired water heater with the flame playing onto the underside of the hot water vessel.

Electrical controls

Although many water heaters are manually switched, time switches ensure that the electric heater or oil burner is switched on at a time that allows for the water to be heated ready for washing. Thermostats are used to control water temperatures up to 82°C (180°F) but for systems using near boiling water temperatures, the common bi-metal domestic type thermostats are not very suitable as they have a wide switching differential. In these instances a time switch only may be used to give a heating period just long enough for the water to be heated to near boiling in time for washing. Many installations however are fitted with both a thermostat and a time switch. The thermostat used in this instance will be the more expensive vapour pressure type which has a very narrow switching differential of approximately 1°C (1.5°F).

Water controls

Water heaters are either filled by a hand operated valve or arranged to fill automatically when the milking machine vacuum pump is switched off. The object is to prevent cold water entering the tank during the cleaning process thus lowering the water temperature and reducing the efficiency of cleaning. The automatic filling system usually consists of a diaphragm operated shut off cold water valve, held closed until cleaning is completed by the vacuum of the milking machine (2 examples are shown in Fig III 50). With a hand operated valve it is usual to arrange for the overflow pipe to discharge in a conspicuous position as a reminder to turn off the cold water supply.

Self contained cleaning units

Cleaning units are available in the UK programmed to circulation clean a pipeline milking installation. They rely on a good hot water supply. The usual cleaning cycle is as follows: (a) a cold water rinse discharged direct to waste; (b) a hot detergent wash which is recirculated for some time (15 min) before discharging to waste; (c) a final chlorinated cold rinse discharged direct to waste.

These automatic cleaning units are effective. Their advantages for cleaning methods involving recirculation are that they relieve the operator of the task of changing from one stage to the next and ensure reproducible cleaning routines.

REFERENCES

(1) Slade, J.R. & Hoyle, J.B. (1974) *International Dairy Congress 19, New Dehli,* 1E, 5.

(2) Hoyle, J.B. & Dodd, F.H. (1970) *Journal of Dairy Research,* 37, 133.

(3) *British Standards Institution* (1975) BS 5226.

(4) *British Standards Institution* (1962) BS 799.

Chapter IV

MAINTENANCE AND MECHANICAL TESTING

H S Hall

MAINTENANCE

Milking machines, like all mechanical equipment, require maintenance to ensure that they continue to function correctly. The manufacturer normally provides detailed information by means of wall charts and handbooks covering the routine mechanical maintenance which is necessary for efficient operation and to reduce the risk of breakdown. The user, in his own interest, should follow these instructions meticulously.

The most important element of effective maintenance is the routine observation and checking of components in the course of use and cleaning. An experienced operator can carry out nearly all of the necessary checks during his normal work without taking extra time. These checks, together with the periodic maintenance specified by the manufacturer, will almost eliminate unexpected failures.

Checks at every milking
1. When the vacuum pump is started observe the vacuum gauge. Working vacuum should be reached in a few seconds. If there is delay, check for air leakage at open stall or drain cocks, misplaced recorder jar or receiver lids and, where a centrifugal milk pump is used, for failure of the non-return valve. If all these items are in order belt slip may be suspected.
2. The working vacuum level should be constant from day to day and as soon as it is reached the sound of air entering the regulator should be obvious. If the vacuum level rises first to a high value and then drops to the working level the regulator needs cleaning.

3. The sound of the pulsators should be regular. Any abnormality must be investigated at least before the next milking.

4. When the vacuum pump is finally switched off after completion of cleaning check that all automatic drain valves in the pipelines, sanitary trap and interceptor have opened and residual water is draining freely. Where no automatic drain is fitted, as in the case of many interceptors, the container must be removed, emptied and replaced. The presence of milk in the interceptor may indicate a split liner.

Other checks at milking time

Certain other checks should be made, not necessarily at every milking but at least several times per week.

5. In handling each cluster look for damage to the rubberware, particularly the short milk and air tubes at the claw nipples. Check that air admission holes are clear: this may be recognized by sound or by applying a finger tip to the hole. If in doubt clear the hole with a stove pricker or other suitable means.

6. Listen for abnormal air leakages particularly at milk pipeline joints and gaskets. Often these can be rectified immediately but if they are not great enough to interfere with the milking in progress investigate the cause before the next milking.

7. When all units are in operation listen at the vacuum regulator. If there is no sound of air entering the regulator a serious shortage of pump capacity is indicated and should be investigated without delay.

8. At the beginning of milking note the precise reading of the vacuum gauge when only the first unit has been put on and compare this with the reading when all units are working. The two values should not differ by more than 2 kPa (0.6 inHg, 15 mmHg) which will only just be noticeable on most vacuum gauges. If the difference is greater than this the regulator is not functioning correctly.

The checks briefly described above must be fitted into the normal routine to suit the operator's convenience and the best sequence depends to some extent on the type of installation. For instance, it is usual with a pipeline machine using jetters to leave the teatcups on the jetters between milkings. The starting-up procedure could then be:

> start vacuum pump
> check rise of vacuum
> check that air is entering regulator
> before removing each unit in turn from the jetters, check for rubber damage, air admission and presence of pulsation by lightly gripping the short pulse tube between thumb and finger. The vacuum levels with all units and with only one unit connected may also be compared.

PERIODIC MAINTENANCE

Periodic maintenance is usually required each week and the necessary time should be allocated on the same day of the week to establish a routine which is easily remembered. This weekly service is best carried out after the morning cleaning operation but any time other than milking is suitable.

The following items should receive attention before the machine is started.

1. Check the vacuum pump oil reservoir and refill to the correct level with the grade of oil specified by the manufacturer.

2. Check the belt tension. With vee belts deflect the belt downwards at a point midway between the pulleys. The movement possible from the rest position should not exceed about 15 mm. If adjustment is needed care should be taken to keep the pulleys in alignment.

3. Dismantle the regulator and clean the valve and its seating and the air inlet screen if one is fitted. Detergent solution and a small stiff brush are useful aids. Metal polish can be used for brass parts.

4. Clean all pulsator air inlets and claw air admission holes using suitable brushes and a pricker if appropriate.

The machine should then be started and the various brief checks described above repeated more carefully. In the course of this the teatcup liners should be examined for damage and maladjustment. The effectiveness of pulsation can be judged by inserting the thumbs into diagonal pairs of teatcups in turn, the other pair hanging in the cut-off position, the controls being set for milking. To make cut-off more certain the pair of teatcups not being checked can be closed with bungs or left in the jetters if they are fitted. Where pinch cocks are used the rubber tubes should be examined for permanent set.

These and various other points of detail specific to the make of machine will be included in the manufacturer's instructions. The user should work out a logical and convenient programme for the weekly maintenance work to reduce the risk of individual items being left unchecked.

MECHANICAL TESTS

The objective in machine milking is to extract milk from the animal efficiently, with the least harm to the animal or to the milk. It is obvious that perfection in meeting this objective cannot be expressed quantitatively nor can the efficiency of a milking installation be expressed in simple numerical terms.

However, standards based on practical experience can be set for the performance of various components and for some of the characteristics of the installation as a whole. This leads to the necessity for mechanical tests in which measurements can be made to compare the actual performance of the component concerned, or the installation, with whatever standard is regarded as acceptable. Such standards have been formulated in a number of countries and an international standard will soon be available. In the UK, British Standard Code of Practice CP 3007 specifies details of construction and performance but this will be superseded by the international standard when it is published. The Ministry of Agriculture, the Milking Marketing Boards and the Milking Machine Manufacturers' Association use an agreed method of test for their advisory and service work.

Mechanical tests have three distinct applications:

1. by the manufacturer, in the development of new designs and in quality control of production;
2. by the manufacturer, in commissioning new installations before the completion of purchase;
3. by a servicing agency (which might be the manufacturer) or advisory body in the course of periodic overhauls or trouble shooting.

If follows that tests under (2) and (3) must be done on site and so are normally limited in scope as they must be conducted under farm, rather than laboratory, conditions. Further, to simplify the procedure they are normally done without liquid flowing through the system. This helps to make the results more reproducible although the effects of milk flow on vacuum stability must be recognized (see Chapter V).

Apart from measuring the speed of the vacuum pump (and of the prime mover if it is desired to calculate belt slip) almost all field tests involve only the measurement of vacuum levels and air flowrates.

THE MEASUREMENT OF VACUUM LEVEL

All milking machine installations are fitted with at least one vacuum gauge as described in Chapter III. While this type of instrument is satisfactory for the routine indication of vacuum level, periodic checking is necessary. Similarly gauges or recorders which are used to measure vacuum level at various points in the installation or in the laboratory in the course of testing must be calibrated from time to time.

The mercury manometer described in Chapter II (Fig II 2) is the most convenient reference instrument. The scale can be read accurately to 1 mm, equivalent to 0.133 kPa, and so is adequate to check a gauge the scale divisions of which will probably be not less than ten times this value. The manometer and the gauge under test must be connected to the test vacuum as nearly as possible at the same point. The test vacuum must be reasonably stable as a mercury manometer responds only slowly to vacuum change.

Note that the accuracy of a manometer or a vacuum gauge is not affected by variation of atmospheric pressure as they are constructed to measure the pressure difference between a vacuum and the prevailing atmosphere at the time and place of measurement. When adjusting a vacuum gauge found to be in error reset the pointer so that it gives a correct reading at the nominal working vacuum. Any error at zero vacuum is unimportant.

In the course of testing an installation the vacuum level should be measured close to the regulator with one unit and with all units in operation, the teatcup liners being stoppered. In most plants having a master pulsator controller all the pulsators will be in operation in both cases so that any difference in vacuum levels is likely to be very small. The mean of these readings is the nominal working vacuum. For the same operating conditions the vacuum level at the inlet to the vacuum pump should be established to obtain the vacuum drop in the connecting pipeline. Measurements may also be made at other points in the installation to establish vacuum drop.

THE MEASUREMENT OF AIR FLOWRATE

Air flowmeters

A variety of instruments is available for measuring air flowrate but the type most suited to measurements on milking machine installations is the multi-orifice meter. This is designed to measure air flowrate at the point of admission to the vacuum system and is based on the fact that for a given applied vacuum the flowrate through the orifice is constant. If the orifice is very small the flowrate of free air will be nearly constant at any vacuum above a certain minimum value.

One example of an air flowrate meter is shown in Fig IV 1. This has 55 precision orifices each of which passes 0.5 ft^3/min (14.16 ℓ/min) free air when the applied vacuum is 15 inHg (50.8 kPa). The orifices are arranged in a circle and can be opened or closed sequentially by a plastics face pad. The number of orifices in use is indicated by a circular dial calibrated in ft^3/min of expanded air at 15 inHg. A ½ ft^3/min expanded air bleed and a connection for checking the vacuum gauge are also provided. The gauge is connected to a plenum chamber into which the air admitted through the orifices passes. The spigot which provides the connection to the vacuum system is a 1¼ in o.d. tube incorporating a flow straightener.

Another type of flowmeter which may be used and is more suitable for laboratory measurements is the variable-orifice meter (e.g. Rotameter, GEC Elliott Process Instruments Ltd). This consists of a glass tube having a tapered bore containing a float of suitable material. The tube is mounted vertically with

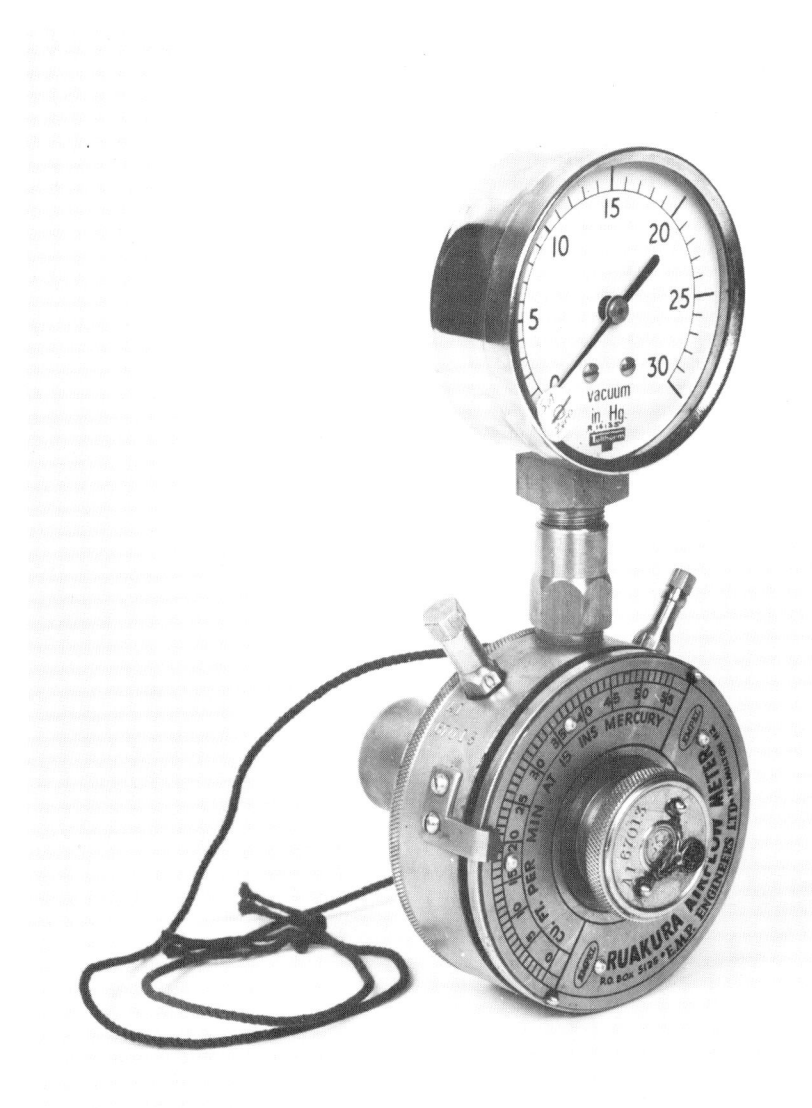

Fig IV 1. The Ruakura air flowmeter

the smaller end at the bottom where the annular space around the float is at a minimum. The air to be measured enters at the bottom and the float rises, increasing the annular space to a point where the pressure drop (which is characteristically small) balances the weight of the float. The float level is read from a scale and indicates air flowrate. The upper end of the tube is connected to the vacuum system through a throttling valve which is set to give the vacuum level required.

Use of the flowmeter

The airflow measurements to be made in testing an installation can be considered under two headings.

1. *Air extraction.* All the air entering the milking machine must be extracted by the vacuum pump. Measurement of pump performance is therefore a first requirement in a mechanical test. To do this the pump is disconnected from the air pipeline and the airflow meter is substituted. The meter orifices are fully opened and the pump is started. The orifices are then closed in succession until the vacuum gauge on the meter indicates the vacuum level, previously measured, at which the pump inlet normally operates. This will be slightly higher than the value measured at the regulator. Assuming this difference is small and a meter of the type described above having small orifices is used, the reading in terms of free air will not have a significant error even though the meter has been designed for a vacuum drop through its orifices of 15 inHg. Meters having larger orifices may be in error and should be calibrated for the higher level of vacuum.

2. *Air admission.* Direct measurement of the air admitted to the milking machine system at each or all of the points indicated in Fig II 3 is not practicable. However, the air admitted to the system as a whole, or through individual sections or components, can be calculated by difference using a simple technique. The flowmeter is connected to the system at the position of the regulator (the point at which the air admitted to the units has expanded to correspond with the working vacuum) and adjusted to admit air to keep the vacuum level constant when the whole plant, or the various components connected in turn, are in operation. For this purpose the vacuum level measured previously when all units are in operation should be taken as the working vacuum.

The steps in this procedure are as follows, but the order in which they are given can be reversed if this is more convenient. Before making these measurements the vacuum pump should have been running for 15 - 20 minutes to reach normal operating temperature.

(a) The regulator, pulsation system, milking units and any other components admitting air are disconnected or sealed off. The pump is started and air is admitted to the system through the flowmeter which is adjusted until the vacuum gauge on the meter indicates working vacuum. The difference between the meter reading (in free air) and the *pump capacity* determined as described above is the *pipeline system leakage*. This can be determined, if desired, section by section.

(b) All the clusters are then connected to the system, the mouth of each teatcup being plugged, and the flowmeter is re-adjusted to the working vacuum. The new reading (which will be less) subtracted from the reading obtained in (a) gives the total *cluster air admission*.

(c) The pulsation system is then connected and all pulsators set in operation. The flowmeter is again adjusted to working vacuum and the new reading subtracted from the reading obtained in (b) gives the total *pulsator air requirement*.

(d) Any components still disconnected, e.g. a milk pump pulsator, are then brought into operation and their consumption calculated as above.

(e) With the whole plant except the regulator in operation the flowmeter is adjusted to an operating vacuum level 2 kPa below working vacuum. The reading then obtained is called *manual reserve* and is the arbitrary surplus of pump capacity over the system requirements assuming no regulator leakage.

(f) The regulator is now reinstated and measurement (e) is repeated. The reading then obtained is the arbitrary surplus of pump capacity under full normal working conditions and is called *effective reserve*. The difference between (e) and (f) is *regulator leakage*.

In many cases it is sufficient to test an installation only for effective reserve (test (f)), omitting the tests (a) to (e) if the reserve is satisfactory and about the same as measured in previous tests.

The relationship between vacuum pump capacity, unit consumption, reserve air and system leakage is shown graphically in Fig IV 2.

Measurement of air admitted by individual components

The technique described above is generally not sufficiently sensitive to measure the air admitted by a single component such as a claw or a pneumatic pulsator. In such cases a direct measurement can be made if the component concerned is enclosed in an airtight box which has an opening to which a variable-orifice flowmeter is attached. The vacuum connection to the component is sealed through a wall of the box and the vacuum applied is controlled to the normal working value. The meter reading should be corrected for the small vacuum drop in the meter which can be measured by a water manometer.

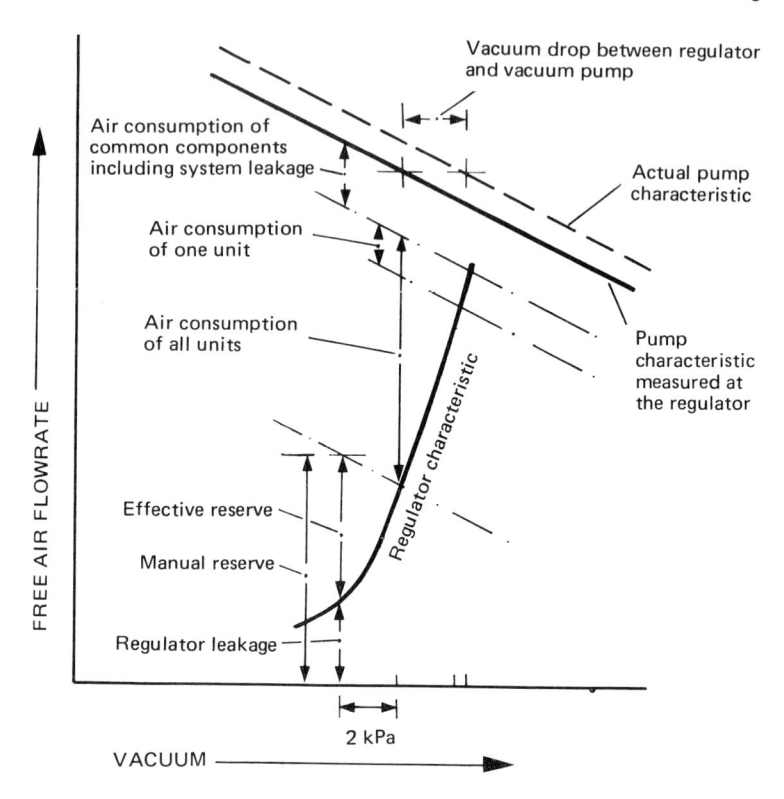

Fig IV 2. Relationship between air consumption and vacuum level

REGULATOR TESTS

The tests described above give important information regarding the behaviour of a vacuum regulator, *viz.* vacuum level when one unit and when all units are operating and regulator leakage, when operating in an installation. However a more comprehensive picture of the performance can be obtained from separate tests to establish maximum capacity, sensitivity and stability.

Operating characteristic

The regulator is enclosed in an airtight chamber in its normal operating attitude. The chamber has an air inlet of adequate size connected to a variable-orifice flowmeter. The regulator is connected to a vacuum system in which the vacuum level can be varied from zero to a suitable maximum.

The applied vacuum is increased from zero in suitable steps, the free air flow being measured at each step. On reaching the maximum the applied vacuum is decreased in similar steps, air flow readings being taken as before. Free air flow is then plotted against vacuum producing a curve similar to Figs II 5 and IV 2. If the air flow readings for rising vacuum are distinctly different from those for falling vacuum, two curves will be obtained showing a hysteresis effect. This is a design fault and if it is excessive the regulator is not suitable for its purpose. Normally the characteristic curve is sigmoid and the working range should be confined to the relatively straight centre portion. The steeper the slope the greater the sensitivity and the smaller will be the vacuum variation in a milking machine system.

Stability

The test rig described above is provided with a flap valve which can be closed rapidly. The vacuum system is adjusted so that closing this valve will change the airflow through the regulator, and hence the vacuum level, from a predetermined minimum to maximum value. A suitable vacuum recorder is connected to the system. When the system is operating steadily at the lower value the flap valve is closed rapidly. Vacuum fluctuations will be induced and should die out within a few seconds, the vacuum level then becoming steady at the higher value. The amplitude and duration of the fluctuations can be measured from the record and compared with standard values where these have been prescribed.

IMPLICATIONS OF BAROMETRIC PRESSURE VARIATIONS

If the machine is operated at constant vacuum the absolute pressure in the system depends on the ambient barometric pressure. Thus the expansion ratio of the air on entering and the compression ratio on extraction will be lower if barometric pressure is above normal and *vice versa*. The measured values for effective reserve and pump capacity must therefore be corrected if they are to be compared with values under standard barometric pressure, say 100 kPa, at which norms and manufacturer's ratings are likely to be expressed. Correction to standard conditions also facilitates the comparison of successive tests on the same installation.

Effective reserve is equal to pump capacity *minus* machine usage. It cannot be corrected by a simple factor but may be calculated from the corrected values for pump capacity and machine usage. Pump capacity can be corrected as indicated in Chapter II but this does not take into account the change in volumetric efficiency resulting from change in compression ration. This may be insignificant for new pumps but for older pumps it is better to measure the maximum vacuum, P_{max}, the pump can produce against a closed inlet under the the ambient barometric pressure, P_B. This test must be done rapidly because the pump temperature will soon rise and damage may result. The overall correction factor K_p by which the measured capacity must be multiplied to give the capacity at 100 kPa barometric pressure (assuming a vacuum of 50 kPa) is:

$$K_p = \frac{100 P_{max} - 50 P_B}{P_B(P_{max} - 50)} \times \frac{P_B}{100}$$

The machine usage (measured pump capacity *minus* measured effective reserve) must also be corrected but as only about half the air is affected by barometric pressure the correction factor K_U is $(100 + P_B)/200$. The predicted effective reserve at a standard barometric pressure of 100 kPa, E_s is therefore:

$$E_s = K_p . C_m - K_U (C_m - E_m)$$

where C_m and E_m are the measured values for pump capacity and effective reserve. The smaller the effective reserve as a percentage of pump capacity the greater the effect of these corrections.

PULSATION TESTS

The rapid changes of vacuum in the pulsation system necessitate the use of a vacuum recorder to examine the characteristics of the applied pulse. Such a recorder must have a rapid response as the vacuum may change from zero to working level in about 0.05s, and must be sufficiently robust for use under farm conditions. Several makes of instrument have been designed specifically for this purpose and most use a bellows which is connected to the pulsation system. Changes in vacuum cause the bellows to expand or contract and this movement is transmitted to a pen arm which marks a trace on a moving paper strip. The example shown in Fig IV 3 has a clockwork motor driving the paper at about 25 mm/s (1 in/s) and that shown in Fig IV 4 has a mains voltage electric motor with a 2-speed drive giving alternative paper speeds of 40 mm/s and 45 mm/min. This recorder uses a stylus sliding on a cross bar to give rectilinear co-ordinates and marks the trace on wax-coated paper. Both recorders have a pen deflection of about 50 mm (2 in) for 51 kPa vacuum.

To obtain a *pulsation characteristic* the vacuum connection is attached to a tee inserted in a short pulse tube and the mouth of the liner is closed by a stopper. The vacuum and pulsation systems are put into operation and the recorder is run to record several pulsation cycles. The curve obtained will be of the general form shown in Fig IV 5 which covers one completed pulsation cycle. For convenience of analysis this is divided into four phases defined as

    a increasing vacuum   b maximum vacuum   c decreasing vacuum
           d minimum vacuum

These are determined by the intersection of the trace with vacuum levels 4 kPa above atmospheric pressure and 4 kPa below working vacuum. By definition the expression

$$\frac{a+b}{a+b+c+d} \times 100\%$$

is called *pulsator ratio.*

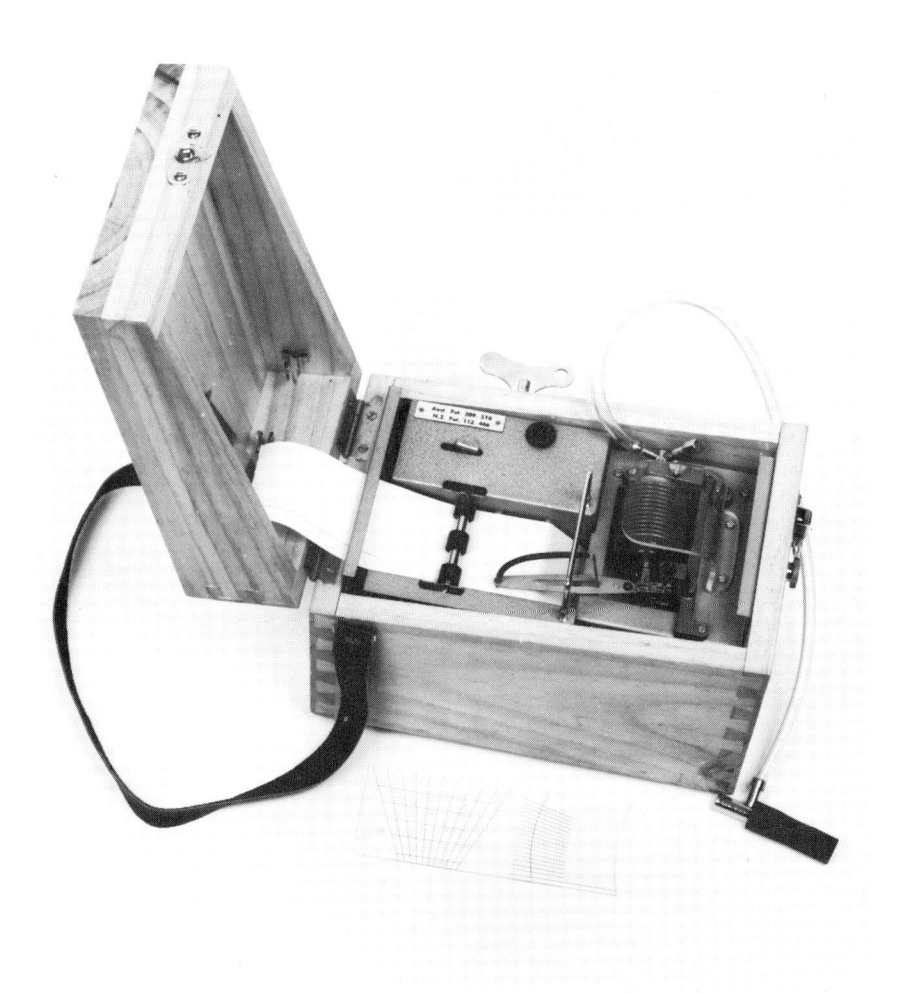

Fig IV 3. The Ruakura vacuum recorder

Fig IV 4. The Alfa-Laval vacuum recorder

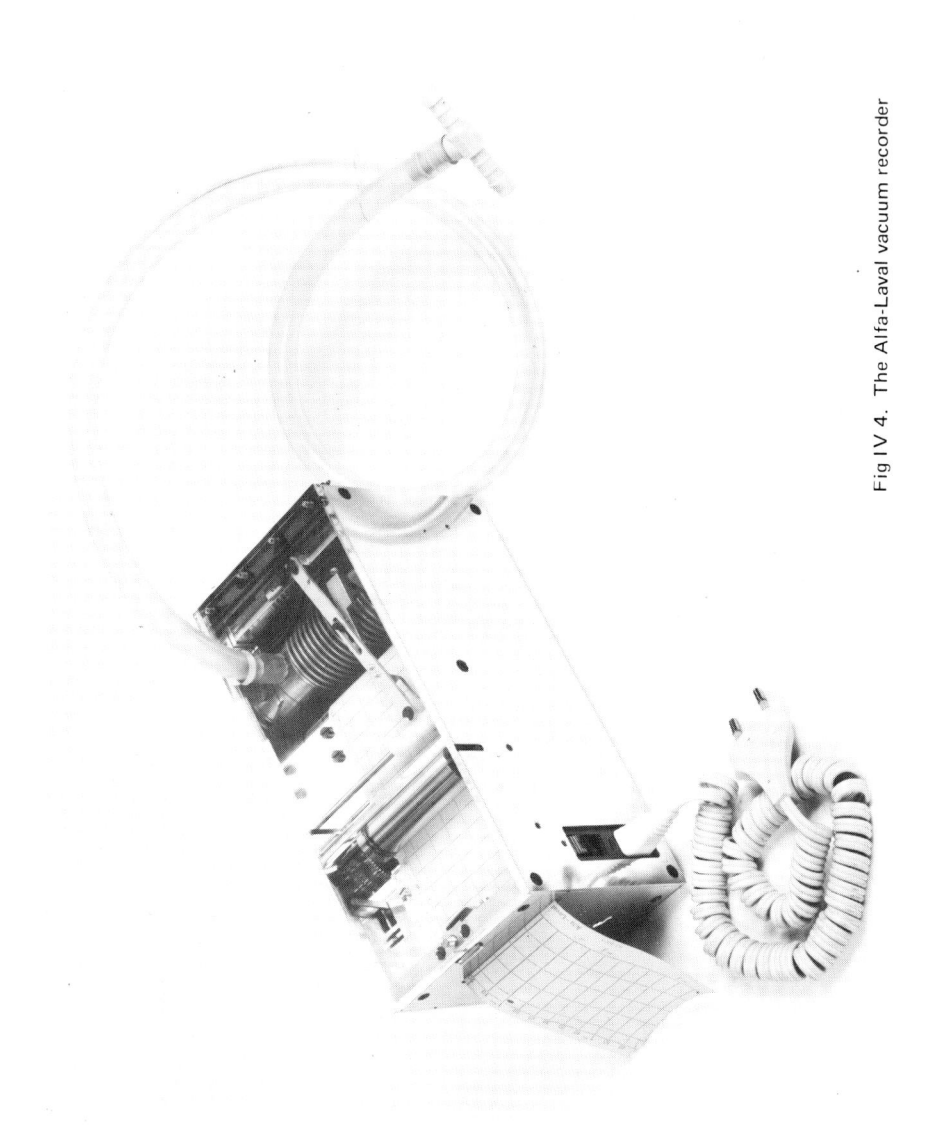

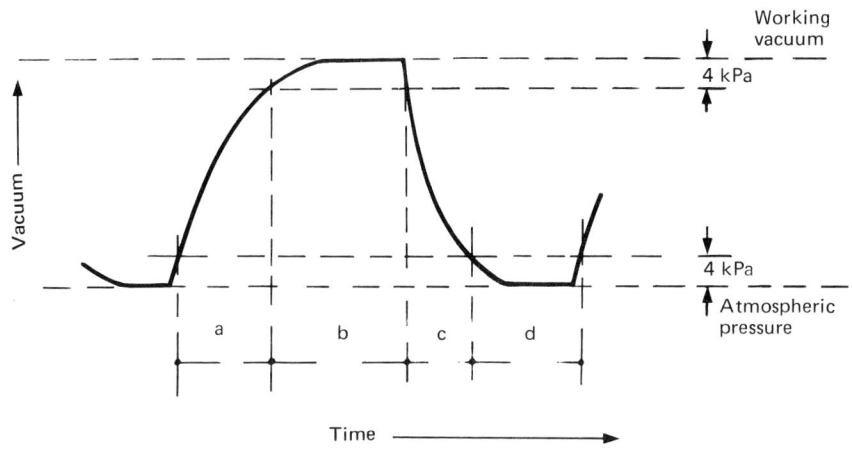

Fig IV 5. Pulsation chamber vacuum record

RECORDING MILKING VACUUM

This type of recorder adapted to give a paper speed of 25 to 50 mm/min can be used to record vacuum in the milk system during milking. Probably the most useful application is recording the vacuum fluctuations in a long milk tube, the recorder being connected to a tee inserted near the claw with the branch of the tee and the connecting tube more or less vertical. The recorder must be placed well above the tee and a transparent connecting tube used so that any accumulation of milk in that tube can be observed. If this occurs the tube must be disconnected and emptied to avoid error in the vacuum record.

TEST RECORDS

The data obtained in the initial commissioning test and in subsequent, possibly annual, service tests form a valuable record of the mechanical history of a milking machine installation and should be preserved, together with a record of component replacements, for future reference. A service history of this kind is invaluable to the person conducting a test at any stage in the life of the installation.

As the report form used by most testing organizations normally contains a great deal of detail it may be an advantage to maintain a summary record showing only a few important data such as pump capacity, manual reserve and effective reserve. Progressive deterioration with time will then be evident.

Chapter V

ACTION OF THE CLUSTER DURING MILKING

C C Thiel and G A Mein

The teatcup liner is the only component of the milking machine that comes into contact with the cow's teat. Consequently, all forces applied by the milking machine to the cow are transmitted to the teat via the liner. When the teatcups are applied to a cow, the teats enter the liners to a greater or lesser extent and then come to rest. The milk in the teat sinus is a little above ambient atmospheric pressure and, with reduced pressure (vacuum) within the open liner beneath the teat, the resulting pressure difference causes the streak canal to open and milk begins to flow. Continuous application of milking vacuum to the exposed tip of the teat in the open liner is said to be uncomfortable or even painful to the cow, so pulsation is used to counter these ill-effects. Pulsation consists of alternate collapse of the liner beneath the teat, when air is admitted to the pulsation chamber of the teatcup, and re-opening when the pulsation chamber is re-evacuated. Milk flowrate from the teat is largely determined by the bore of the open streak canal, but it is influenced by machine factors such as design of the liner and vacuum level, and it is increased by faster pulsation rates and wider pulsation ratios. Flowrate from the udder declines as the end of milking approaches. When finally milk flow stops, some trapped milk can almost always be obtained by manually assisting the milking machine (machine stripping). In most existing milking machines, the milking vacuum has the dual function of causing milk to flow from the teat and of transporting it for varying distances within the machine. The presence of milk in the tubes and pipelines of the machine causes the liner vacuum to fluctuate in several distinct ways.

This chapter attempts to add more substance to this brief description of the action of the cluster during milking, using our present incomplete knowledge of the various aspects of the mechanics involved. There are six main parts: (1) the patterns of flowrate during milking; (2) the patterns of flowrate within a single pulsation cycle; (3) vacuum conditions in the cluster and liner wall movement; (4) the action of the liner in relation to the teat; (5) vacuum stability and (6) summary. For readers not interested in the detailed mechanics of milk extraction, the first two parts on patterns of flowrate cover general aspects and the summary gives an outline of liner action covering the range of current knowledge.

PATTERNS OF FLOWRATE DURING A MILKING

The pattern of milk flowrate during the course of milking a cow with a conventional cluster is influenced by the different milking out characteristics of the individual quarters, the effects of variable distribution of cluster weight between the quarters, and by the design and action of some machine components, notably those in the cluster. Components other than those in the cluster cannot affect the pattern of milk flow from the teat unless they affect either liner vacuum or liner wall movement.

PATTERN OF MILK FLOW FROM THE WHOLE UDDER AND FROM SEPARATE QUARTERS

Flow pattern from the udder

A typical graph of cumulative milk yield plotted against time for the whole udder of one cow at a milking is given in Fig V 1. It shows the usual 3 regions: (a) a period of increasing slope, however brief, during which flowrate increases to a maximum; (b) a period of more or less constant slope indicating constant flowrate, often referred to as maximum or peak flowrate and usually assessed from cumulative milk yield graphs as the maximum yield in any 1 minute of milking; and (c) a period of declining slope as the flowrate falls. This low flowrate period may be quite long and involve a substantial proportion of the total yield, until the graph becomes horizontal as milk flow ceases.

A fourth period (d) is also shown. It represents a period of machine stripping when increased weight may be applied to the cluster, after milk flowrate has slowed or stopped, to extract the last of the available milk in the udder.

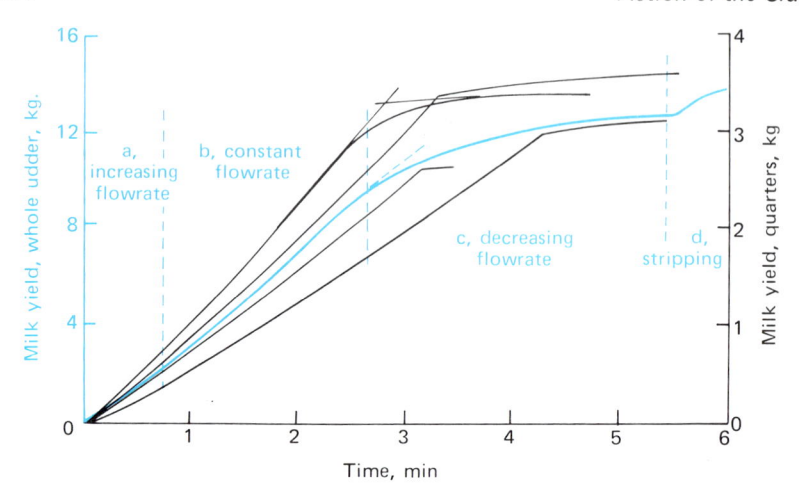

Fig V 1. Typical curves of cumulative milk yield plotted against time for a whole udder at a milking, and for the separate quarters plotted on a four times more open scale. Note the long period of declining flowrate of the whole udder curve, and the sudden change from peak flowrate to low flowrate for three of the separate quarter curves[1].

Flow pattern from individual quarters

As can be seen from the curves of yield versus time for the separate quarters, also plotted in Fig V 1, much of the decline in flowrate from the whole udder occurs simply because the different quarters milk out at different times. Although the curves for the separate quarters also typically show increasing flowrate for a time, the period of constant flowrate usually ends abruptly and is followed by a period of reduced flow at constant or declining rate.

The extent to which the pattern of milk flow varies between the quarters of udders is illustrated by some NIRD results in which a total of 24 Friesian cows were quarter milked with one type of liner (Alfa-Laval 99220003-02) in 4 experiments[1]. Mean machine time (to an end-point of 0.05 kg/min) for milking the slowest quarter of each of the 24 udders was 60% longer than that of the quickest milking quarters. The mean machine time for hind quarters was 15% longer than the mean time for fore quarters, hind quarters yielded 30 - 40% more milk than fore quarters, and mean peak flowrate from hind quarters was 15 - 20% higher than from fore quarters. Mean strip yields from hind quarters were about 60% higher than those of the fore quarters when cows were milked with separate independent teatcup assemblies (each with its own long milk and pulse tubes), and about 90% higher when milked with a cluster made up by clamping together the 4 small claw bowls of the independent teatcup assemblies.

Data from these 4 experiments confirmed an earlier conclusion[2] that most of the milk from a quarter is obtained at almost constant flowrate. However, after the main period of constant flowrate, during which about 90% of the available milk was obtained, there was usually a second period of greatly reduced flowrate. Only about 5% of the available milk was obtained during this period which represented 25% of the mean machine time of quarters. The remaining 5% of available milk was removed by machine stripping.

Milk flowrate data from an experiment comparing the milking characteristics of twelve types of transparent and conventional liners assembled in complete clusters similarly showed that the reduced flow time of the slower quarters considerably prolonged machine time[3]. In all these experiments, had the mean peak flowrate of the slower quarters been maintained for 10 - 20 s longer, then all the milk including strippings would have been obtained in about 80% of the machine times of the slowest quarter of the udders (i.e. in 80% of the mean machine times of the cows).

Low flowrate time, low flowrate yield and strip yield of quarters can be regarded as inter-related measures of inefficiency in the milking properties of the

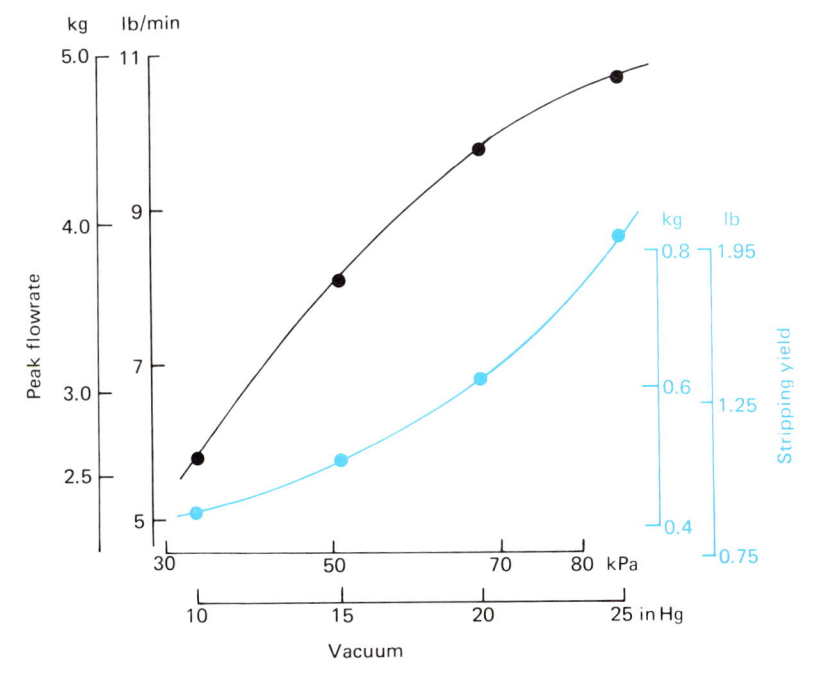

Fig V 2. The effect of vacuum level on peak flowrate and machine stripping yield (from Ministry of Agriculture, Fisheries and Food (1959) Bulletin 177. London: HMSO).

machine. They result from the phenomenon known as teatcup crawl, some aspects of which are considered later.

EFFECT OF MACHINE FACTORS ON THE PATTERN OF MILK FLOW

Much of the published information on the effect of various machine factors on milking has been brought together in a review by Labussiere & Richard[4]. Very little new information has appeared since.

Vacuum level

It is well known that increasing the vacuum level increases peak flowrate and decreases machine time, but increases strip yield[4]. The progressive increase in peak flowrate with increasing vacuum level, accompanied by a marked rise in strip yield at the higher vacuum levels, is shown in Fig V 2. In practice, more economical use of labour is obtained by adjusting machine conditions to reduce strip yields rather than to give faster milking with high strip yields.

Vacuum stability

With currently accepted standards of milking machine design, vacuum fluctuations should not be severe enough to cause teatcups to fall, so interrupting the work routine. Increased milking time has been reported, although one detailed comparison of "low" and "high" cyclic fluctuations in vacuum, in combination with reduced vacuum level due to elevation of milk, showed no practical effect on milking performance[5]. Incomplete closing of the liner (liner walls below the teat not touching) may occur if there is no air admission at the claw and considerable elevation of the milk in the long milk tube. With less extreme machine conditions, however, the liners will open and close fully in each pulsation cycle. It seems therefore that a high degree of vacuum stability is not essential either for normal liner movement or for high milking performance of an installation. The main interest in vacuum stability is in relation to mastitis, as discussed later.

Pulsation characteristics

Smith & Petersen first showed that widening the pulsation ratio increased milking rate[6]. This has been confirmed many times since and is now accepted practice. Increasing the pulsation rate also increases milking rate[7]. This too has been confirmed in a number of subsequent experiments although some results have been inconclusive[4]. The effects, on peak flowrate, of varying pulsation rate and ratio in one of these experiments[7] are shown in Table V 1.

When pulsation rate was increased from 40 to 160 c/min while ratio was held constant at 50%, or when pulsation ratio was widened from 50 to 75% while the rate was held constant at 40 c/min, peak flowrate increased by 30 to

40%. Pulsation rates between 45 and 70 c/min, and pulsation ratios (proportion of each pulsation cycle during which the liner is more than half open when tested with the liners stoppered) between 50 and 75%, are recommended in the UK.

The rate of change of pressure in the pulsation chamber has been shown to have only a small influence on peak flowrate[8]. Cine-radiographic and cinephotographic studies (described later) indicate why this may be so.

Table V 1 The effects of varying pulsation rate and ratio on peak flowrate

	Pulsation rate, c/min			
Pulsation ratio, %	40	80	120	160
50	100	108	127	137
67	123	136	142	141
75	134	142	141	140

Comparative peak flowrates for a group of cows milked at a vacuum of 51 kPa (15 inHg). The results are expressed as percentages of the peak flowrate obtained at a pulsation rate of 40 c/min and a pulsation ratio of 50%, i.e. liner more than half open for 50% of each pulsation cycle when tested with the liners stoppered[7].

Liner characteristics

Ideas and information as to the features of a good liner are conflicting. In the UK, over 100 different types of liners, some in alternative materials, are currently available. Present-day liners tend to show a relatively small range of differences in their observed milking characteristics. It is therefore not surprising that these small differences in the performance of different liners have failed to indicate the relative importance of various design features. The barrel is thought to influence milking rate while the mouthpiece is said to influence the stripping characteristics of a liner, but good evidence for these relationships is not available. Liners under tension give noticeably faster milking, possibly because the proportion of the pulsation cycle during which milk flows is increased.

As liner design is still highly empirical, comparative tests provide the usual means of judging liners in relation to the criteria outlined below. It is usual to make such tests only with new liners under uniform experimental conditions such as common pulsator settings, vacuum level and cluster weight, irrespective of the manufacturers' recommendations. Although comparative tests may provide a useful indication of those liners that perform best under a given set of experimental conditions, such trials are of little value in quantifying the relative importance of various physical factors of liners such as their length, bore, wall thickness, mouthpiece characteristics and frictional properties. Analytical, rather than comparative, experiments are needed for this purpose.

The criteria used to assess liner design have included, or should include, the following.

1. Milking characteristics. Low strip yields, infrequent slipping or falling teatcups, and short machine times are the desirable milking properties of practical consequence. Although peak flowrate is a sensitive and repeatable measurement, it is seldom of practical importance because a substantial increase is needed to give a useful decrease in machine time. Low flowrate time should be as short as possible because it can account for such a large part of the machine time. Strip yield should be low, to reduce its influence on lactational yield if stripping is not practised and to reduce stripping time if it is practised.

2. Mastitis. Although it is now clear that liner design and action influence new infection rates, little information is available on the relative influence of different makes of liner. According to current opinion, liners with hard mouthpieces or wide-bore barrels should be avoided. The way in which liner design and action might affect mastitis is discussed more fully in the chapter on mastitis.

3. Effect on cow behaviour. Little or no information is available on liner design in relation to the comfort of the cow and the consequent effect on her lactational yield. Presumably, cows with small teats should not be milked with large bore liners. Perhaps soft liners are more comfortable than liners made from hard rubbers, but evidence is lacking. Teats often penetrate so deeply into some short liners that the liner is unable to collapse fully beneath the teat. It seems reasonable to expect that a liner should be long enough to avoid such failure of pulsation.

4. Physiological response of the cow. Again there is little information on the influence of different liners on the stimulation of cows and the consequent effects on their yield. The production increases ascribed to good hand stimulation of Jersey cows in New Zealand are well known. Very large breed differences in the amount of stimulus required might exist, however, because similar work with Friesians in the Republic of Ireland and other experiments elsewhere have not shown statistically significant increases in production. Attempts to make the teatcup itself stimulate cows more effectively have been made in New Zealand[9] and in East Germany[10]. Results claimed suggest that further research on such methods of mechanical stimulation would be worthwhile.

Weight of the cluster

The main effect of increasing cluster weight is to reduce strip yields both with conventional liners[11,12] and transparent liners[3]. This effect is shown graphically in Fig V 3. The effective weight of clusters varies from about 1.5 - 3.5 kg (3 - 8 lb). Increasing weight of the cluster is accompanied by increased slipping and falling off and due consideration must be given to this in cluster design. Some manufacturers put a relatively high proportion of the weight in the teatcups and others in the claw. When much of the weight is in the claw, distribution of the weight is very dependent on udder shape and the way in which the long milk and long pulse tubes affect the claw. Often most of the weight of the claw is taken by the front quarters. Ideally, most of the weight should be in the teatcups to provide more equal distribution between the four quarters.

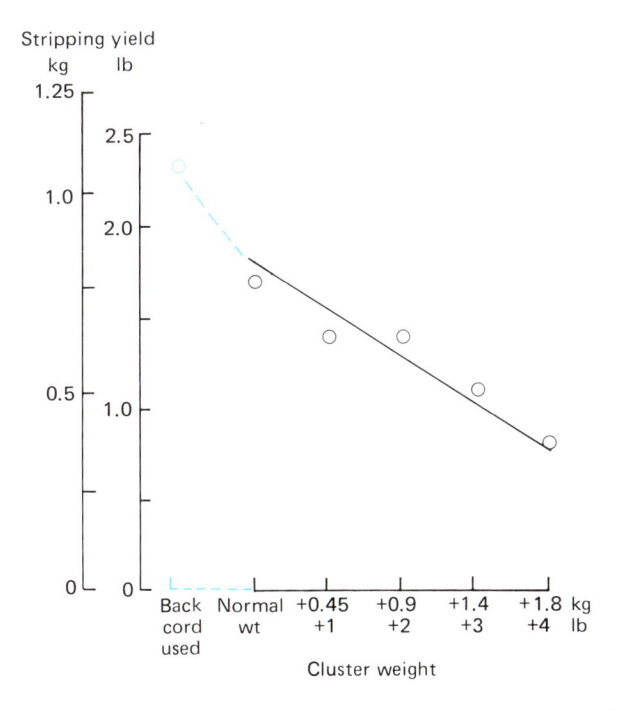

Fig V 3. Effect of cluster weight on yield of machine strippings[12].

PATTERNS OF FLOWRATE WITHIN A SINGLE PULSATION CYCLE

The ways in which various machine adjustments may affect the general pattern of flowrate during milking can be predicted or explained from studies of flowrate within a single pulsation cycle. Two methods have been used for visual observation of the teat and liner during milking; cine radiography using conventional teatcup liners[13] and cine photography using transparent teatcup liners and shells[14]. A method of determining the pattern of milk flowrate from a teat within a pulsation cycle has also given useful results[15].

MILK FLOW TIME AND LINER WALL MOVEMENT: INFORMATION FROM CINE X-RAY STUDIES

Many of the effects of the liner on the teat within a single pulsation cycle have been shown by cine radiography[13]. In this work most of the milk was removed from the quarter, mixed with aqueous barium sulphate suspension, and returned to the quarter, because milk is insufficiently radio-opaque to distinguish it within the teat. Tracings from one sequence of cine X-ray pictures, reproduced in Fig V 4, show the main events in one pulsation cycle during peak flowrate of milking. The camera was operated at 50 frames per second. The tensioned liner, pulsated at 40 c/min, was more than half open for 50% of the cycle. There were 75 pictures covering the complete cycle, each representing 0.02 s. They are numbered consecutively in Fig V 4. Frames 8 - 42 showing the liner fully open, which were similar to frames 7 and 43, have been omitted together with frames 49 - 75 with the liner closed, which were similar to frame 48. The pictures show some distortion and it is helpful to remember that the glass teatcup shell and the barrel of the open liner were cylindrical.

The sequence of events starts with the liner fully closed beneath the teat in frame 1. The teat appears rather triangular, being most compressed near the tip. It does not quite fill the space above the completely collapsed part of the liner, however, as shown by the small clear zone immediately below the teat. As the pulsation chamber is re-evacuated, the first apparent changes are a widening of the liner in contact with the teat and an increase in the size of the teat sinus (frame 2), the liner beneath the teat remaining closed. In the next picture (frame 3) the liner has begun to open beneath the teat. There is some indication that milk is present in the streak canal, and the teat sinus has expanded further. Continued expansion of the teat sinus and opening of the liner can be seen until frame 7 when the liner is fully open. Milk can be seen flowing from the teat in frame 4, i.e. when the liner is about half open. The liner remained fully open up to and including frame 43, in which milk can be seen backing up in the liner. In the next frame, 44, the liner is already about half closed and it appears that milk flow has almost stopped. The teat sinus has already become considerably smaller. In the subsequent stages of closing of the liner the teat sinus becomes progressively smaller (frames 43 - 48) and the region of the streak canal appears elongated, due to complete collapse of the teat sinus for some distance. As the liner closes, milk flow ceases long before complete collapse beneath the teat, indicating clearly that cessation of milk flow is not caused by the liner cutting off the vacuum to the end of the teat but by the force exerted by the closing liner itself.

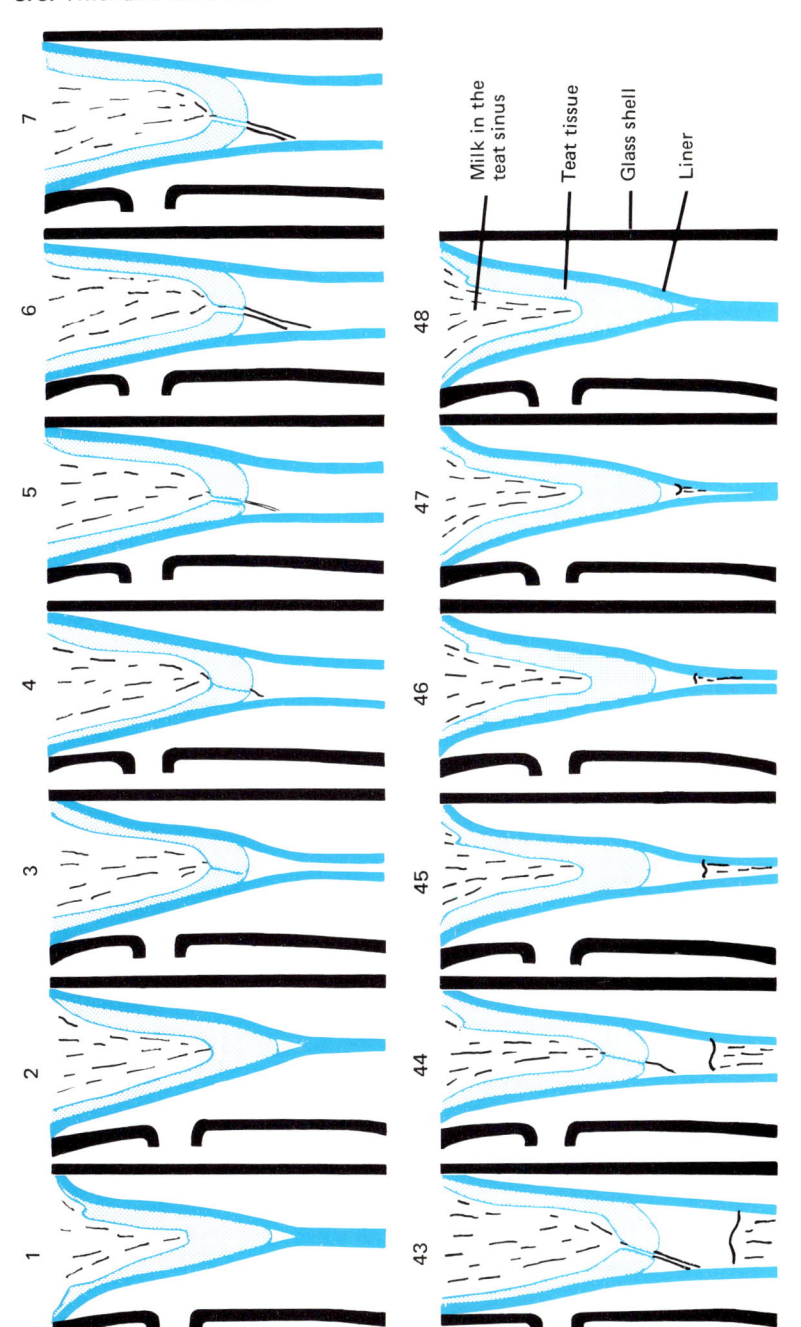

Fig V 4. Tracings of two sequences of frames from a cine X-ray film taken at 50 frames/s, showing the liner opening and closing in one pulsation cycle. Pulsation rate, 40 cycles/min; fast liner wall movement(13).

From other cine radiographs in the same study as those shown in Fig V 4, it appeared that the streak canal might be closed progressively upwards from the teat orifice which may be of importance as a mechanism in mastitis infection. The reverse process seemed to occur when the liner opened, but these cine films did not have sufficient resolution to show such detail with reasonable certainty.

MILK FLOW TIME AND LINER WALL MOVEMENT: INFORMATION FROM CINE PHOTOGRAPHIC STUDIES

As conclusions derived from studies with transparent teatcup liners seemed applicable to conventional liners[3], additional information on milk flow times in relation to liner wall movement was obtained using high speed cine photography[14]. A mirror was mounted vertically on the transparent teatcup at a horizontal angle of 45 degrees to the cine camera (operated at 100 frames per second), to provide simultaneous views of the teat and liner both parallel and perpendicular to the plane of collapse of the liner.

The main results of this study are given in Table V 2 for events within a pulsation cycle during the period of peak milk flowrate of four Friesian cows.

It is apparent that milk flow started in about half the time taken for the liner to open halfway and stopped in about twice the time needed to half close the liner again. It is also evident that the liner closed faster than it opened.

Table V 2 Mean times from the start of liner movement, and mean pressure differences across the liner wall beneath the teat

	Time s	Pressure difference kPa	(inHg)
Increasing vacuum phase			
Liner started to open	0	33	(10)
Milk flow started	0.03	14	(4)
Liner walls separate beneath teat	0.04	9	(3)
Liner half open	0.06	7	(2)
Liner fully open	0.11	2	(0.5)
Decreasing vacuum phase			
Liner started to close	0	5	(1.5)
Liner half closed	0.02	17	(5)
Liner walls touch beneath teat	0.03	23	(7)
Milk flow stopped	0.04	33	(10)
Liner fully closed	0.08	46	(14)

Mean results for three types of transparent liner (liners A, B & D shown in Plate 1 of [3], used at different tensions (13, 37 or 70 N) and different rates of pressure change in the pulsation chamber (0.05, 0.1 or 0.2 s duration of both the increasing and decreasing vacuum phases) at a pulsation rate of 60 c/min.

Increasing liner tension, pulsation rate and rate of pressure change did not alter the relative sequence of events although the times and pressure differences at which events occurred varied with rate of pressure change as might be expected. In Fig V 5, the various events have been superimposed on the records of pressure change in the pulsation chamber for one of the liners operated at slow and fast rates of pressure change. An interesting feature of this figure is that the time of milk flow within a pulsation cycle is the same in both cases, 0.64 s.

The results of this study are similar to those of the cine radiographic study just described except that milk flow started sooner and ended later in the pulsation cycle. This might reflect physical differences between the types of liners used or cows, but more likely is related to better visibility with the cine photographic technique and a wider range of measurements.

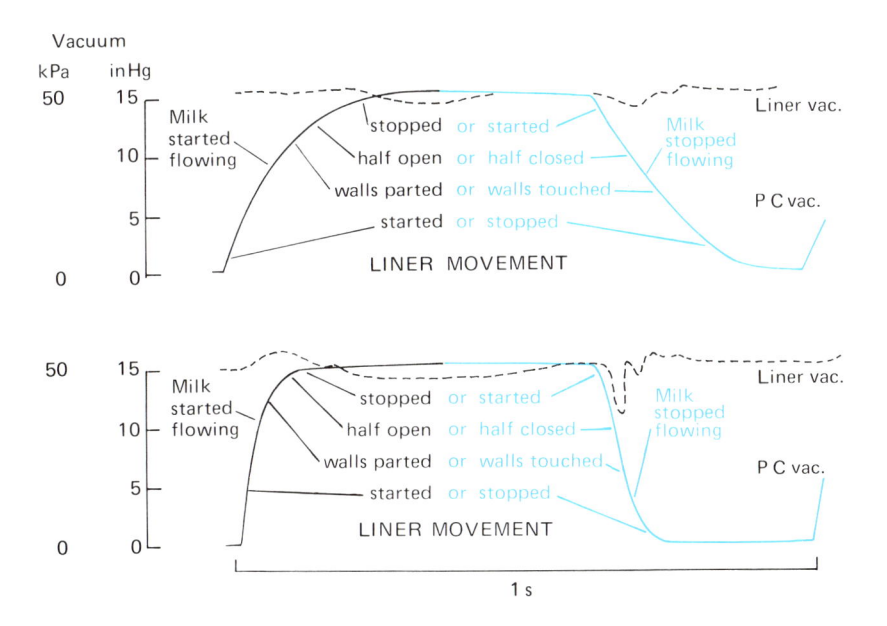

Fig V 5. Milk flow time in relation to liner wall movement within a single pulsation cycle for slow (upper record) and fast (lower record) rates of pressure change in the teatcup pulsation chamber (PC). The measurements were made on the same cow during her period of peak milk flowrate using a Zero PVC transparent liner mounted at a tension of 70N. The cine camera was operated at 100 frames/s in synchronism with a UV oscillograph used for simultaneous recordings of liner and pulsation chamber vacuum (G.A.M. not published).

MEASUREMENT OF MILK FLOWRATE WITHIN A SINGLE PULSATION CYCLE

If the proportion of each pulsation cycle during which milk flows from a teat is kept constant, the total time during which milk flows in a minute will be constant whether pulsation rate is fast or slow. Nevertheless milking rate increases with increasing pulsation rate, which implies some change in the pattern of flowrate during the brief milk flow period of each pulsation cycle. A summary of the main data from an investigation of flowrate patterns within single pulsation cycles[15] is given in Fig V 6.

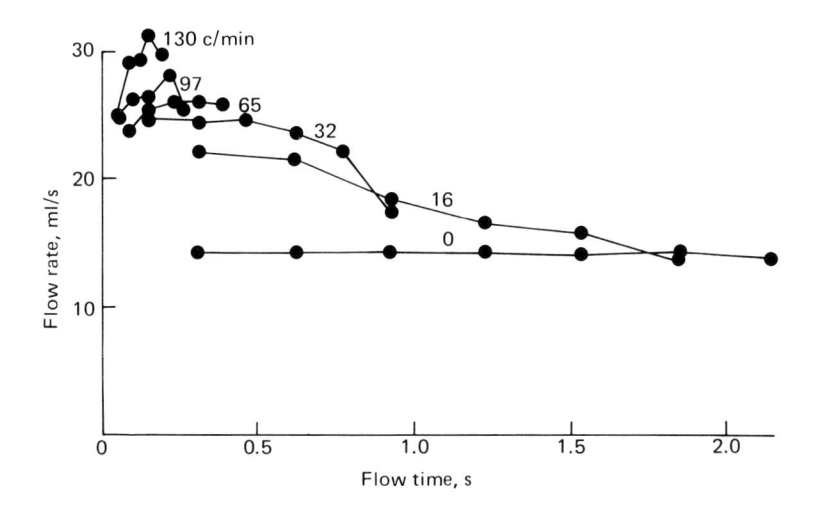

Fig V 6. Mean flowrate curves (hind teats of six cows) during the milk flow period of a single pulsation cycle at six pulsation rates. Milk was collected during one cycle only in a circle of cups rotating just below the liner at one revolution per pulsation cycle. The apparent rise in flowrate for the first 0.1 - 0.2 s for the three higher pulsations rates was thought to be due to measurement errors[15].

With the experimental technique used, only the highest pulsation rates gave any indication of just how rapidly maximum rate of flow in each cycle could be attained. It was not more than 0.04 s, and possibly less than 0.02 s, that is, about the time taken to open the liner in these experiments. As can be seen from Fig V 6, flowrate steadily increased with increasing pulsation rate. At high pulsation rates (above about 60 c/min) flowrate was constant within the limits of experimental error for the entire flow period. At 32 and 16 c/min flowrate began to decline after about 0.5 s, and at 0 pulsation, flowrate was again constant but at the lowest value of the 16 c/min curve.

This investigation has led to some interesting conclusions about the mechanism, presumably muscular, which normally keeps the streak canal closed. Once the pressure difference between the teat sinus and liner vacuum caused the streak canal to open, the observed flowrate in that cycle presumably resulted from the intrinsic bore of the streak canal and the elasticity of its walls, the pressure difference, and residual force of the muscle normally holding the streak canal closed. The fact that flowrate early in each cycle steadily increased as pulsation rate increased implies that the level of flowrate in a cycle depends on the length of preceding cycles. This is consistent with the view that the sphincter muscle increasingly loses tone as it is more frequently forced open. The slow decline in flowrate, beginning after about 0.5 s, appeared to be complete on average after a further 1.5 s since the flowrate after 2 s (16 c/min) had fallen to about the same flowrate as that measured after 30 s of continuous flow (0 c/min). This suggests that the control system of the closure mechanism is very slow acting (about 0.5 s on average) and that the closure mechanism itself takes a long time (more than a second) to establish its full closing force. A possible alternative explanation for these results might be that fluids in the teat wall accumulate in the tip of the teat when the liner is open and they are squeezed out of this region each time the liner closes. Perhaps the return and progressive accumulation of fluid in the tissues surrounding the streak canal gradually reduces the flowrate by reducing the diameter of the streak canal.

VACUUM CONDITIONS IN THE CLUSTER AND LINER WALL MOVEMENT

The methods of measurement which have been employed in research on the relationships between liner wall movement and vacuum conditions, are not suitable for routine use on farms. Liner wall movement has been measured by means of cine photography, in conjunction with a transparent shell. An alternative method uses a linear transducer mounted in the shell. The principle is that movement of a probe in contact with the liner produces a change in electrical resistance. For measuring vacuum, pressure transducers with an electrical output are preferred. Bellows-operated recorders, specially developed for field testing of milking machines, lead to problems of interpretation when used for measuring vacuum in confined spaces such as in the liner, where both milk and air are present. When these instruments are used with connecting tubes of small bore, a mixture of milk and air will not be able to move in and out of the system fast enough for the recorder to follow fast vacuum changes. If the connection is large and an air-liquid separator is used, volume of the measuring system will be so large that it is likely to modify the pressure changes in the confined space.

In a cluster equipped with two-chambered teatcups, liner vacuum does not normally remain steady. In describing the relationship between liner vacuum, pulsation chamber vacuum, and liner wall movement, it is an advantage to begin with a milking machine in which the vacuum to the pulsator and at the outlet of a downward-sloping long milk tube are identical and steady. It is also convenient to take first the simple case of no milk present in the cluster and then to add the complication of milk in the system.

VACUUM CONDITIONS AND LINER WALL MOVEMENT: NO MILK IN THE CLUSTER

When the liners are stoppered, or during milking when no milk is flowing from the teats and the cluster is substantially free from milk, vacuum measured inside the liners (liner vacuum) is usually quite stable during each pulsation cycle. Vacuum conditions and consequent liner wall movement for conditions of moderately fast and slow rates of change of vacuum in the pulsation chamber with the liner stoppered, are shown in Fig V 7. Liner vacuum remained close to the controlled level throughout the cycle in each case. It is usual to find that liner vacuum will remain at least as stable as this, irrespective of the internal dimensions of the components forming the cluster. This is because, with no liquid in the system, there is so little resistance to movement of air into and out of the liners as their internal volume changes on opening and closing that little change in liner vacuum occurs in a pulsation cycle.

It can be seen from Fig V 7 that the liner begins to close very soon after pulsation chamber vacuum begins to decline, i.e. when air pressure in the pulsation chamber is only slightly higher than in the liner. The two sides of the collapsing liner come into contact long before pulsation chamber vacuum declines to zero. Conversely, opening of the liner does not begin until pulsation chamber vacuum level has risen considerably and is completed only shortly before full vacuum is attained.

Rate of liner wall movement

Both the rising and falling curves of vacuum change in the pulsation chamber begin with the fastest rate of change (because pressure difference driving air into the pulsation chambers or exhausting air from them is then at a maximum) and they end with the slowest as the pressure difference declines to zero. Consequently, since the liner opens and closes when the pulsation chamber vacuum is nearer to liner vacuum than to atmospheric pressure, the liner closes faster than it opens, provided the resistance to flow of air is the same in both directions (as it was when Fig V 7 was recorded; angles marked on the records of liner wall movement indicate rate of movement).

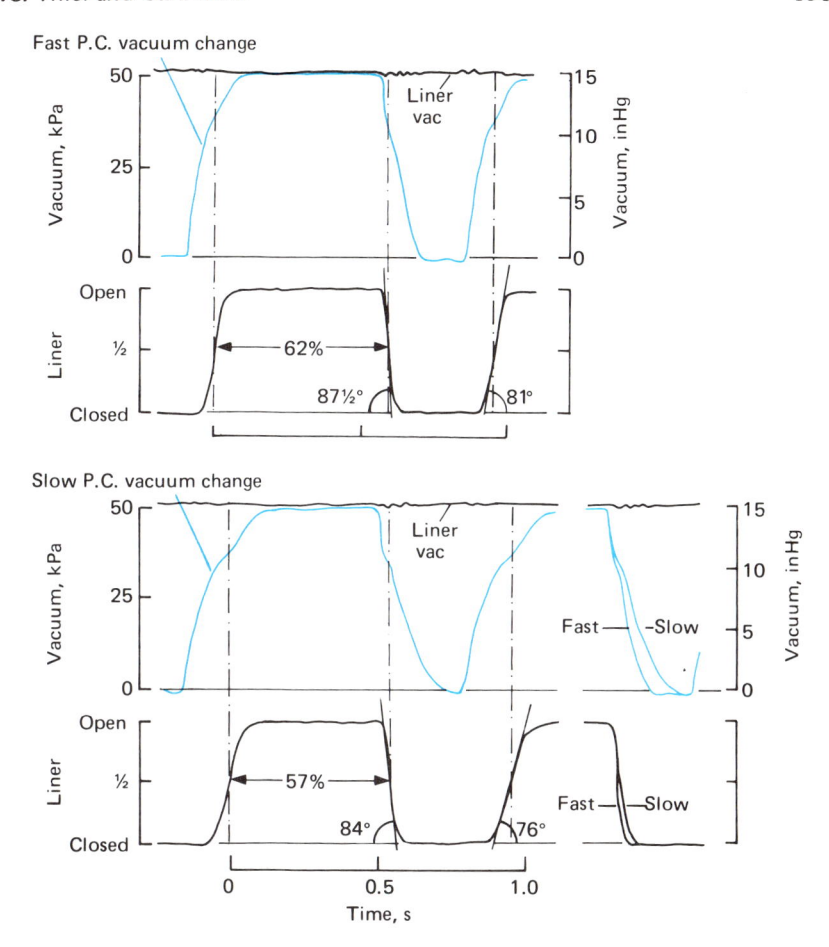

Fig V 7. Fast and slow vacuum changes in the pulsation chamber (P.C.) of a teatcup and consequent liner wall movement, liners stoppered. Extruded natural rubber liners (Alfa-Laval 9922999-2) mounted under a tension of 70N (16 lbf), downward sloping long milk tube.

A further consequence of the basic shape of the pulsation chamber vacuum record is that the rate of collapse of the liner, for the faster of the two vacuum changes shown, cannot be greatly slowed at the same pulsator ratio by throttling the air inlet. The record of the slower vacuum change shows the slowest possible rate of decreasing vacuum which still allows the pulsation chamber to reach atmospheric pressure. The two partial waveforms of decreasing vacuum have been drawn in close proximity to the right in Fig V 7, together with the corresponding curves of liner wall movement. Although the records of declining vacuum as the air is admitted diverge substantially, the slower closing took only 0.02 s compared with 0.01 s for the fast closing – twice as long but still fast. It is also noticeable that the same degree of throttling (imposed by a restrictor in the long pulse tube) had a much larger effect on the time taken for the liner to open, because of the smaller pressure difference exhausting air from the pulsation chambers as the liners opened.

Similar conclusions are apparent from the data given in Table V 2 and Fig V 5.

Pulsation ratio

The curves of changing vacuum level in the pulsation chamber in Fig V 7 show distinct irregularities in the region of 38 kPa. The reason for these is volume change of the pulsation chambers as the liners close and open, which slows down the rate of vacuum change. Vertical lines drawn from the graphs of liner wall movement at the half-open point intersect the regions of irregularities on the curves of pulsation chamber vacuum. Thus, on these particular pulsation chamber vacuum records the times when the liner was half open and half closed may be defined either from the positions of the irregularities, or by drawing a horizontal line at a vacuum level of 38 kPa to intersect the record. The vacuum level corresponding to the half open positions of the liner will vary with different liners and would, for example, be nearer to liner vacuum if the same liner were mounted slack in the shell. Such methods of inferring liner wall movement from pulsation chamber vacuum records can be used to extract a value for pulsation ratio.

Pulsation ratio is in effect an expression of the proportions of each pulsation cycle which are devoted respectively to the purposes of extracting milk and of pulsation. A figure denoting these proportions is useful for characterizing a pulsation system. It is convenient to take the percentage of the total pulsation cycle during which the liner is more than half open since this may be expected to approximate to the period of milk flow. There is some evidence from cine X-ray and other studies that milk flow begins and ends at about these stages of liner wall movement[13, 14]. Even if flow begins and ends at other than the halfway position, the walls of a half open or half closed liner usually move rapidly so that the time error will be small.

Pulsation ratio inferred from pulsation chamber vacuum records taken in field testing with the liners stoppered, will predict actual behaviour of the liner during milking reasonably well only if the presence of milk in the system does not interfere with liner wall movement. It is usually not possible to check liner opening and closing during milking by examining the record of pulsation chamber vacuum because the types of irregularities which occur with the liners stoppered do not appear. This is mainly due to the smaller volume changes with teats present (see for example Fig V 8a and d).

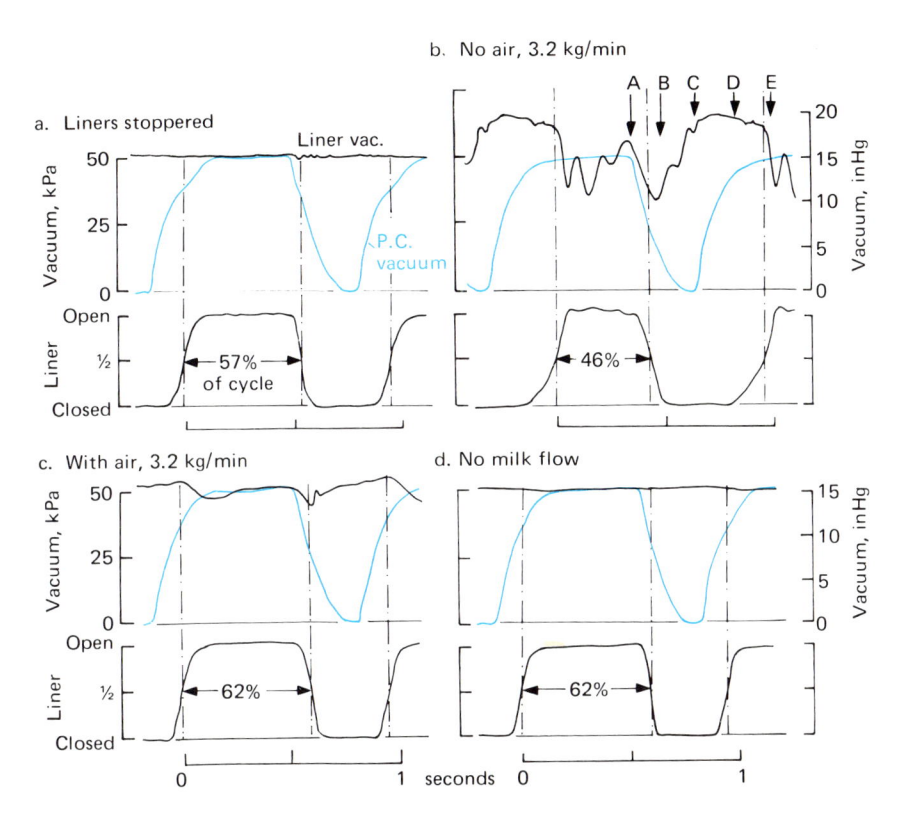

Fig V 8. Vacuum conditions and consequent liner wall movement (a) before milking; (b) & (c) during milking at 3.2 kg/min without and with air admission at the claw (7 ℓ/min free air); and (d) after milk flow had stopped. Cluster as Fig V 7 slow pulsation chamber vacuum change. A - E, upper right, see text.

VACUUM CONDITIONS AND LINER WALL MOVEMENT DURING MILKING

Violent fluctuations in liner vacuum can occur when milk is flowing, particularly if there is no air admission at the claw. An example of extreme liner vacuum instability within a pulsation cycle is shown in Fig V 8b. It was recorded during milking with no air admission at the claw, a downward sloping long milk tube with stable vacuum at the outlet, and stable vacuum at the pulsator. At the same flowrate with air admission, cyclic vacuum fluctuations were much less (8c). They were almost absent with no milk flowing at the end of milking (8d) or with the liners stopped before milking (8a). Note that liner wall movement changed markedly in the presence of large cyclic vacuum fluctuations (8b).

It is clear why liner vacuum was stable during these measurements when no milk was present in the cluster, but could be remarkably unstable when milk was flowing[16]. When air was admitted to the pulsation chamber forcing the liners to begin closing (at point A in Fig V 8b), there was a sharp fall in liner vacuum. This was because milk in the system did not move away quickly enough to accommodate the decreasing volume of the liners so that air in them, trapped behind the milk, was compressed. Hence, liner vacuum declined. Thus, by the time the liners were about half closed a substantial pressure difference, forcing milk out of the cluster, was established from the high pressure in the liners (low vacuum) to the low pressure (high vacuum) at the outfall of the long milk tube. This increasing force rapidly accelerated milk in transit so that the liners could continue to close without further decline in liner vacuum. However, the liners were soon as fully closed as the small difference in vacuum across the liner walls would allow (point B, 8b). At this time the liners were almost closed, but milk was leaving the cluster with considerable speed and hence considerable kinetic energy. Again a force acting on the milk, this time to decelerate it, rapidly developed. The milk in transit acted as a piston once the liners were closed, generating a high vacuum in the liners. In this particular case, increase in liner vacuum was augmented (from point C in 8b) by re-evacuation of the pulsation chamber. This re-evacuation added elasticity of the liner as an additional force acting to open the liners. Consequently, liner vacuum increased. The net effect was a high final vacuum within the liner, considerably higher than either the pulsation chamber vacuum or vacuum at the outfall of the long milk tube. The milk in transit took some time to stop and then begin to flow back towards the liners. Only at this stage did liner vacuum decline sufficiently to allow the liners to begin to open (point D, in 8b) and milk to flow from the teats (point E).

This brief description of cyclic fluctuation in liner vacuum illustrates the complex sequence of events which occur in a pulsation cycle when there is sufficient milk in the cluster to obstruct any of the passage-ways. The cluster is very likely to become substantially filled with milk when there is no air admission at the claw. Air originally present and leakage air are readily entrained in milk and removed with it: air lost to the system in this way can be replaced only by milk. One of the main effects of air admission at the claw is to reduce substantially the quantity of milk in transit in the cluster. This can greatly reduce cyclic fluctuations in liner vacuum, as can be seen by comparing Fig V 8b and c. Only when the cluster is extremely free draining, that is with all tubes of large cross-section and the long milk tube either completely absent or sloping downwards all the way to a milk vessel, is it likely that fluctuations in liner vacuum will be reduced to, say, 10 kPa with a fast-milking cow.

Vacuum conditions and liner wall movement in the teatcup when there is little or no milk in the cluster at the end of milking are similar to those with the liners stoppered (Fig V 8a and d). There are two changes, however. Rates of change of vacuum in the pulsation chamber with teats present are appreciably faster (and therefore the liner in d is more than half open for a longer proportion of the cycle than in a) because of the smaller volume change. For the same reason, irregularities on the vacuum records at about 38 kPa are absent in d, as already discussed.

PRESSURE IN THE TEAT AND UDDER SINUSES DURING MILKING

Measurements have been made of pressures occurring simultaneously in the teat sinus, the udder sinus, the pulsation chamber, and in the liner beneath the teat[17]. These and additional measurements of teat sinus vacuum[18] are of great value in understanding the behaviour of the teatcup during milking. The main findings were as follows.

When milk ejection is complete, pressure measured in the gland sinus varies from about 3 to 10 kPa (1 - 3 inHg) above atmospheric pressure. This pressure declines steadily during milking to about ambient atmospheric pressure. Pressure in the teat sinus accurately follows pressure in the gland sinus up to the stage in milking when milk flow from the teat ceases (or declines, as later work[20] showed). At this point there is an abrupt change, with vacuum appearing in the teat sinus in each pulsation cycle when the pulsation chamber is evacuated. This vacuum increases and decreases in phase with vacuum in the pulsation chamber, attaining maximum values of about half to three-quarters of the pulsation chamber maximum value (Fig V 9). Both teat sinus vacuum and mouthpiece

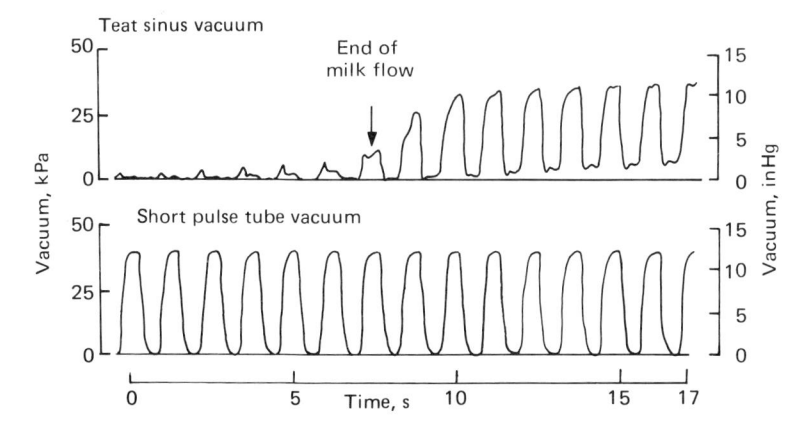

Fig V 9. Vacuum in the short pulse tube and in the teat sinus before and after milk flow from the teat stopped. As peak milk flow came to an end, vacuum changes in phase with those in the pulsation chamber were rapidly established in the teat sinus. (Data kindly supplied by Dr. J.S. McDonald, National Animal Diseases Laboratory, Ames, Iowa.)

chamber vacuum probably derive from liner vacuum[19]. Typical changes in mouthpiece chamber vacuum are described later in the chapter.

ACTION OF THE LINER IN RELATION TO THE TEAT

TEAT MOVEMENT IN THE LINER DURING MILKING

When the teatcup is applied at the beginning of milking the teat moves rapidly into the liner, coming to rest when there is an equilibrium of all the forces acting on it. The teat nearly always moves further into the liner during milking, sometimes in a series of small movements, sometimes in one or more sudden large movements[3]. In independent teatcup assemblies, the teat usually remains stable in the liner while most of the milk is withdrawn at nearly constant flowrate[1]. A sudden increase in the depth of teat penetration (even up to 60 mm) often marks the end of this period (Fig V 10a). Occasionally, however, when the initial depth of penetration of the teat is small, the teat may move deeper into the liner early in milking. Sometimes this caused an immediate increase in milking rate (Fig V 10b). The reason why flowrate may increase with greater teat penetration is not clear.

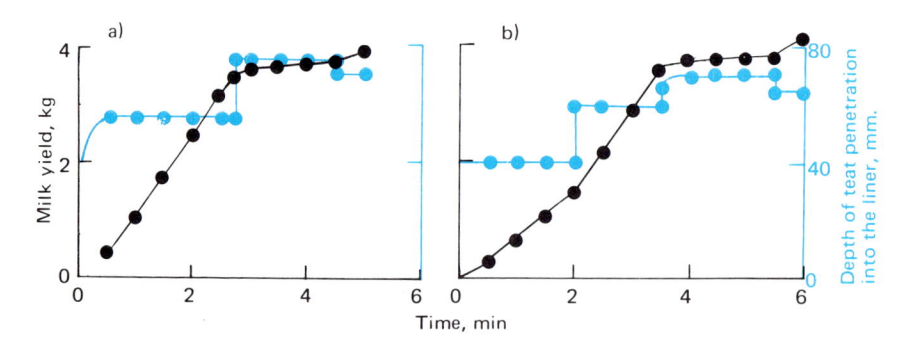

Fig V 10. Relationships between the pattern of milk flowrate from a single quarter and the pattern of teat movement in an independent teatcup assembly (single teatcup not attached to a claw.)

(a) From its initial position, the teat moved further into the liner during the first 0.5 min of milking. No further movement occurred until, after 2.75 min of milking, the teatcup 'jumped' up the teat. The jump coincided with the end of the peak flowrate period. The teatcup moved back about 5 mm when extra weight was applied to it for machine stripping.

(b) The same teat in the same liner at another milking. When the teat moved deeper into the liner during the peak flowrate period, flowrate increased to that in (a), an increase of about 70%[1].

Radiographic studies[20] show that deeper penetration of the teat into the liner as milking progresses is due to more of the teat entering at the top and not to increased stretching of teat already in the liner (see Fig V 11 for example). Teats stretch 35 - 50% on entering the liner and further stretching throughout milking is negligible.

Cyclic movement of the teat relative to the teatcup shell

The radiographs in Fig V 11 illustrate the familiar, rhythmic, up and down movement of the teatcup that occurs in synchronism with pulsation. By comparing the positions of the reference marks on the teat and barrel with the positions of the mouthpiece lip and the top of the shell, it can be seen that the shell moved relative to the liner while the liner remained stationary with respect to the teat. Therefore the shell moved up and down with respect to the teat, so producing the normal cyclic vertical movement of the teatcup. The rhythmic movement occurs because the upper part of the liner moves upwards and inwards with atmospheric pressure in the pulsation chamber. A definite floor to the mouthpiece chamber forms and the mouthpiece lip is carried higher above the top of the shell at this stage of the cycle.

Within a pulsation cycle no regular movement of the teat relative to the mouthpiece lip or the barrel of the liner occurs during the period of peak milk flow. Some cyclic sliding of the teat in the liner barrel may occur, however, after the end of peak flowrate. The reference marks on the teat and barrel of the liner in Fig V 11 show that some cyclic sliding occurred during the periods of restricted flow (nos. 45, 46 and 47) and no flow (nos. 48 and 49), reflecting the lower frictional forces at these stages of milking.

INFORMATION FROM STILL X-RAY STUDIES

Compared with cine radiography, radiography is a simpler but nonetheless useful technique for studying the action of the liner in relation to the teat. Tracings of a particularly informative series of radiographs, from the studies referred to previously[20], are shown in Fig V 11. The radiographs were made in pairs at different stages of milking, first with the liner closed and then, after a few pulsation cycles, with the liner open. These tracings may be used to illustrate several more aspects of the action of the teatcup.

Factors limiting milk flowrate

The odd numbered radiographs made with the liner open show marked changes in the appearance of the teat and size of the milk stream as milking progressed from peak flowrate to the beginning of the reduced flowrate period, then to the end of that period, and finally when flow had stopped. The liner used (an experimental silicone rubber liner by Surge) was particularly soft and easily

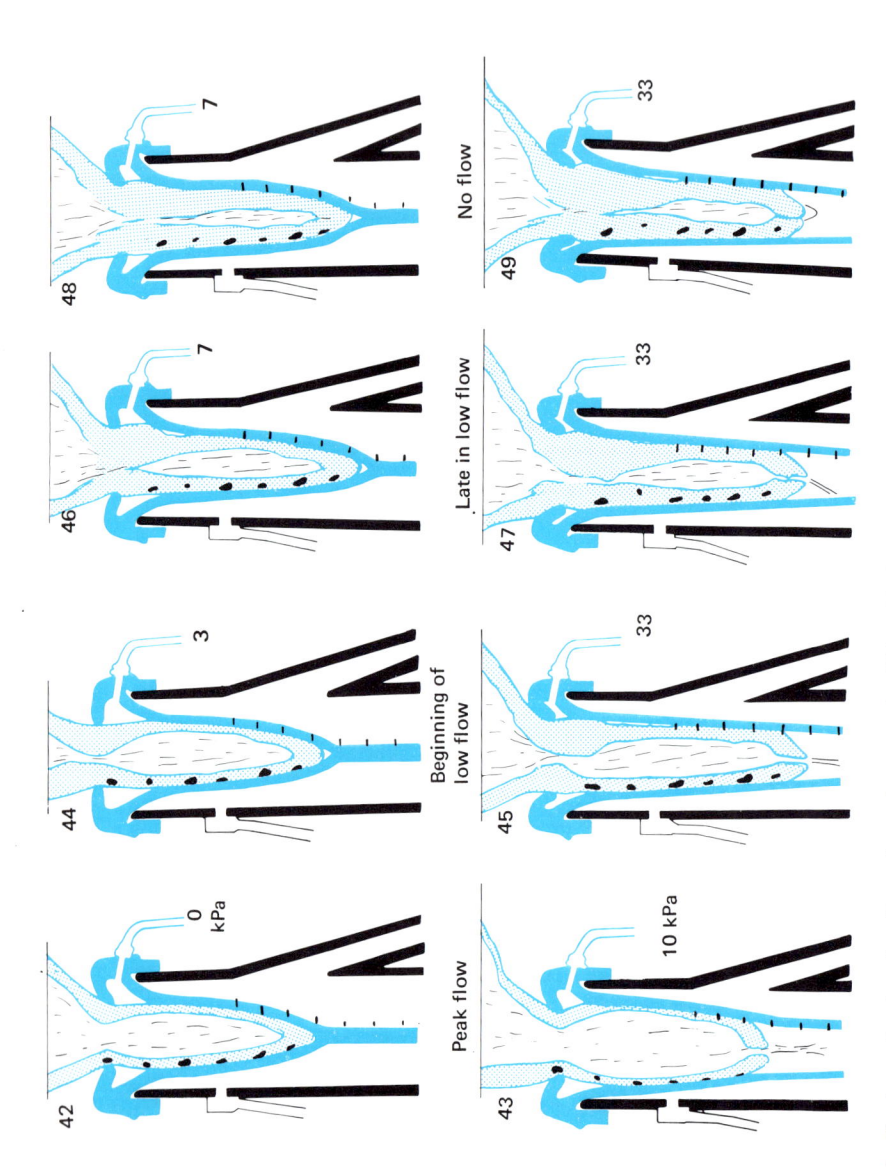

Fig V 11. Tracings of pairs of radiographs at 4 stages of milking. Marks on left of teat were originally 10 mm apart and 10 mm scale marks are on right of liner. Values of mouthpiece chamber vacuum are shown.

distended and in radiograph no. 43 it can be seen to be considerably distended by pressure exerted on its inner surface by the teat. The wall of the teat is thin and the volume of the teat sinus enormous with a wide connection to milk in the udder sinus. The stream of milk in the streak canal is substantial, and vacuum level in the mouthpiece chamber is low. Clearly, the rate of milk flow from the teat is governed by the physical dimensions of the streak canal and the pressure difference present. This is consistent with the observation that the main reason for differences in peak flowrate between cows is the variation in size of the streak canal[21].

In radiograph no. 45, taken only eight pulsation cycles after the beginning of the reduced flowrate period, all these properties have changed. The liner is not much distended (indicating little pressure exerted on it by the teat), flowrate is low and the volume of the teat sinus is much reduced. The obvious explanation is that the connection to the udder sinus is restricted, as indeed it appears to be. This causes a fall in pressure in the teat sinus and consequently lower flowrate, less distension of the teat and liner, less friction between the teat and liner barrel (the teat has moved in about 20 mm as shown by the reference marks on the teat and liner) and less restriction to passage of air past the teat, causing a rise in mouthpiece vacuum (values are given on each radiograph). In the two subsequent stages there is little further change in the outer shape of the teat, but further thickening of the wall occurs as the sinus becomes smaller. The connection to milk in the udder sinus appears even more tenuous. When extra weight was applied to the teatcup the teat was seen to swell against the liner and milk started to flow from the teat immediately. The radiographic appearance during this type of machine stripping was similar to nos. 46 and 47.

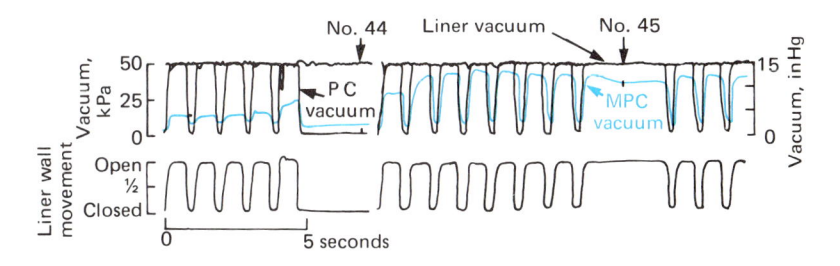

Fig V 12. Recordings of vacuum in the pulsation chamber (PC), in the liner, and in the mouthpiece chamber (MPC), and liner wall movement, at the time radiographs nos. 44 and 45 in Fig V 11 were made. The pulsator was stopped for about 3 s with the liner closed for no. 44 and, after a few cycles of milking, with the liner open for no. 45. The small change in MPC vacuum in each pulsation cycle, shown at the beginning of the record, is typical of peak flowrate milking. The rapid transition to a large cyclic change in MPC vacuum, beginning just before 44 and completed a few cycles after, typically occurs at the onset of the low flowrate period. It persists, normally at about the same level, until the teatcups are removed[20,22].

Fig V 12 shows part of the continuous records of vacuum and liner wall movement made in conjunction with the radiographs in Fig V 11. The records, associated with radiographs nos. 44 and 45, span the change from peak flowrate to low flowrate. The patterns of cyclic change in mouthpiece chamber vacuum before radiograph no. 44 was exposed and at the time radiograph no. 45 was taken are very similar to the cyclic changes throughout milking for the peak flowrate and reduced flowrate periods respectively. Fig V 12 also shows that the large increase in amplitude of the cyclic changes in mouthpiece chamber vacuum occurred over a very few pulsation cycles. This abrupt change in amplitude was common to a variety of liners (although the amplitude was characteristic of individual liners) and coincided with the transition from peak flowrate to reduced flowrate[22]. The small spike on the record of mouthpiece chamber vacuum as the liner opened two cycles before radiograph no. 44 was made (Fig V 12) is the first sign of the end of peak flow and was frequently observed in this work.

The changes initiating the period of reduced flowrate seem to have occurred above the teatcup. Compared with its appearance in no. 42, the teat above the teatcup in no. 44 is more parallel-sided. In addition, the walls of the udder and base of the teat are thicker, presumably because they contracted as the volume of milk in the udder diminished. Both of these changes would tend to close the connection between the udder and teat sinuses. The deduction that the subsequent decline in teat sinus pressure is the cause of reduced flowrate of milk is consistent with the pioneer findings of Witzel & McDonald[17] in relation to pressure in the udder and teat sinuses during milking. However, the radiographs indicate a change to vacuum conditions in the teat sinus from the beginning of the reduced flowrate period rather than at the end of milk flow, and in this respect are somewhat more precise. However, no detailed description is available of the factors which decide peak flowrate and reduced flowrate from a particular teat at milking, nor is it clear in detail how pressure difference from teat sinus pressure to liner vacuum causes the streak canal to open.

Teat congestion

Thickening of the wall of the teat was not apparent during the peak flowrate part of milking, as shown by radiographs nos. 42 and 44 taken near the beginning and at the end of that period. From then on the volume of the teat remained more or less constant but the volume of teat tissue increased as the volume of the teat sinus shrank. The degree of congestion was calculated as the percentage increase in tissue volume between the marks on the teat originally 10 and 60 mm from the tip. Compared with no. 43, the volume of tissue had increased by 20% in no. 45, 40% in no. 47, and 50% in no. 49. It is likely that these changes occurred because fluid (blood and serum) was able to move into the teat wall more easily than milk could pass from the udder sinus to the teat sinus. If so, teat congestion seems unavoidable once flow of milk into the teat becomes restricted.

Force exerted on the teat by the closing or closed liner
The main purpose of pulsation is said to be to maintain circulation of blood in the teat, so avoiding discomfort or more severe pain to the cow. Absence of exact information as to what constitutes effective pulsation is well illustrated by contrasting views of the way in which the collapsing liner achieves its main purpose. A common view is that force exerted on the teat by a relatively stiff liner is necessary. On the other hand, the use of a thin-walled liner has been advocated[23], implying that bringing the surface of the teat to atmospheric pressure is all that is needed. Another view, that pressure above atmospheric pressure in the pulsation chamber is necessary to squeeze the teat, has some support[10]. Alternatively, it is possible that all the benefits of pulsation are achieved by the force applied while the liner is collapsing rather than by the lower force operating while the liner is collapsed.

It is also possible that pulsation may be unnecessary. Recent work in New Zealand and Germany has revived interest in the use of single-chambered teatcups[24, 25]. Results on these linerless teatcups are not yet available, however.

In most conventional teatcups the liner will be about half closed below the teat for a pressure difference of 10 - 15 kPa (3 - 4 inHg), i.e. if the liner vacuum is 51 kPa (15 inHg) the liner will be about half closed when the pulsation chamber vacuum has declined to 35 - 40 kPa (10 - 12 inHg). The liner continues to show some movement for the whole time that the pressure difference is increasing to its maximum. In the final stages, however, movement is confined to a tighter bending of the walls of the liner around the end of the teat.

The force applied by the closing and closed liner will vary with different teats and different liners but no measurements have been reported so far. With the same teat, a slack liner will show a more gradual curvature than the same liner under tension because the tensioned liner is stiffer. A wide-bore liner can more readily accommodate the volume of tissue of a normal sized teat and will therefore show less curvature than a small bore liner containing the same teat. Normally, radius of curvature of the collapsed liner is not a useful means of measuring force exerted by the liner on the teat. A small force gives a measurable change in curvature compared with the transition shape in the absence of a teat, but the liner soon stiffens so that further increments of force have a rapidly decreasing effect on curvature. Nevertheless, degree of curvature of the collapsed liner indicates the region of maximum force. The even-numbered radiographs in Fig V 11 show clearly that the pressure exerted on the teat by the closed liner is greatest near the end of the teat. Moreover, decreasing radius from radiographs nos. 42 to 48 suggests that movement of milk from the teat sinus back into the udder[26] becomes more difficult late in milking.

Air leakage past the teat and mouthpiece chamber vacuum

An interesting feature of pairs of radiographs such as those in Fig V 11 is that the mouthpiece chamber is larger and its pressure higher (vacuum lower) when the liner is closed than when open. As the larger volume is accompanied by the higher pressure it follows that there is a greater quantity of air in the mouthpiece chamber of the closed liner. In most instances pressure in the mouthpiece chamber does not rise at any part of the cycle to ambient atmospheric pressure, as shown by the vacuum record in Fig V 12. It is therefore clear that air lost from the mouthpiece chamber as the liner opens travels between the teat and the barrel of the liner in the direction of the lower pressure beneath the teat. Moreover, the record of mouthpiece chamber vacuum in Fig V 12 shows that the pressure change can occur quite rapidly in both directions — air must enter the mouthpiece chamber past its lip for a brief period as the liner closes and must leak past the teat for a brief period as the liner opens. Rates of air leakage, calculated from the shapes of the mouthpiece chamber and vacuum levels shown in Fig V 12 and multiplied by four to represent air leakage rate in a cycle for a whole cluster, were 0.5 ℓ/min of free air during peak flow and 1.1 ℓ/min during reduced flow at a pulsation rate of 60 cycles/min. However, there was no way of telling from the radiographs whether an additional continuous flow of air takes place. An elegant method of direct measurement of leakage air using a sensitive gas meter gave similar results indicating that a continous flow of air past the teat was unlikely. Overall mean values, when the teatcups were quite stable on the teat, for four types of liner and eight cows were 0.2 ℓ/min of free air at peak flow for a cluster; 0.8 ℓ/min during the period of reduced flow; and 0.4 ℓ/min at the end of this period[22]. Air flowrates were much higher as the teat penetrated further into the liner and when the teatcups slipped down the teat.

The level of vacuum in the mouthpiece chamber at any stage is determined not so much by the rate of air leakage as by the relative restriction to the flow of leakage air past the teat and mouthpiece lip compared with the restriction to flow past the teat and barrel. Thus, mouthpiece chamber vacuum changes when the site of main force between the teat and liner changes simply because the relative restriction to flow of leakage air past the mouthpiece lip and the barrel changes. The rise in maximum mouthpiece vacuum at the end of the peak flowrate period follows a reduction in the force between the teat and barrel, due to the fall in teat sinus pressure. There is also, probably, an increase in the force between the teat and mouthpiece lip due to filling of the mouthpiece region of the liner with teat tissue. This means that the characteristic rise in maximum mouthpiece vacuum, which coincides with the abrupt fall in milking rate and often with the increase in depth of teat penetration in the liner, is one of the changes resulting from (rather than a cause of) the restriction to the free flow of milk from the udder sinus to the teat sinus.

The vacuum level in the teat sinus near the end of milking seems to depend on the average ambient pressure surrounding the teat in the teatcup. The level is often remarkably similar to that in the mouthpiece chamber and it is likely that the pressures in these two sites are mutually interdependent. If so, it is unlikely that a reverse pressure gradient (i.e. liner vacuum lower than teat sinus vacuum) could be sustained long enough to allow retrograde flow through the teat canal from the liner at any stage in the pulsation cycle or at any stage of milking.

FRICTION BETWEEN THE TEAT AND TEATCUP LINER

A teat in an open liner is analogous to a piston in a cylinder. With the space under the teat at the milking machine vacuum there will be a force, proportional to the vacuum level and the cross-sectional area of the teat filling the liner, tending to thrust the teat (the piston) further into the liner (the cylinder). Weight of the teatcup assembly offsets part of this force, but only about half the force when the liner is open is usually accounted for in this way. The remainder must be accounted for by friction between the liner and the teat, otherwise the liner would continue to slide up the teat until it came to rest on the sole of the udder or until the mouth of the liner was jammed with tissue.

Coefficient of friction

Measurements of the coefficient of friction, μ, between teats and liners under conditions of steady load indicated large differences in the frictional properties of liners[27]. The method of making these measurements is shown in Fig V 13. Values measured for the friction between teats and liner materials under a range of conditions, varied from $\mu = 0.2$ to 1.9 with most values falling between 0.5 and 1.0. By comparison, friction between tyres and most road surfaces is of the order of $\mu = 0.5$ but on wet roads it may be as low as 0.2[28].

Nature and influence of friction between the teat and liner

In the series of studies reported in[27] the influence of friction on milking characteristics was investigated by lubricating specific regions of the teat or liner with petroleum jelly. In addition, estimates of the total frictional force between the teat and liner were made by measuring the force required to pull a teatcup off the teat with the liner either open or closed. From the wealth of information recorded, the following points are of greatest immediate relevance. Friction between the teat and liner follows the general frictional behaviour of elastic solids[29]; that is, friction is associated mainly with adhesion and partly with deformation. Even when water and milk were used as lubricants adhesion dominated in these studies. Thus, friction under these conditions is approximately equal to adhesion As, where A is the true area of contact and s is the shear strength of the interfacial junctions. Since s remains more or less constant, friction is simply dependent on the way A varies with the experimental conditions. For example, friction between the liner barrel and a teat increases with time as the soft tissue of the surface of the teat moulds to the comparatively hard rubber surface, so increasing the area of true contact. It is common experience that teatcups are less likely to slip or fall after a short time than when first placed on the teats.

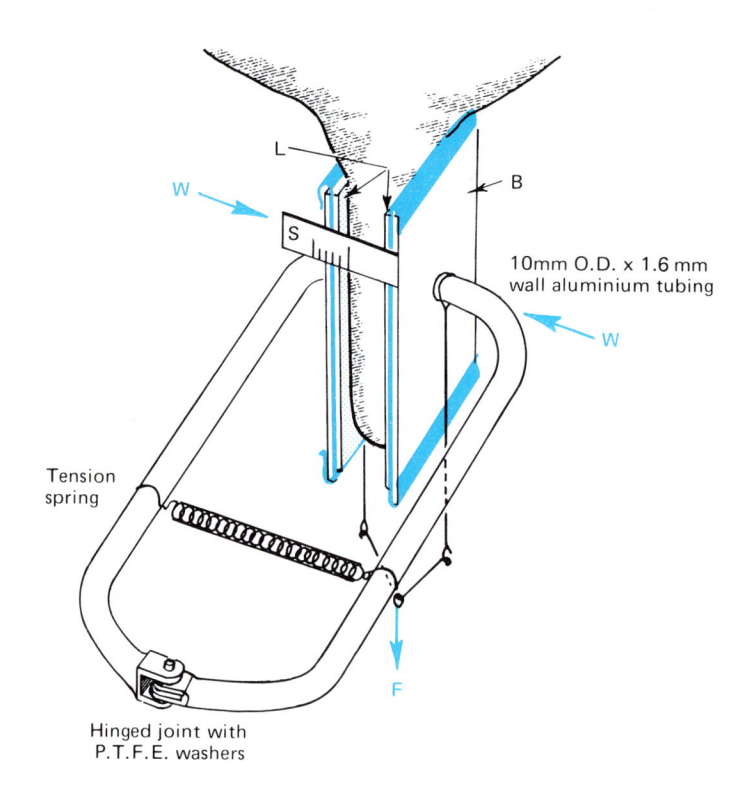

Fig V 13. Technique for measuring the coefficient of friction between teats and liner materials[27]. Pieces of liner L, 75 x 50 mm were glued to a pair of metal backing plates which slid onto base plates B. A spring balance was used to measure force F required to initiate movement of the plates down the teat. Force W was noted from the previously calibrated scale S. Because the teat was squeezed between two plates the coefficient of friction is given by the equation   u = F/2W

Barrel friction was more or less doubled during peak flowrate when the apparent area of contact was doubled by inducing deeper penetration of the teat into the liner. Thus, the large area of contact between the teat and liner barrel was the main source of friction at this stage of milking, and frictional force was fairly uniform over the surface in contact.

Barrel friction early in milking is nearly always more than sufficient to keep the liner stationary on the teat with the liner either open or closed. Barrel friction with the liner open suddenly falls at the onset of reduced flowrate and is usually insufficient to keep the liner stationary. The reduction in barrel friction reflects the lower horizontal force exerted by the teat on the barrel when the teat sinus pressure falls at the end of the peak flowrate period, that is as soon as vacuum appears in the teat sinus. The mouthpiece lip then becomes the main site of friction, but the nature of friction in this area is less well understood than barrel friction. Both barrel friction and mouthpiece lip friction are increased by increasing vacuum level in the mouthpiece chamber. Both were about doubled by increasing the mouthpiece chamber vacuum from a controlled level of 16 kPa (5 inHg) to one of 33 kPa (10 inHg).

VACUUM STABILITY

Unstable vacuum in machine milking may be detrimental because it reduces the milking performance of an installation, or perhaps because pulsation becomes ineffective enough to affect the cow's comfort, but the main preoccupation is with its influence on udder health. The current trend is to improved vacuum stability, with considerable emphasis for example on adequate levels of vacuum pump reserve capacity and air admission at the claw even with bucket units where formerly it was not thought important.

Some vacuum fluctuations are generated in the cluster as discussed previously, others elsewhere in the machine. Three types of vacuum fluctuation, affecting liner or pulsation chamber vacuum, have been distinguished[30].

1. Irregular vacuum fluctuations

These are vacuum fluctuations in a large part or all of the vacuum system of an installation, resulting from changes in air flow during milking. They are influenced by such factors as reserve vacuum pump capacity, sensitivity and stability of the regulator, and frictional losses in the pipe system particularly with increased air admission during cluster changing.

Given that the air extracting capacity of the pump is not exceeded, vacuum stability will depend on the response of the regulator to changes in the quantity of air admitted elsewhere in the machine and additional frictional losses in driving this extra air through the pipe system. With extra air admitted when a cluster is changed, for example, some local decline in vacuum (increase in pressure) must occur to increase rate of flow of air away in the direction of the vacuum pump. The greater the rate at which additional air is admitted, the longer the pipe system and the smaller its diameter, the greater will be the decline in vacuum. When air in excess of reserve vacuum pump capacity is admitted, as often happens in cluster changing and when a cluster falls, these irregular fluctuations will of course be accentuated. Irregular vacuum fluctuations of this type (Fig V 14) have been associated with increased new mastitis infections[18].

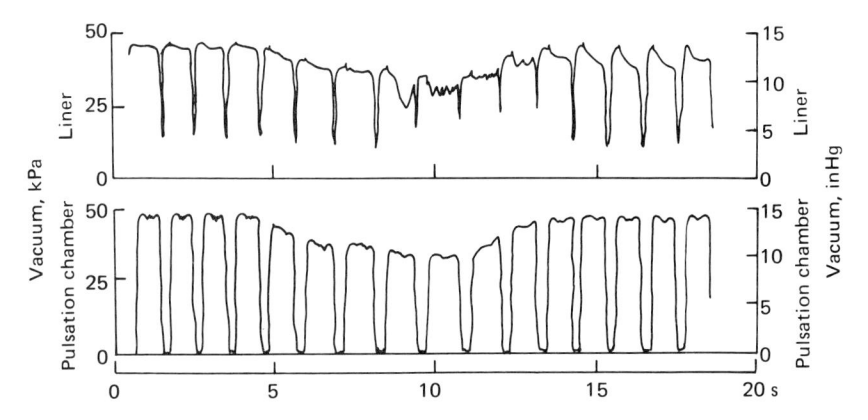

Fig V 14. The type of irregular fluctuation in vacuum, affecting both liner and pulsation chamber vacuum, which has been associated with increase in mastitis infections[18].

2. *Liquid flow vacuum fluctuations in dual purpose milk tubes and pipelines*

These are vacuum fluctuations caused by the presence of milk in tubes and pipelines used for the dual purpose of providing vacuum at the cow's teats and for transporting milk. Loss of vacuum is attributed to friction and varying hydrostatic head when milk is elevated above the teats.

Maximum vacuum fluctuations in a dual purpose milk plus air pipeline occur when the milk travels as distinct slugs, separated by pockets of air. Under these conditions surging is prevalent, a state of flow required for in-place cleaning, but inevitably leading to fluctuating vacuum. Ideally, milk should flow in the lower part of the pipeline with a clear continuous space above for the much larger volume of air to pass over it. The effect of air velocity is critical. Above about 1 m/s, the drag of the air on the milk surface raises it into waves, the first stage of slug formation[31]. For stable vacuum such pipelines should be short, slope in the direction of milk flow to prevent flooding, have as few bends as possible and no risers, and have carefully designed milk entry ports.

3. Cyclic vacuum fluctuations

These are vacuum fluctuations generated within the cluster as a consequence of pulsation, milk in transit impeding free movement of air into and out of the liners as they open and close.

VACUUM FLUCTUATIONS ASSOCIATED WITH THE CLUSTER

In describing the effects of various aspects of cluster design on liner vacuum it is not always possible to separate the contributions of inertia of milk in transit, frictional losses, and vacuum drop due to hydrostatic head, i.e. the contributions of the second and third types of vacuum fluctuations outlined above.

Factors originating in the cluster which influence vacuum changes within the liners may be grouped into (a) those influencing both the mean milking vacuum within the open liner and the amplitude of vacuum fluctuation in a cycle; (b) those predominantly influencing the amplitude of cyclic vacuum fluctuation, and (c) those predominantly affecting mean milking vacuum[5].

(a) Factors influencing both mean milking vacuum level and amplitude of vacuum fluctuation

Increasing milk flowrate, so giving more milk in transit to impede free movement of air, greatly increases cyclic vacuum fluctuation and, by requiring more energy for milk transport, decreases mean milking vacuum. Admission of air at the claw markedly reduces milk in transit and has a dramatic effect in moderating cyclic vacuum fluctuation, sometimes preventing liner vacuum from exceeding the nominal machine vacuum. In addition, the mean milking vacuum is increased by air admission when milk is elevated above the outlet of the claw by reducing the density of the fluid in transit. Increasing the rate of admission over the range 3 - 7 litres per min (0.1 - 0.25 ft^3/min) of free air may, however, have only a small effect on mean milking vacuum within a pulsation cycle. Presumably the beneficial effects of minimizing the quantity of milk in transit through the unit, and of reducing the mean density of fluid in transit, are offset by the general increase in fluid velocities which occur with increasing rate of air admission.

(b) Factors predominantly influencing the amplitude of cyclic vacuum fluctuations

Next to admitting air at the claw, increasing the bore of the short milk tube has the greatest effect in decreasing cyclic vacuum fluctuations. Volume change within the milk system of the cluster as the liners open and close is the cause of vacuum fluctuation, so smaller volume changes when liners pulsate in pairs or when teats penetrate deep into the liners, markedly decrease the amplitude of vacuum fluctuation. Pulsating liners in pairs, with one pair opening as the other closes, also substantially reduce cyclic fluctuations. Once pressure differences between the interior of liners and the bowl of the claw are reduced to negligible proportions by increasing the bore of the short milk tube, volume of the claw bowl appears to be of secondary importance.

(c)   Factors mainly influencing mean milking vacuum within the liner

The mean milking vacuum within the open liner is depressed by increasing length of the long milk tube and decreasing bore (both of which increase frictional losses) and by increased elevation of milk above the outlet of the claw.

VACUUM FLUCTUATIONS AND MASTITIS

Extensive changes have been made to the design of milking machines in the last decade. They include the provision of larger vacuum pumps and pipelines, better regulators, narrow-bore liners and large bore milk tubes, and the adoption of low level milking machines or machines with independent air and milk transport. These changes have resulted from the accumulating evidence that vacuum instability, in particular the occurrence of irregular fluctuations in vacuum, is associated with increased risk of udders becoming infected. However, the reasons why higher mastitis infectivity accompanies vacuum fluctuations are still little understood.

A series of studies to measure the value to mastitis control of suppressing one particular effect of unstable vacuum has been made at the National Institute for Research in Dairying. The identification of an impact mechanism associated with irregular fluctuations in vacuum led to a technique of evaluating the importance of that particular milking machine effect (there may well be others, not yet identified, associated with irregular fluctuations) on commercial farms[19].

Impact effect

Experiments using half–udder milking machines applying either minimal or severe vacuum fluctuations to cows' teats that were artificially contaminated with mastitis organisms showed large differences in numbers of intramammary infections. Frequent infections occurred only when the design of the half-cluster caused marked cyclic fluctuations in vacuum (average 40 kPa (12 inHg) at peak flowrate), and only then if irregular vacuum fluctuations (40% fall in vacuum and recovery artificially imposed 7 times/min) also occurred. Further experiments showed that the cow was at greater risk towards the end of a milking.

The machine-induced infections in these experiments were found to be due to force of milk striking the end of the teat (Fig V 15). This suggested the following mechanism. Milk returning to the liner as it opens in the pulsation cycle can strike the end of the teat with sufficient force of impact to implant any bacteria present deeply enough into the teat canal to increase the risk of infection. The later an implantation occurs in a milking the smaller is the chance that the bacteria will be washed out by milking. Impacts may also occur at cluster removal when air entering the first teatcup to be detached sweeps any milk present back into the other liners.

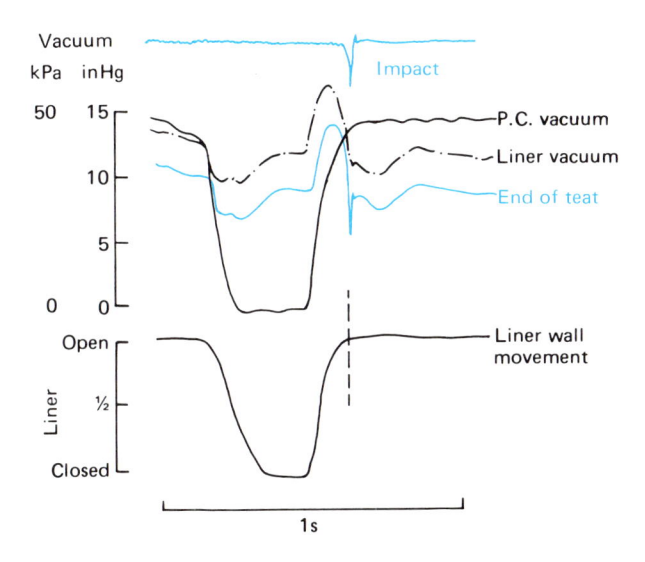

Fig V 15. Occurrence of an impact on the end of an artificial teat under simulated milking conditions like those used in experimental milkings. A pressure transducer in the end of the teat responded to liner vacuum + impact force. A separate pressure transducer measured liner vacuum only (record displaced from end of teat record for ease of reading). The difference between the signals from these two transducers is represented by the impact record. It can be seen from this record and the records of pulsation chamber vacuum and liner wall movement that impacts occur at about the instant when the liner becomes fully open[19].

The milking machine characteristic most closely associated with high impact force is high velocity in the short milk tube as liners open. Many machine characteristics which lead to large cyclic and irregular fluctuations in vacuum will also increase velocity in the short milk tube. Thus this high velocity may well be the link between increased risk of infection and vacuum fluctuations (with impact as the operative event). Nevertheless, the occurrence of impact may not always be accompanied by high vacuum fluctuations. Indeed there was an indication from impact studies using an artificial teat in simulated milking that a machine factor which markedly reduced cyclic vacuum fluctuations (pulsating the liners in a cluster in pairs so that the opening of one pair coincided with the closing of the other, thus favouring quick air transfer) led to the generation of substantial impact force.

Shields for farm trials

In the experiments at NIRD, a horizontal disc or deflector plate fitted centrally in the liner above the opening of the short milk tube appeared to shield the teat orifice from impacts and give good protection to cows against the entry of bacteria. By fitting shields in two teatcups in the same diagonal positions on all milking clusters on a commercial farm, it was expected that the prevailing hazards of infection by pathogens would remain for unprotected quarters. If fewer new infections occurred during lactation in quarters protected by shields the difference would be a measure of a machine effect based on impact. The shields consisted of discs about 4.5 mm larger in diameter than the bore of the short milk tube, supported about 10 mm above its opening into the liner. The diagonal placing of shields in a cluster ensured that the same quarters were protected whether the cow was milked from her left or right side.

The results of farm trials suggested that only a proportion of UK farms would benefit from the use of shields. The implications for mastitis control of these studies are put in perspective in the chapter on mastitis.

SUMMARY

The foregoing account indicates that a clearer picture of the action of the cluster has emerged since the foundation cine X-ray studies of Ardran, Kemp, Clough and Dodd in 1958[13] and pressure measurements in the teatcups, and the teat and gland sinuses by Witzel and McDonald in 1964[17].

Action of the teatcup within a pulsation cycle

The use of cine radiography showed that, within a pulsation cycle, milk flow starts when the liner is about half open and ends when the liner is about half closed. Cine photographs using transparent teatcups showed the same general pattern although milk flow appeared to start before the liner was half open and end after the liner had half closed. Studies of the milk flowrate pattern within a single pulsation cycle indicate that the control system of the closure mechanism of the streak canal is very slow acting and that the closure mechanism itself takes more than 1 s on average to establish its full closing force. Therefore, cessation of milk flow as the liner collapses is not caused by the liner cutting off the vacuum to the teat end, nor by the slow-acting closure mechanism, but by pressure exerted on the teat by the closing liner. Pressure measurements have shown that teat sinus pressure is higher than pulsation chamber pressure and yet, when the liner collapses, the teat is seen to change shape. It does so because the liner itself exerts pressure in bending around the teat. Radiographs show clearly that force exerted on the teat by the closing or closed liner is greatest near the end of the teat.

Action of the teatcup during the course of a milking

When the teat first enters the liner it comes to rest when friction between the teat and liner is at least equal to the difference between the weight of the assembly and upthrust of the teatcup due to liner vacuum. The frictional force is derived from pressure of milk in the teat sinus and is no doubt influenced by air pressure, probably at some intermediate level between liner and mouthpiece chamber vacuum, in pockets and channels where the teat is not in adhesive contact with the liner. The teat stretches longitudinally 33 - 50% on entering the liner, but deeper penetration which almost invariably occurs is due to more teat entering the liner rather than further stretching. In spite of the high level of friction during peak flowrate of milking between the teat and barrel of the open liner the teat often moves deeper into the liner. This suggests that barrel friction can be considerably lower at some other stage in the pulsation cycle. Cine photography using transparent liners showed that movement occurred only as the liner opened, which seems a likely time for low friction associated with low sinus pressure as the teat refills with milk.

Early in milking flowrate rapidly builds up to a steady value (peak flowrate) which is maintained until most of the milk available in the quarter has been obtained. With a large connection between the teat and udder sinuses flowrate is influenced mainly by the dimensions of the streak canal.

Eventually, failing milk supply from the udder and thickening of its walls on emptying seem to initiate restriction of the connection to the teat sinus at the base of the teat. With pressure difference from the udder sinus to liner vacuum now shared between the restriction formed at the base of the teat and the restriction of the streak canal, teat sinus pressure falls considerably. Other changes promptly follow from this low pressure (typically half to two-thirds of liner vacuum) induced in the teat sinus. Milk flowrate declines because there is now less force to open the streak canal and less pressure difference to force milk through it. However, no detailed description is available of the way pressure difference from teat sinus pressure to liner vacuum causes the streak canal to open, and what factors decide peak flowrate and reduced flowrate. Friction between the teat and liner barrel is much reduced as this is derived from teat sinus pressure. At this stage the mouthpiece lip becomes the main site of friction maintaining the stability of the teatcup on the teat. Less force between the teat and liner barrel also means less restriction to movement of air from the mouthpiece chamber to the liner beneath the teat. Mouthpiece chamber vacuum is therefore markedly increased during the time the liner is open, which greatly increases friction at the mouthpiece lip.

Flow continues during the reduced flowrate period at a steady or a declining rate, equal on average to about 15% of peak flowrate, until eventually the connection to the udder sinus is blocked. During this time the quantity of milk in the teat sinus usually diminishes although the total volume of the teat remains more or less constant, the difference being accounted for by swelling of the teat tissue by as much as 50%.

Vacuum fluctuations

Three types of vacuum fluctuations are generated in the milking machine.

1   Irregular vacuum fluctuations in a large part or all of the vacuum system of an installation, resulting from changes in air flow during milking. They are influenced by such factors as reserve vacuum pump capacity, sensitivity and stability of the regulator, and frictional losses in the pipe system particularly with increased air admission during cluster changing. With increase in air admission at any point in the machine, some local decline in vacuum (increase in pressure) must occur to increase rate of flow of air away in the direction of the vacuum pump. When air in excess of reserve vacuum pump capacity is admitted, as often happens in cluster changing and when a cluster falls, these irregular fluctuations will of course be accentuated by a general decline in vacuum below the regulated level.

2   Vacuum fluctuations in dual purpose milk and air tubes and pipelines resulting from friction and varying hydrostatic head when milk is elevated above the udder. Fluctuations in vacuum due to transport of milk in dual purpose pipelines are reduced by larger bore, shorter length, and slope in the direction of flow to avoid flooding, by few bends and no risers, and by carefully designed milk entry ports. The influence of elevation of milk on vacuum level in the cluster, which depends on height of elevation may be reduced by optimizing the bore of the long milk tube and rate of air admission to the cluster.

3   Cyclic vacuum fluctuations generated within the cluster as a consequence of pulsation. Cyclic fluctuations are initiated when atmospheric air is admitted to the pulsation chambers, and will occur if air at milking vacuum within the liners cannot escape from the cluster as the liners collapse because milk gets in the way. The characteristic pattern of changing liner vacuum begins with a fall in vacuum as the liners close because air trapped by milk in transit is

compressed, the increase in pressure (fall in vacuum) eventually pushing milk out of the cluster. The rush of milk out of the cluster may cause a sharp increase in vacuum when the liners become suddenly fully closed (as with water hammer when a tap is closed quickly, but in the reverse direction in a milking machine). In any case, as the pulsation chambers of the teatcups are re-evacuated, liner vacuum will increase. This may precede, augment, or follow the increase due to outflow of milk, depending on the duration of the pulsation cycle. The severity of vacuum fluctuations due to pulsation is governed by many features of the milking cluster, but the characteristic pattern described is always discernible.

Practical value of this research

Better understanding of teatcup action may lead to practical improvements in three areas.

1. Improvements in extracting milk more quickly and completely. If the connection between the teat and udder sinuses could be kept open for an average of 10 - 20 s longer, then the last of the available milk could be withdrawn from the udder at peak flowrate. Under these conditions all of the milk, including strippings, could be obtained in about 80% of the mean machine times at present required to milk the slowest quarter of an udder.

2. Higher lactational yields because of greater comfort and better stimulation. The small amount of information available to date indicates that further research in this area would be worthwhile.

3. Reduction or elimination of any direct effect of the machine on mastitis infectivity. On the basis of present knowledge this is likely to be the greatest benefit from an increased understanding of the action of the cluster during milking.

REFERENCES

(1) Mein, G.A., Thiel C.C. & Clough, P.A. (1973) *Australian Journal of Dairy Technology,* **28,** 26.

(2) Jennische, L.M. (1969) *Proceedings of the Symposium on Machine Milking 1968,* Reading: National Institute for Research in Dairying, p.42.

(3) Mein, G.A., Clough, P.A., Westgarth, D.R. & Thiel, C.C. (1970) *Journal of Dairy Research,* **37,** 535.

(4) Labussiere, J. & Richard, P. (1965) *Annales de Zootechnie,* **14,** 63.

(5) Thiel, C.C., Clough, P.A., Westgarth, D.R. & Akam, D.N. (1968) *Journal of Dairy Research*, **35**, 303.

(6) Smith, V.R. & Petersen, W.E. (1946) *Journal of Dairy Science*, **29**, 45.

(7) Clough, P.A. & Dodd, F.H. (1956) *Agriculture*, **63**, 334.

(8) Phillips, D.S.M. (1963) *Proceedings of the Ruakura Farmers' Conference Week*, p.219.

(9) Phillips, D.S.M. (1970) *Proceedings of the Ruakura Farmers' Conference Week*, p.177.

(10) Wehowsky, G. (1969) *Proceedings of the Symposium on Machine Milking 1968*, Reading: National Institute for Research in Dairying, p.134.

(11) Dodd, F.H. & Henriques, E. (1949) *Agriculture*, **56**, 212.

(12) Clough, P.A. (1965) *Report National Institute for Research in Dairying*, 1964, p.26.

(13) Ardran, G.M., Kemp, F.H., Clough, P.A. & Dodd, F.H. (1958) *Journal of Dairy Research*, **25**, 154.

(14) Mein, G.A., Maguire, P.D. & Sangiorgi, F. (1974) *International Dairy Congress*, 19, New Dehli, 1E, p.14.

(15) Thiel, C.C., Clough, P.A., Westgarth, D.R. & Akam, D.N. (1966) *Journal of Dairy Research*, **33**, 177.

(16) Thiel, C.C., Clough, P.A. & Akam, D.N. (1964) *Journal of Dairy Research*, **31**, 303.

(17) Witzel, D.A. & McDonald, J.S. (1964) *Journal of Dairy Science*, **47**, 1378.

(18) Nyhan, J.F. (1969) *Proceedings of the Symposium on Machine Milking 1968*, Reading: National Institute for Research in Dairying, p.71.

(19) Thiel, C.C. (1974) *Biennial Reviews National Institute for Research in Dairying*, p.35.

(20) Mein, G.A., Thiel, C.C. & Akam, D.N. (1973) *Journal of Dairy Research*, **40**, 179.

(21) Baxter, E.S., Clarke, P., Dodd, F.H. & Foot, A.S. (1950) *Journal of Dairy Research*, **17**, 117.

(22) Mein, G.A., Thiel, C.C., Fulford, R.J. & Hoyle, J.B. (1973) *Australian Journal of Dairy Technology*, **28**, 31.

(23) Phillips, D.S.M. (1965) *Proceedings of the Milking Machine Summer School and Conference*, Hobart, Launceston: Department of Agriculture, p.132.

(24) Tolle, A. & Hamann, J. (1975) *Bulletin, International Dairy Federation*, Document 85, p.193.

(25) Phillips, D.S.M., Woolford, M.W., Millar, P.J. & Phillips, E.M. (1975) *Bulletin, International Dairy Federation,* Document 85, p.200.

(26) Thompson, P.D. & Miller, R.H. (1974) *Journal of Dairy Science,* 57, 1489.

(27) Mein, G.A., Thiel, C.C., Westgarth, D.R. & Fulford, R.J. (1973) *Journal of Dairy Research,* 40, 191.

(28) Conant, F.S. & Liska, J.W. (1960) *Rubber Chemistry and Technology,* 33, 1218.

(29) Bowden, F.P. & Tabor, D. (1950 & 1964) *The Friction and Lubrication of Solids,* I & II. Oxford: University Press.

(30) Cowhig, M.J. (1969) *Proceedings of the Symposium on Machine Milking 1968.* Reading: National Institute for Research in Dairying. p.15.

(31) Kosterin, S.J. (1949) *Izvestiya Akademii Nauk SSSR, Otdel Tekhnologicheskikh Nauk,* 19, 1824.

CHAPTER VI

ANATOMY AND PHYSIOLOGY OF THE UDDER

A T Cowie

To understand the processes of milk formation and milking it is necessary to know something about the structure of the udder, and so before dealing with physiological mechanisms, a brief description will be given of the anatomy and microscopic structure of the udder of the cow.

ANATOMY OF THE UDDER

The udder is made up of four 'quarters' or mammary glands, each quarter being a separate milk-secreting gland drained by its own teat. The four quarters are closely packed together within the skin of the udder and held suspended under the pelvis or hip bone and muscular wall of the belly by strong ligaments. On the skin surface the line of demarcation between the two right and the two left quarters of the udder is indicated by a median groove of varying depth. There may occasionally be a slight groove, particularly in old animals, between the hind and fore quarters, but this is always much less distinct.

SUPPORTING STRUCTURES OF THE UDDER

The udder of a lactating cow is a relatively large organ weighing over 50 kg (110 lb) inclusive of milk and blood, and so it must be adequately supported and securely attached to the skeleton. This is achieved by strong flat suspensory ligaments which are attached to the pelvic bone and to the strong tendons of the abdominal muscles in the region of the pelvis (Fig VI 1). The fibres of the lateral

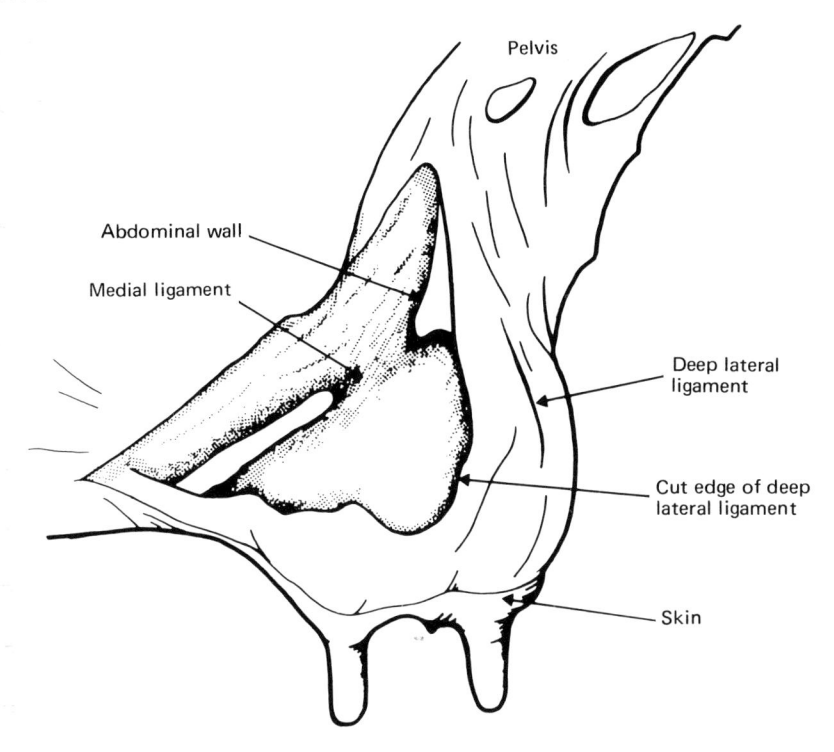

Fig VI 1. Diagram of ligaments supporting the udder.

or side suspensory ligaments fan over the left and right sides of the udder and join with the median or central ligaments which pass between the two halves of the udder. The ligaments are connected to the deeper connective tissues of the udder by fibrous bands or webs of tissues. The right and left halves of the udder are thus, in effect, suspended from the pelvic bone in strong ligamentous slings. The opposing surfaces of the two median ligaments are knit together by connective tissue. The median ligaments are elastic, but the lateral ligaments are not. As the udder fills with milk and its weight increases the median ligaments stretch, resulting in the outward protrusion of the teats characteristic of the well-filled udder. (The cut edges of these suspensory ligaments can be seen in Fig VI 2).

The mammary glands are connected with the abdominal cavity only by way of the inguinal canals. These canals are two narrow oblique passages through the abdominal wall, one on either side, through which pass blood and lymph vessels and some of the nerve trunks to the udder (Figs VI 10, 11 and 12).

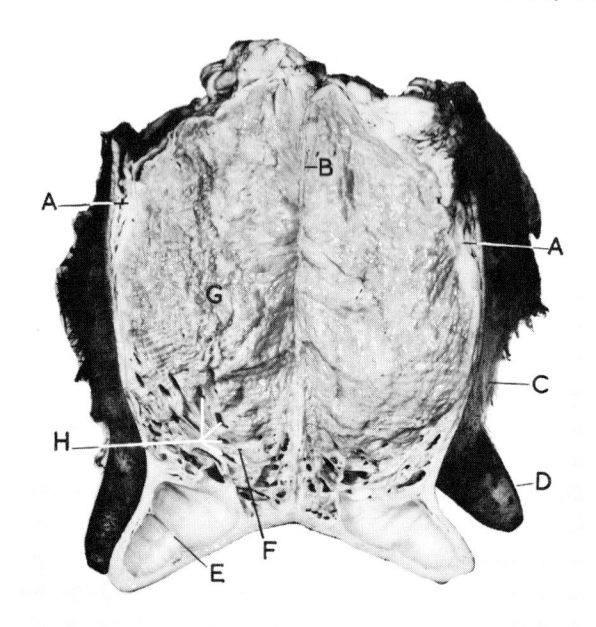

Fig VI 2. Cross-section of the rear quarters of a cow's udder

2 cm

A — lateral suspensory ligament
B — median suspensory ligament
C — skin
D — front teat
E — teat cistern
F — gland cistern
G — glandular substance
H — ducts opening into gland cistern

(From Ministry of Agriculture, Fisheries and Food (1959) Bulletin 177. London: HMSO).

STRUCTURE OF THE MAMMARY GLANDS

The mammary glands are made up of two main types of tissue. There is the glandular or secretory tissue which produces the milk and the connective and fatty tissues which protect and support the glandular tissue and carry the blood vessels supplying the glandular tissue. The glandular tissue of each of the four mammary glands is functionally quite separate and under no normal circumstances can milk formed in one gland pass into another gland. While it is a relatively easy matter to dissect apart the left and right halves of the udder — it has already been noted that the median suspensory ligaments lie between the two halves — the fore and hind quarters of each side are anatomically so closely related that it is not possible to distinguish the two quarters. The limits of a quarter, however, can for anatomical purposes be determined by injecting special dyes into the quarter.

A.T. Cowie

Since the milk-producing capacity of the udder depends ultimately on the quantity of the glandular tissue present within the udder, the size of the udder is not always a reliable criterion for assessing milking capacity, for a large udder may owe its bulk to the presence of much connective and fatty tissues.

Little of the structure of the glandular tissue can be determined by the naked eye. On cutting through a teat and its quarter, the teat canal (or streak canal) and cistern are exposed. Above, and opening into the teat cistern, is the irregularly shaped gland cistern with its numerous pocket-like recesses into which open the large ducts. Above the gland cistern lies the gland substance which beyond the cistern region becomes less spongy and more solid in appearance (Figs VI 2 and 3).

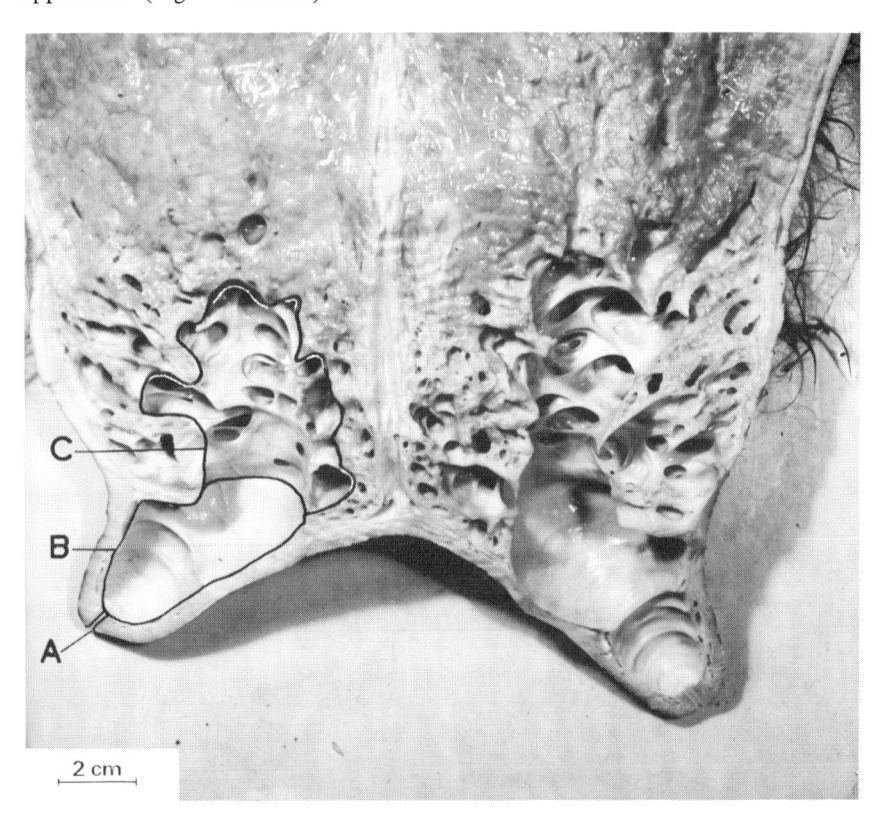

Fig VI 3. Cross-section of cistern region of a cow's udder.

A — teat canal   B — teat cistern   C — gland cistern

(From Ministry of Agriculture, Fisheries and Food (1959) Bulletin 177. London: HMSO).

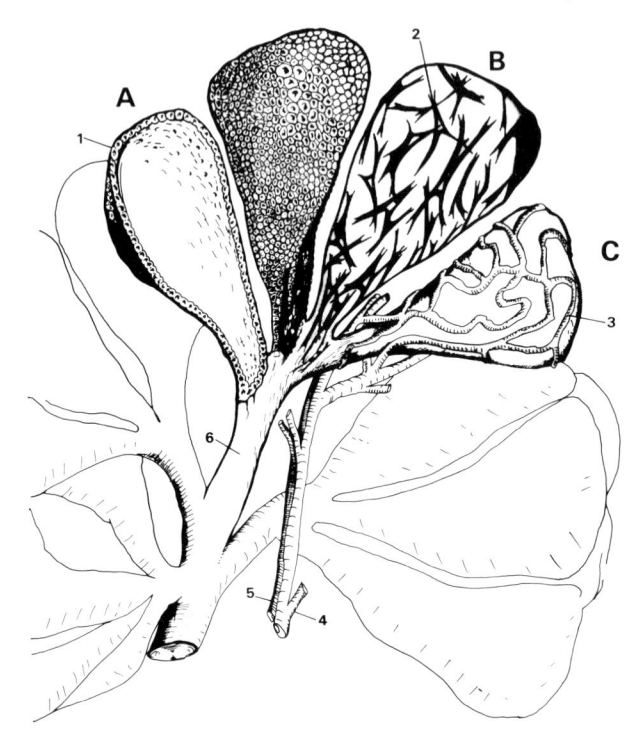

Fig VI 4. Diagram of a cluster of alveoli

  A — alveolus with part of wall removed. Wall is made up of a single layer of alveolar cells, 1

  B — alveolus showing myoepithelial cells, 2, lying on outer surface of alveolar cells

  C — alveolus showing network of capillaries, 3, which lie on top of the myoepithelial cells

      4 — small artery   5 — small vein
      6 — small duct

(The diagram is not drawn to scale and the branching and arrangement of the alveoli have been simplified for diagrammatic purposes).

The detailed structure of the udder can be determined only by microscopic examination using specialized histological techniques. The glandular tissue is then seen to be composed of alveoli and ducts. The alveoli (Fig VI 4) are small, pear-shaped or egg-shaped sacs whose walls are composed of a single layer of cells — the alveolar cells — resting on a fine membrane — the basement membrane. The sections cut for microscopic examination are so thin that the alveoli are cut into slices which explains their net-like appearance in photographs of such preparations (e.g. Figs VI 5 and 6). When the alveoli are distended their cross-section measures just under 0.5 mm (Fig VI 5). The alveolar cells are responsible for manufacturing the milk from precursor substances in the blood. The blood is brought to the alveoli by a fine network of capillary vessels which surround each alveolus (Fig VI 4). When the alveolus is empty its walls are contracted and are often collapsed into folds (Fig VI 6). As the milk is formed by the alveolar cells it is passed into the lumen or central cavity of the alveolus where it is stored; as the walls of the alveolus become stretched the shape of the alveolar cells is altered. In the empty alveolus the alveolar cells are tall and elongated (Fig VI 7), but as the alveolus becomes distended the cells become greatly flattened (Fig VI 8), and their milk-producing activity will, eventually, be much decreased.

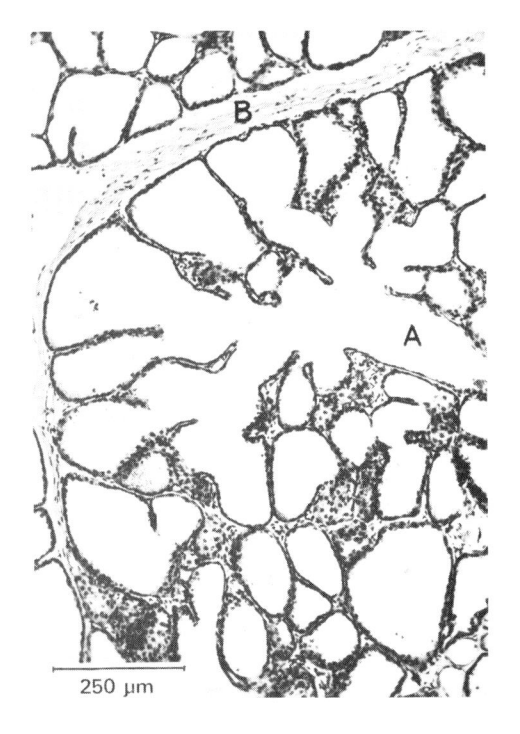

Fig VI 5. A section from part of a lobule from the left half of a goat's udder which was full of milk

    A — cluster of alveoli opening into a duct
    B — connective tissue septum

(From Ministry of Agriculture, Fisheries and Food (1959) Bulletin 177. London: HMSO)

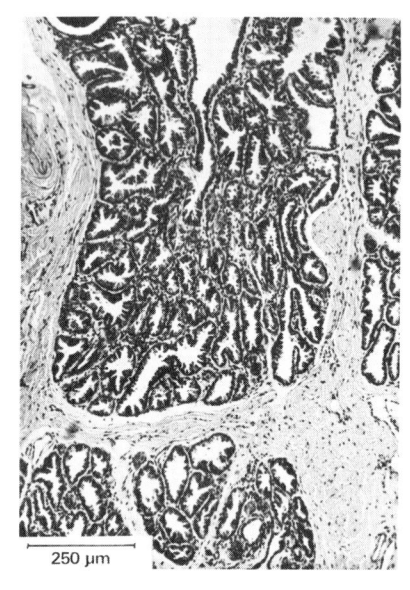

Fig VI 6. A section from part of a lobule from the right half of the udder of the same goat within an hour of milking, i.e. the udder was relatively empty. Compare with Fig VI 5.
(From Ministry of Agriculture, Fisheries and Food (1959) Bulletin 177. London: HMSO).

The alveoli are arranged in lobules or small clusters. This arrangement is well illustrated in the wax model constructed from drawings of microscopic sections as shown in Fig VI 9. These lobules are surrounded and bound together by bands or septa of connective tissue. Parts of these bands of connective tissue are seen in Figs VI 5 and 6. These septa link up with the bands of connective tissue which spring from the suspensory ligaments and form a strong supporting meshwork for the glandular tissue. The mouth of each alveolus opens into a tube or duct. Sometimes a cluster of alveoli may have a common opening into a duct (Fig VI 5) or an alveolus may open directly into another alveolus. The ducts from adjacent alveoli join up and eventually large ducts are formed which are clearly visible to the naked eye. The ducts within the lobules have thin walls like the alveoli, but as the ducts increase in diameter their walls become thicker and are composed of two layers of cells resting on a basement membrane. The large ducts converge and open into the gland cistern which, as we have already seen, is an irregularly-shaped cavity at the lower portion of the quarter communicating directly through an opening in its base with the teat cistern. It is by way of the duct system that the milk, formed by the alveolar cells, ultimately reaches the gland and teat cisterns. The mechanisms involved in the transfer of the milk from the minute alveoli to the gland and teat cisterns will be described later.

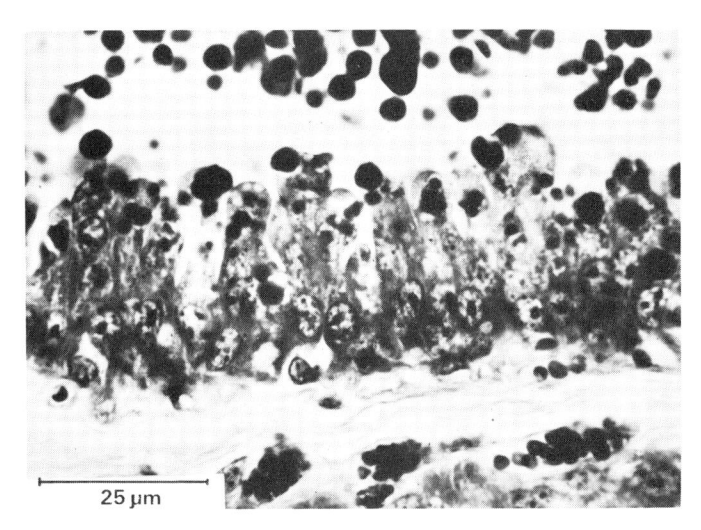

Fig VI 7. Alveolar cells of a goat recently milked (from right half of udder — see Fig VI 6). The alveolar cells are tall and contain dark-stained fat droplets. There are also fat droplets lying free in the lumen of the alveolus.
(From Ministry of Agriculture, Fisheries and Food (1959) Bulletin 177. London: HMSO).

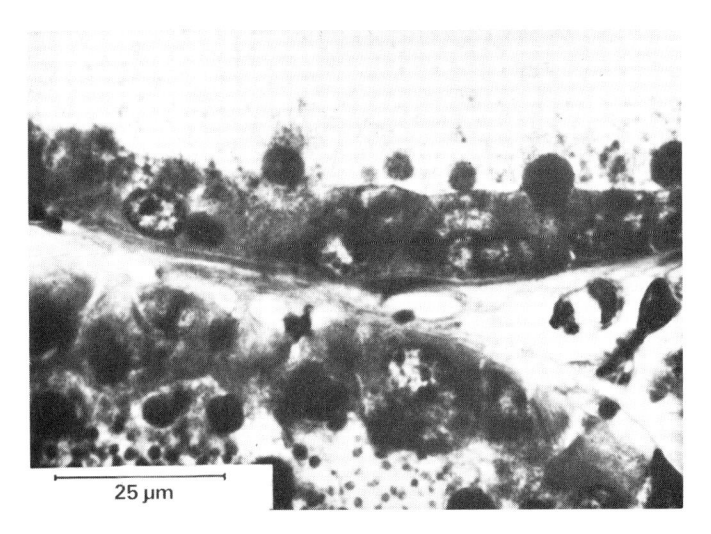

Fig VI 8. Alveolar cells of a goat just before milking (from left half of udder — see Fig VI 5). The alveoli are full of milk and so the cells are stretched and flattened. A few fat droplets are visible.
(From Ministry of Agriculture, Fisheries and Food (1959) Bulletin 177. London: HMSO).

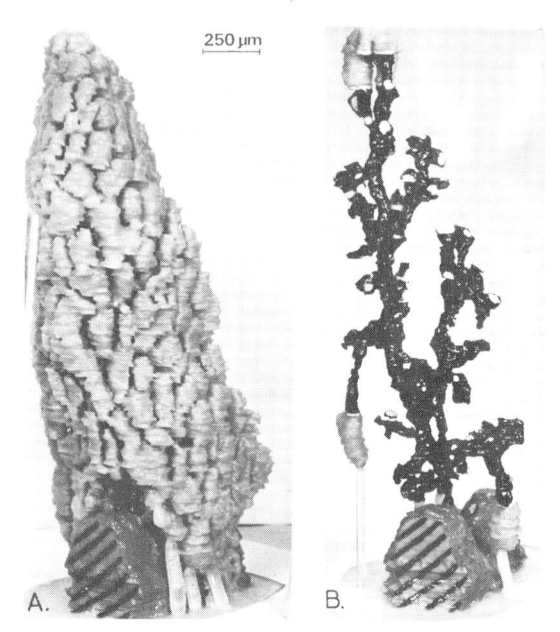

Fig VI 9
 A. Wax reconstruction of a mammary lobule from a cow's udder.
 B. Same model with alveoli removed to show duct system within the lobule.

(By courtesy of Professor A.F. Weber and American Journal of Veterinary Research; see ref. 6).

The gland and teat cisterns are lined by a delicate membrane — an epithelial membrane — which is made up of two layers of cells. This membrane is similar to, and is in continuity with, the membrane of the ducts. The lining membrane of the teat cistern may be thrown into folds and there is sometimes a fold at the junction of the teat and gland cisterns. The teat canal is lined with quite a different type of membrane. It is composed of many layers of flattened cells and is virtually a continuation of the skin of the teat. The abrupt change in the character of the membrane occurs just where the teat canal opens into the teat cistern. The teat canal is kept closed by a network of smooth muscle fibres in the wall of the tip of the teat.

The wall of the teat is rich in large blood vessels. These lie under the basement membrane of the lining epithelium, and when they become engorged with blood the teat becomes turgid.

A.T. Cowie

BLOOD SUPPLY OF THE UDDER

Since milk is derived from substances present in the blood stream, the routes by which the blood reaches and leaves the udder are worthy of some consideration. Some idea of their importance will be gained when it is realized that in a cow giving 20 kg (44 lb) of milk per day, about 9000 kg (20 000 lb) of blood will circulate through the udder each day. Arterial blood from the heart reaches the udder mainly by way of large paired arteries (the external pudic or mammary arteries), about 10 mm (0.4 in) in cross-section, each of which passes through the inguinal canal on its own side. The connection of these arteries with the heart is shown in Fig VI 10. On emerging from the canal, the artery (and also the associated vein) shows an S-shaped curve which probably provides for the lengthening of these vessels when the median ligaments stretch as the udder fills up. Before entering the udder, the artery sometimes gives off a small branch (subcutaneous abdominal artery) which sends off some side branches into the udder substance. On entering the udder the mammary arteries divide up into smaller and smaller branches which finally become capillaries. These, as we have already seen, surround each alveolus (Fig VI 4). The left artery supplies the left

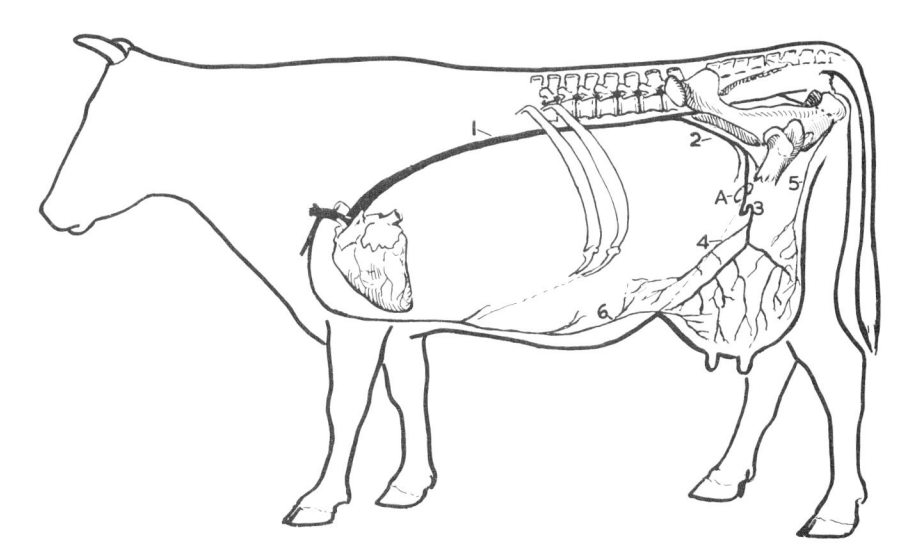

Fig VI 10. Diagram of the arteries supplying the udder with blood.

1. Aorta
2. External iliac artery
3. External pudic or mammary artery
A. Inguinal canal

4. Subcutaneous abdominal artery
5. Perineal artery
6. Cranial epigastric artery

(From Ministry of Agriculture, Fisheries and Food (1959) Bulletin 177. London: HMSO).

half of the udder and the right artery the right half but there are some small arterial connections which pass from one half of the udder to the other, so that some of the blood passing into the left half of the udder may find its way into the right half and vice versa. There are two other routes (the cranial epigastric and perineal arteries) by which blood may reach the udder. These, shown in Fig VI 10, are of minor importance. The blood in the capillaries is brought into intimate contact with the alveolar cells, for through the walls of the capillaries there is an interchange between the fluid constituents of the blood and the fluid in the tissue spaces which bathes the alveolar cells.

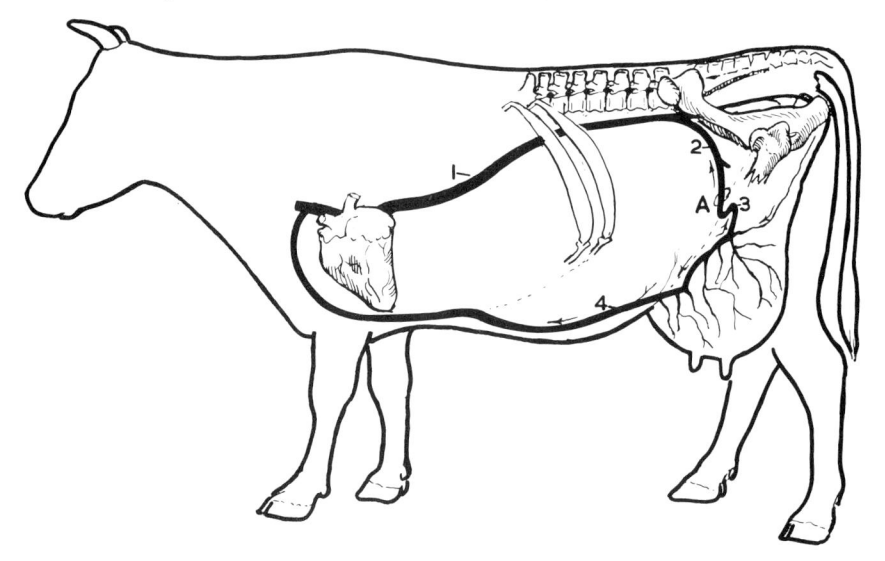

Fig VI 11. Diagram of the veins draining the blood from the udder

1. Vena cava
2. External iliac vein
3. External pudic vein.
4. Subcutaneous abdominal vein
A. Inguinal canal

The arrows indicate direction of blood flow in the lactating cow.

After passing through the capillaries the blood reaches the small veins. The small veins unite to form larger veins that eventually link up at the base of the udder to form a venous circle. In the lactating cow the blood leaves the venous circle by two routes (Fig VI 11). There is a large vein (external pudic vein) which follows the course of the artery through the inguinal canal and there is the subcutaneous abdominal vein or milk vein which passes along the abdominal wall just under the skin until it pierces the muscles of the abdominal wall at a depression (the milk well) near the xiphoid cartilage (breast bone). Either of these veins is large enough to carry all the blood from the udder and perhaps undue importance has in the past been attached to the size of the milk veins in the belief that large milk veins indicated a large volume of blood passing through the udder. As long ago as 1838 Youatt[1] emphasized that "the subcutaneous or milk vein has nothing to do with the udder, but conveys the blood from the fore part of the chest and sides to the inguinal vein". Only recently has the venous drainage of the udder been carefully investigated[2]. In the heifer the situation is much as Youatt described it, the flow of blood in the milk vein in the neighbourhood of the udder being towards and into the external pudic vein but in the lactating cow the valves in the milk vein become permanently incompetent and the blood flow is then in the opposite direction (Fig VI 11) so that the milk vein does indeed carry much of the blood leaving the udder when the animal is standing.

LYMPHATIC SYSTEM OF THE UDDER

Not all the fluid leaving the capillaries re-enters them. Some of the fluid returns eventually to the blood stream by way of the lymphatic system which is well developed in the udder. The lymph vessels begin as small capillary-like vessels which join up to form ultimately one or several large vessels which pass through the inguinal canal to join the lumbar lymph trunks thence into the thoracic duct which ends by draining into a vein in the neck. Interpolated in the course of the lymph vessel are the lymph nodes or lymph glands which, among other functions, serve as a very efficient filter mechanism for removing and destroying micro-organisms. Just before calving and in early lactation the lymph flow from each half of the udder may exceed 1 litre per hour[3].

NERVE SUPPLY TO THE UDDER

The main nerves of the udder are the first and second lumbar nerves, the external spermatic and the perineal nerves (Fig VI 12). While these nerves are mainly sensory i.e. they carry sensory impulses from the udder to the brain, they also contain some fibres from the sympathetic nervous system which are motor fibres to the muscular tissue in the walls of the arteries and to the scanty muscle fibres in the connective tissue of the udder. These sympathetic nerve fibres are responsible for controlling the rate of blood flow through the udder by regulating the diameter of the arteries. There is no direct nervous control of the secretory (i.e. milk producing) activity of the alveolar cells, but the activity of the cells may be influenced indirectly by alterations in the rate of blood flow. The glandular tissue itself is poorly supplied with sensory nerve endings; the teat, on the other hand, has a rich sensory innervation although the highly specialized sensory nerve endings said to be present in the nipples of some species have not been found in the teat of the cow[4].

Further information on the anatomy of the udder will be found in references [5] and [6].

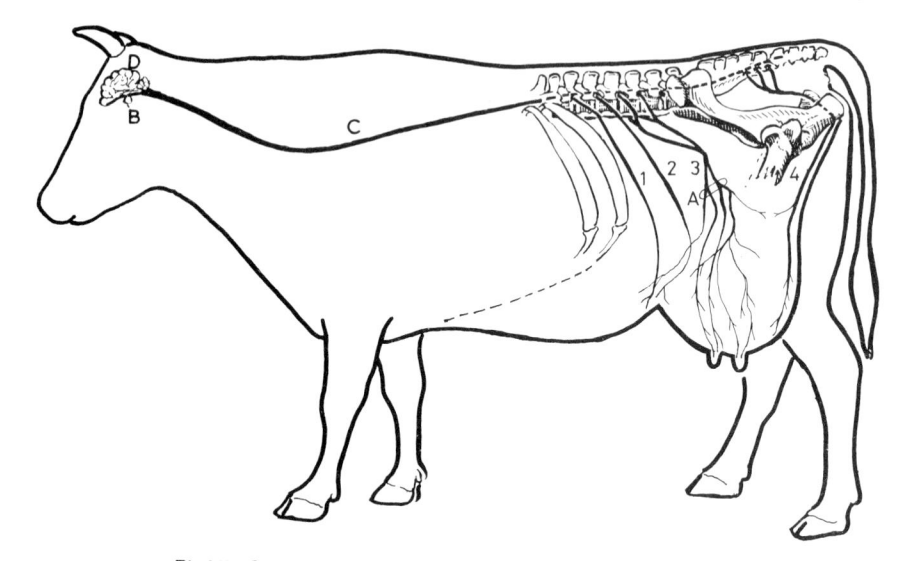

Fig VI 12. Diagram of the nerves supplying the udder.
1. Ventral branch of the first lumbar nerve
2. Ventral branch of the second lumbar nerve
3. External spermatic nerve
4. Perineal nerve
    A.   Inguinal canal   B.   Pituitary gland
    C.   Spinal cord   D.   Brain

DEVELOPMENT OF THE MAMMARY GLAND

The various stages of development of the udder need not be described here in any detail and it will suffice to say that evidence obtained from transplantation experiments has established that the growth of the mammary gland is under the control of hormones* arising from certain endocrine† glands. Before pregnancy the mammary gland of the heifer consists of a simple duct system embedded in fat and connective tissues. There are probably few alveoli present. When pregnancy ensues this duct system grows rapidly under the stimulus of hormones from the ovary (oestrogen and progesterone) and other

---

* A hormone is a specific chemical substance, produced by an endocrine gland, which circulates in the blood to all parts of the body and which, in small amounts, influences the rate at which cellular functions proceed in other organs without itself contributing significant quantities of matter or energy.

† An endocrine gland is a ductless gland which pours its elaborated product (a hormone) into the blood or lymph.

hormones from the anterior-pituitary gland and the placenta. This growth of the duct system is accompanied by the appearance of alveolar tissue. The building up of the glandular tissue is particularly rapid in the second half of pregnancy and its growth probably continues into the early part of lactation.

As lactation declines there is, in most mammals, a slow disappearance or involution of the lobules of alveoli in the mammary gland; after weaning these changes in the mammary gland are speeded up and in some species the alveolar tissues completely disappear so that the mammary gland consists again of a simple collection of ducts as existed before pregnancy. In the dairy cow, however, the sequence of events is probably somewhat different in that early in lactation the cow usually again becomes in-calf and lactation and pregnancy are concurrent. The interval at the end of lactation when the cow is dried off, and when rapid involution of the mammary gland might be expected, is thus the last two months of pregnancy when the mammary gland is normally becoming fully developed for the next lactation. Under these circumstances it might be expected that the atrophy of the alveolar tissue would be prevented by the hormonal changes of late pregnancy. There have been but few relevant studies on the structure of the cow's udder at this period but what evidence there is indicates that little glandular involution occurs in the usual dry period between lactations.

Although no major structural changes have been observed in the mammary tissues during the dry period, this period is important for the success of the next lactation. It has been known for some time that if the dry period is unduly short (less than 6 weeks) then the milk yields in the subsequent lactation (i.e. after calving) tend to be reduced; indeed if the dry period be eliminated then, as has been shown in experiments with identical twins, cows not allowed a dry period will give only 70% of the milk yield of their partners who were dried off for a two-month period. It has been customary to believe that the beneficial effects of the dry period were due to the improvement in the nutrition of the cow when the drain of lactation ended. Evidence, however, that nutritional factors were not the essential cause of reduced yields has been provided in studies in which two quarters of a cow were dried off while the other two were regularly milked to the end of pregnancy; when the cow calved the two quarters which had been dried off gave considerably more milk throughout the lactation than the two quarters which had not been rested. Clearly, since all the quarters were receiving blood of the same composition so far as milk precursors and hormones were concerned, the reduced yields were not directly due to nutritional or hormonal factors. That the continued milking of the two quarters by maintaining the alveolar cells in a functional state prevented the usual renewal and replacement of the cells could well be a reasonable explanation since the evidence all points to some local alteration in the functional capacity of the alveoli. (For further information and references, see reference [7].

PHYSIOLOGY OF LACTATION

In discussing the physiology of lactation it is convenient to consider this process as consisting of two distinct components. First there is *milk secretion*, this is the manufacture of milk constituents by the alveolar cells and the passage of the constituents from the cells into the lumen or cavity of the alveolus. Milk secretion is a continuous process, proceeding at a steady rate so long as the alveoli do not become excessively distended with milk. Secondly there is the *ejection of milk* from the alveolar cavity into the large ducts and cisterns where it can be readily withdrawn by the suckling calf or by the milking machine. As will be noted below milk ejection occurs in response to suckling or milking and is therefore an intermittent process. In the past these two stages in lactation have been confused and considerable misunderstanding of the physiology of lactation has thereby arisen.

MILK SECRETION

Hormonal control

The secretory activity of the alveolar cells is controlled not by nerves as in the case of certain other glands (e.g. salivary glands) but by hormones. Of major importance in maintaining the production of milk are the hormones of the anterior pituitary. (The pituitary is a small endocrine gland attached to the base of the brain; it has two parts — an anterior lobe and a posterior lobe — the hormones of the anterior lobe regulate the secretory activity of the alveolar cells of the udder while a hormone (oxytocin) of the posterior lobe is concerned with milk ejection (see below).) Prolactin was the first anterior-pituitary hormone shown to be concerned with milk secretion but research soon demonstrated that, in the cow, pituitary growth hormone was equally important in this process. In addition to prolactin and growth hormone the thyrotrophic and the adrenocorticotrophic hormones of the anterior pituitary are also necessary as they control the secretory activities of the thyroid and adrenal glands respectively whose hormones must be present in the blood to ensure full milk production. The injection of growth hormone or thyroid hormones to cows in declining lactation generally increases the milk yield which suggests that the decline in yield may be associated with a reduction in the production of these hormones. The effects of the surgical removal of the pituitary gland from the lactating cow have not, so far as we are aware, been reported but in the lactating goat this operation causes the milk yield to fall to negligible amounts within a few weeks; yields can be restored to normal by the regular administration of suitable hormones — prolactin, growth hormone, thyroid hormone and adrenocortical hormones.

It is generally believed that the production of the anterior-pituitary hormones necessary for maintaining milk secretion is regulated by the nervous impulses arising from stimulation of the teat during suckling or milking, these impulses being carried by way of the spinal cord to the hypothalamic region of the brain where the centres controlling anterior-pituitary activities are situated. This may well be the usual mechanism of control but in some species, e.g. the goat and sheep, it is not the only mechanism since the udder of the goat or sheep may have its nerve supply completely disrupted by surgically cutting the nerve, or in the case of the goat, by transplanting the udder to a different region of the body, yet the milk yields in such animals will remain virtually normal despite the fact that no nervous stimuli from the teats or udder can then reach the brain[7]. The mechanisms co-ordinating mammary and pituitary functions in such circumstances are at present unknown.

Milk

Milk is for a time the sole nourishment of the new-born and must of necessity be a complete and balanced diet supplying all the essentials for energy requirements and growth. Most of its constituents — the milk fat, lactose and most of the protein — are manufactured in the alveolar cells from precursor substances in the blood but other constituents — the water, the antibodies, vitamins and salts — are directly transferred from the blood. When secreted, milk contains two liquid phases, fat and water, between which are partitioned the many substances it contains. Dissolved in the fat are numerous compounds such as phospholipids, sterols, carotenoids and fat-soluble vitamins while the watery phase holds in solution lactose, the water-soluble vitamins and some of the minerals; it also carries in the colloid state the proteins and the rest of the minerals.

The secretory process

In the past there has been controversy over how constituents of milk pass out of the alveolar cells into the lumen of the alveolus; recently a good deal of information on this process has been obtained with the help of the electron microscope. The milk fat first appears as small droplets in the basal portion of the alveolar cell, these droplets migrate upwards within the cell increasing in size as they do so until they lie at the cell apex. They then push out the cell membrane which clings closely to them. As the degree of protrusion of the droplet increases, the cell membrane constricts behind it and the droplet eventually becomes pinched off and, completely encased in an envelope of cell membrane, it falls free into the alveolar lumen. The process of the pinching-off of the fat droplet permits the cell membrane at the apex of the cell to regain its continuity with minimal loss of the cytoplasm of the cell.

The protein granules first appear in small spaces or vacuoles within the cytoplasm of the cell. These vacuoles move towards the apex of the cell and finally burst open into the lumen thus shedding their granules.

There is little hindrance to the transfer of water from the blood into milk for it has been shown that changes in the osmotic pressure of the blood plasma, as for example after a large drink of water, are rapidly followed by a similar change in the osmotic pressure of the milk. Lactose is the main contributor to the osmotic effect in milk, potassium, sodium and chloride playing lesser roles. The rate of production of lactose thus plays a major role in regulating the volume of milk produced. Milk is isotonic with blood and so it has much lower concentrations of sodium and chloride ions than blood to compensate for the presence of lactose. The maintenance of a low concentration of sodium in milk in comparison with that in the blood means that a mechanism must exist for the active transfer of the sodium ion. Recent studies indicate that this mechanism is present in the alveolar cells since the duct walls are permeable only to water. Damage to the alveolar cells as occurs in infections of the udder (mastitis) interferes with the transport mechanisms with the result that sodium and chloride ions leak into the milk from the blood to offset the osmotic deficiency caused by the lowered lactose levels. (Further information on milk secretion will be found in references [7] and [8].

MILK EJECTION

It has long been recognised that before the full yield of milk can be obtained from a cow at any one milking the milk has to be 'let down'. This phenomenon is characterised by a sudden increase in the fullness and tension of the udder due to an increase in the pressure within the gland and teat cisterns. It was once believed that this sudden increase in intramammary pressure was the result of a very rapid production of milk by the alveolar cells in response to stimulation of the teat at the beginning of suckling or milking. This idea is quite wrong for it is now well established that all the milk obtained at a milking is present as milk in the alveoli, ducts and cisterns before milking begins. The let-down, or as it is now termed, milk ejection, is the forceful expulsion or ejection of milk from the alveoli and fine ducts into the larger ducts and cisterns so that the milk stored in the alveoli can become available to the suckling calf or to the milker.

As milk secretion proceeds the lumen of the alveolus fills up and the pressure within increases, some of the milk passes into the duct system and eventually reaches the cisterns. The milk in the larger ducts and cisterns can be readily removed from the udder, e.g. by simply draining it off through a teat cannula. The milk within the alveoli and fine ducts cannot be so removed since it is retained there by capillary forces. To empty the alveoli of their stored milk the alveoli must contract so that the milk is squeezed out into the duct system.

Milk ejection reflex

A good deal is now known of the physiological mechanisms which bring about milk ejection. The act of milking or suckling stimulates nerve endings in the teat causing nervous impulses to reach the brain by way of the mammary nerves and the spinal cord where they bring about the release of the hormone oxytocin from the posterior pituitary. This hormone is carried in the circulating blood to the mammary gland where it acts on certain specialized cells, the myoepithelial cells, which surround the alveoli, causing them to contract thereby squeezing the alveoli and ejecting or expelling the milk. Milk ejection is a reflex act and being so it is involuntary and is not under the conscious control of the cow, hence the old terms 'let-down' and 'holding up of milk' are rather misleading as they imply conscious acts on the part of the cow. The milk-ejection reflex, unlike most other reflexes in which both the routes to and from the central nervous system are nervous pathways, has a hormonal pathway from the central nervous system and is thus a neurohormonal reflex.

Nervous pathways of reflex

The precise pathways within the spinal cord and brain are still not completely defined although studies on goats, sheep and laboratory animals are helping to clarify them. There can be little doubt that the impulses reach the hypothalamus, a region of the brain overlying the pituitary gland where there are specialized collections of nerve cells, the supraoptic and paraventricular nuclei, in which the hormone oxytocin is produced. The hormone, attached to a carrier substance, passes down the extensions or axons of these cells which lead through the pituitary stalk into the posterior lobe of the pituitary where the oxytocin is stored. The precise method by which the impulses on reaching the hypothalamus bring about the release of oxytocin into the blood is as yet not fully understood, but the essential role of the hypothalamus is clear for it can be demonstrated that if small areas of the hypothalamus are destroyed the milk-ejection reflex no longer occurs; also electrical stimulation of the nuclei in the hypothalamus will bring about milk ejection.

Oxytocin in the blood

It is now possible to measure accurately the concentration of oxytocin in blood. An assay procedure based on the milk-ejection response in the lactating guinea pig or rat has proved to be very sensitive and highly specific. Under conditions which avoided any stress to the cow serial samples of blood were collected from the external jugular vein from the start to the end of the milking procedure. (The blood from the posterior pituitary drains into the external jugular vein so this is a very suitable blood vessel from which to sample the blood.) Oxytocin was detected in the jugular blood in most instances just after the application of the teat cups, a second release sometimes occuring during milking; release of oxytocin was also detected in response to the approach of the milker, the washing of the udder, and to foremilking.

Oxytocin rapidly disappears from the blood; its 'half-life' in the cow is just over a minute. As the oxytocin level in the blood is reduced the alveoli relax and if milking has not been completed some milk will re-enter the alveoli and the intramammary pressure falls. For milking out to be thorough it is therefore desirable that milking should be completed as quickly as possible after the onset of the milk-ejection reflex. It may be noted, however, that normal milking never empties the udder of milk and if injections of oxytocin are given at the end of milking a further quantity of milk (about 15% of the milk present before milking started) of very high fat content will be obtained – this is usually termed the residual milk.

Myoepithelial cells

For years the nature of the mechanism causing the contraction of the alveoli was uncertain. Some research workers believed it was smooth muscle fibres which surrounded the alveoli while others had described special contractile cells. However a detailed description of the contractile elements based on light and electron microscope studies are now available. The contractile cells are termed myoepithelial cells. They are flattened and somewhat star-shaped with long arms or processes and they lie on the outer surface of the alveoli (Fig VI 13). Not only are these cells present on the alveoli but they also lie along the ducts and extend on to the cisterns. The orientation of myoepithelial cells varies with the structure on which they lie; on the ducts, particularly the smallest, they are arranged lengthwise, while on the alveolus their distribution is lengthwise or in spirals around the alveolus. These arrangements are well adapted not only for compressing the alveolus but also for shortening the ducts thereby increasing their cross-section and allowing the milk to flow through them more rapidly. These myoepithelial cells are quite separate structures from the plain muscle fibres with which they were at one time confused. Indeed there is little plain muscle in the udder and that is mostly associated with the blood vessels and teat. Careful morphological studies and the observed effects of local application of oxytocin to the mouse mammary gland leave little doubt that the myoepithelial cells are the active contractile agents which respond to oxytocin.

The significance of the milk-ejection reflex

The milk-ejection reflex does not have the same physiological significance in all species. It can be shown experimentally that in the goat and sheep the reflex can be abolished yet milk production will remain normal if the animals are carefully milked by hand; in the rat, on the other hand, lactation comes to an end if the reflex is abolished since the milk can no longer be removed from the gland and the resultant build-up of pressure within the gland eventually destroys the alveolar cells so ending their secretory activity. It is probable that in the goat and sheep the anatomical structure of the udder facilitates the drainage of the alveoli into the ducts and cisterns; it is also possible that the alveoli may contract, at least partially, in response to direct stimuli such as arise during massage of the udder. The cow would appear to hold an intermediate position in that up to 40%

of the milk can be removed passively from the udder but for the removal of the remainder the proper functioning of the milk-ejection reflex is necessary.

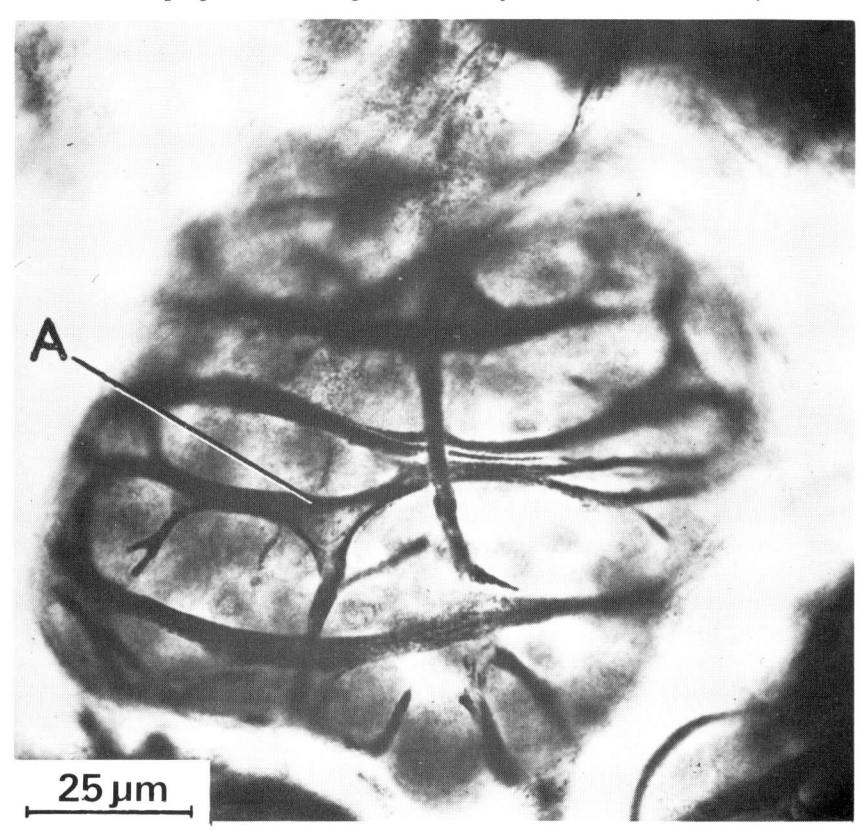

Fig VI 13. A small contracted alveolus in surface view showing myoepithelial cell with nucleus (A). Note the branching processes or arms.
(From Ministry of Agriculture, Fisheries and Food (1959) Bulletin 177. London: HMSO).

Inhibition of the milk-ejection reflex

It has long been recognised that the milk-ejection reflex can be blocked or inhibited under conditions of excitement, fear, stress or pain so that when the cow is milked the full yield is not obtained but only that milk which is present in the cisterns and large ducts. Injections of adrenaline will block the reflex. (Adrenaline is a hormone from the medulla of the adrenal gland which is poured into the blood under stress conditions.) It was believed that adrenaline acted by shutting down much of the vascular system of the mammary gland thus reducing the blood flow through the gland and thereby preventing sufficient quantities of oxytocin from reaching the myoepithelial cells but there is also evidence that adrenaline can act directly on the myoepithelial cells and inhibit their contraction. It is doubtful, however, whether this is the mechanism normally inhibiting milk ejection; subsequent research strongly suggests that the main factor in emotional disturbances of the milk-ejection reflex is a partial or complete block of the release of oxytocin from the posterior pituitary.

Conditioning of the milk-ejection reflex

The milk-ejection reflex occurring in response to the suckling stimulus is an inborn reflex, i.e. it is an immediate and involuntary response to the stimulation of the teat and one which is inherent and requires no training. It has been well established by Pavlov and others that animals can be trained to respond reflexly to stimuli other than those which evoke the inborn reflex. If two stimuli, one of which evokes the inborn reflex, and the other a neutral stimulus, are applied in close association a number of times then the neutral stimulus acquires the ability to arouse on its own the same response as that produced by the unconditioned original stimulus. The neutral stimulus now becomes a conditioned stimulus and the reflex it induces is a conditioned reflex. Feeding a cow concentrates just before milking does not normally cause milk ejection but if the feeding is regularly carried out immediately before milking starts then the act of feeding will become an involuntary signal to trigger off the release of oxytocin, that is the feeding of concentrates becomes a conditioned stimulus. Similarly the appearance of the milker may become a conditioned stimulus and as a result oxytocin can be detected in the jugular blood.

Until recently it was considered that the occurrence of milk ejection to the presentation of the calf was a conditioned reflex but a recent study in Belgium[9] has revealed that milk ejection may occur in newly calved animals (which have not previously been milked or suckled) on seeing their calf for the first time and before the calf has touched the teats: the sight of the calf must therefore be an unconditioned reflex.

The importance of establishing conditioned stimuli for milk ejection lies in the fact that the conditioned reflex in conjunction with the inborn reflex gives a very effective milk ejection permitting efficient and rapid milking out of the udder, particularly if for any reason the unconditioned stimulus has been unusually weak. (The economic importance of rapid milking and the saving of labour time is discussed in Chapters VII and VIII.)

In establishing conditioned reflexes certain conditions must be observed: first, the stimulus which is to become the signal in the conditioned reflex must precede and overlap in point of time the action of the unconditioned stimulus (if this order be reversed no conditioned reflex will be established); secondly, the animal must not be under any sort of stress or be disturbed or distracted by unusual events; thirdly, the animal must be in a good state of health; and lastly the new stimulus must not be of such a nature that the animal would regard it as upsetting or unusual.

Once a conditioned reflex has been established it can be readily 'extinguished', for example if the routine becomes irregular and the conditioned stimulus does not regularly precede the unconditioned one. There are thus very good physiological reasons for keeping to a regular routine in the milking shed, for the quiet handling of animals and for the avoidance as far as possible of unusual or disturbing events. (Further information on milk ejection will be found in reference [7].)

MECHANISM OF SUCKLING

In the past there has been controversy about how the suckling obtains milk from the teat. The argument has now been settled by the use of cineradiography[7]. Sterile radio-opaque material was injected into the gland cistern of a lactating goat and its kid was allowed to suckle while cineradiographs were taken at a speed of 25 and 50 frames per second. The kid compresses the neck of the teat between its tongue and hard palate and displaces the contents of the teat cistern into its mouth by compressing the teat between its tongue and palate from the base of the teat towards the tip, the jaws and tongue are then lowered to allow the teat cistern to fill up again with milk and the cycle is repeated. The cineradiographic examination of a calf feeding from the cow's teat shows that the cow's teat is compressed and refills in a similar manner. Sucking, that is the production of a vacuum within the mouth cavity, is not an essential feature in suckling, although it aids the process, since the young kid or calf can still readily obtain milk from the teat even if it is prevented experimentally from creating a vacuum within the mouth. The act of suckling is thus analogous to hand milking in that the milk in the teat cistern is trapped there by compression of the base of the teat and then forced out through the teat canal by positive pressure on the teat.

Calves and kids very soon learn to obtain milk by suction alone if they are fed through hard rubber teats which they cannot compress or strip but this is not the natural method.

The mechanics of machine milking has also been elucidated by cineradiography (Chapter V).

REFERENCES

(1) Youatt, W. (1838) *Cattle: their breeds, management and diseases.* London: Baldwin & Craddock p.244.

(2) Linzell, J.L. (1974) In: *Lactation,* vol 1, B.L. Larson & V.R. Smith, (Editors) New York and London: Academic Press, p.143.

(3) Lascelles, A.K., Cowie, A.T., Hartmann, P.E. & Edwards, M.J. (1964) *Research in Veterinary Science,* 5, 190.

(4) Linzell, J.L. (1971) In: *Lactation,* I.R. Falconer, (Editor) London: Butterworths, p.41.

(5) Turner, C.W. (1952) *The Mammary Gland I. The Anatomy of the Udder of Cattle and Domestic Animals.* Columbia, Missouri: Lucas Brothers.

(6) Weber, A.F., Kitchell, R.L. & Sautter, J.H. (1955) *American Journal of Veterinary Research,* 16, 255.

(7) Cowie, A.T. & Tindal, J.S. (1971) *Physiology of Lactation.* London: Arnold.

(8) Linzell, J.L. (1972) *Journal of Dairy Science,* 55, 1316.

(9) Peeters, G., De Buysscher, E. & Vandervelde, M. (1973) *Zentralblatt fur Veterinärmedizin, Reihe A,* 20, 531.

Chapter VII

MILKING ROUTINES

F H Dodd and T K Griffin

In their efforts to achieve maximum economic milk production, farmers influence lactation in dairy cows in a number of ways. Cattle are selected and bred for high yields of milk of desired chemical composition. They are then managed to reach their economic potential production by control of reproduction, nutrition and disease. These three methods are not independent and the degree of control exercised over any one of them varies with the skill of the farmer, and the herd circumstances. In addition to these self evident limits on milk production, milk secretion is also influenced by the methods of milk removal. The effect of milking on milk secretion is still difficult to understand but, in simple terms, over a period a cow cannot secrete more milk than is removed from the udder by milking. Thus maximizing milk removal in ways which are economic will take fullest advantage of secretion potential.

Lactation consists of the formation or secretion of milk, followed by milk storage, ejection and removal (see Chapter VI). Milk is secreted within the alveolar cells, and stored in the alveolar lumen and ducts. Secretion is continuous during normal lactation, but the rate of milk formation may not be constant. Once formed most of the milk is held by capillary forces in the fine ducts of the udder though some is expressed or drains to the sinuses of the udder and teat where it is retained by the sphincter of the streak canal. When milk ejection occurs milk is forced from the alveoli and small ducts towards the udder sinus. As a result it is possible for most but not all of the milk to be removed from the udder when external forces open the streak canal of the teat. The milk retained in the udder at the end of milking is called residual milk. The composition of the milk secreted by the alveoli throughout normal milking intervals is probably constant. This holds for all the main components of milk even though the concentration of milk fat in the milk stored in the udder is always much higher than that secreted, or that obtained at individual milkings. This is due mainly to the greater retention of milk fat globules, compared with the aqueous phase of the milk, in the alveoli and fine ducts.

This chapter is concerned with the direct and indirect effects of the management of milking on milk removal, and on the amount of secretory tissue in the udder and its rate of activity.

MILKING METHODS AND RATE OF MILK SECRETION

There is now conclusive proof that milk is secreted continuously in the udder[1], though it has been the conventional view that within normal milking intervals the rate of formation falls from the start to the end of the period as the intramammary milk pressure increases. This decline in rate of secretion was also believed to be much greater for fat than for the other components of milk. These views appeared to be supported by the direct evidence of Ragsdale, Turner and Brody[2] who measured the milk yields after milking intervals of 3 to 18 h and concluded that the rate of secretion of total milk declined by 5%/h and that of fat secretion by a much higher figure. It was also supported by the indirect evidence of milk recording data[3]. The latter show the apparent rates of secretion of both milk and fat to be much greater in the short day interval compared with the longer night interval.

In the interpretation of these studies it was assumed that virtually all the milk obtained at a milking had been secreted since the last milking and that any carryover of milk from a previous interval could be ignored. We now know that this is not correct. Johansson demonstrated that residual milk and particularly the residual fat not removed by milking are considerable and cannot be ignored in measuring the rate of secretion[4]. In later work various methods have been used to eliminate or correct for residual milk[5,6,7,8,9]. These studies gave similar results and indicated that while the rate of secretion of all the main components of milk is curvilinear with time, the relationship is virtually linear for the first 12 h and in some studies for 20 h even with relatively high yielding cows. Furthermore contrary to the conventional view, the rate of fat secretion declines less quickly than that of the aqueous phase as time passes[5,7] but the effect is small (Fig VII 1). This may indicate that the apparent decline in rate of secretion of most of the constituents of milk as the milking interval increases is due to their reabsorption during storage rather than a true decline in secretion rate, the fat being less easily reabsorbed. There are considerable differences between cows in the decline in the rates of secretion after extended milking intervals and this variation is not due primarily to differences in milk yield[6]. The later experiments also demonstrated that once the rate of milk secretion has been depressed by delayed milking it does not immediately revert to normal when milking is resumed[7,9]. For example the yield of milk obtained in an 8 h interval following a 24 h interval will be 25% less than if the 8 h interval had followed on another interval of 8 h (Fig VII 2). These patterns of secretion have been well established by a series of experiments but on their own they do not explain the variability of the fat percentages of the milk obtained from an individual cow at consecutive milkings.

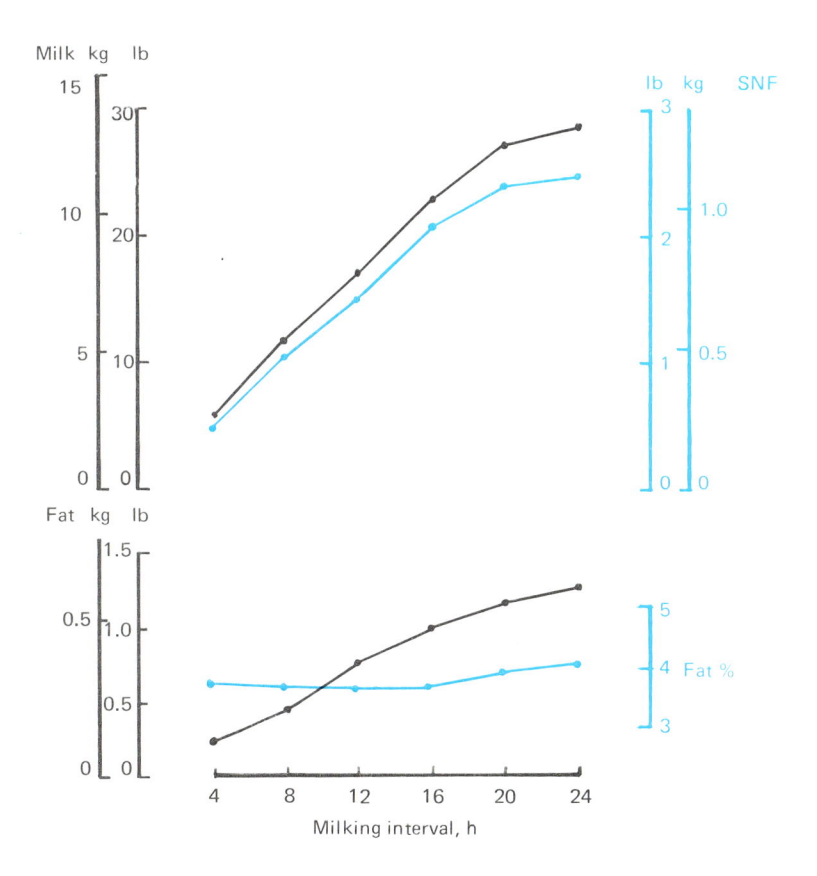

Fig VII 1. The quantities of milk, fat and solids-not-fat (SNF) secreted and the changes in fat content during milking intervals of 4, 8, 12, 16, 20 and 24 hours. Residual milk removed for all milkings[7].

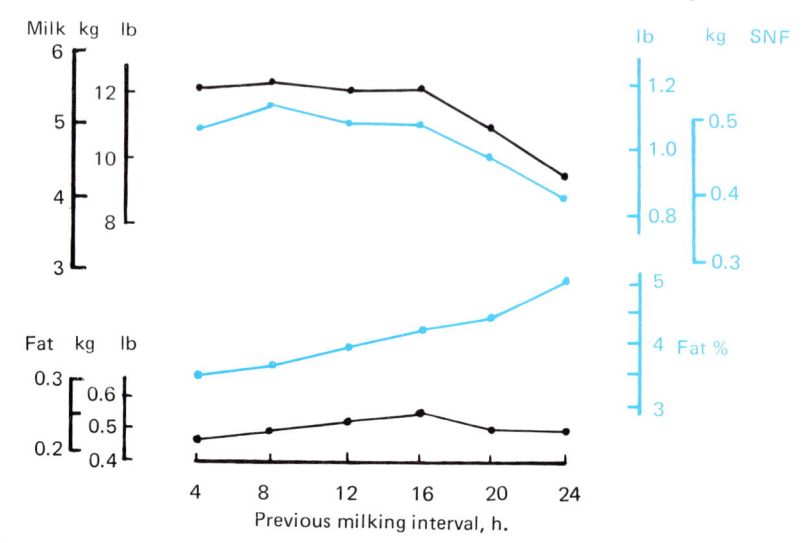

Fig VII 2. The quantities of milk, fat and solids-not-fat (SNF) secreted and the changes in fat content during 8 hours after previous intervals of 4, 8, 12, 16, 20 and 24 hours. Residual milk removed for all milkings[7].

The diurnal trends in the fat content of milk are due to two factors. First, because of their size, the fat globules pass less readily along the fine ducts and become concentrated in the upper udder. Therefore the fat percentage of the first drawn milk may be only 1 - 2% whereas at the end of milking it will be 5 - 10% and in the residual milk 10 - 20% (Fig VII 3). The greater the yield of milk the wider the range in the fat percentage from first to last drawn milk[4]. The second factor is that the quantity of milk retained as residual milk is not constant, partly because of variation in the effectiveness of milk ejection[10] but also because the quantity of residual milk is directly related to milk yield[4]. Thus the apparent random day to day variations in the fat content of milk of individual cows are mainly due to chance variations in the effectiveness of milk ejection affecting the completeness of milking, whereas the regular diurnal variations in fat content at morning and evening milkings are due to a net carryover of fat from the long night interval to the short day. The carryover effect on milk and fat yields is evident in Bartlett's 1929 data[11] shown in Fig VII 4, although the reason was not understood at the time. The hypothetical data in Fig VII 5 illustrates the way carryover effects operate for yields of milk and fat with uneven intervals. The data are based on a uniform secretion rate of 1 kg/h in alternating intervals of 8 and 16 hours, residual milk a constant 11% of total milk present at the beginning of milking, and fat a constant 12% in residual milk. These values are typical. It will be seen from the calculations in the legend to Fig VII 5 that the effect of residual milk with high fat content being carried over in intervals in the ratio of 1 : 2 results in milk yields in the ratio of 1 : 1.7, fat yields in the ratio of 1 : 1.2, and change in fat percentage from 4.89 to 3.47. There are some indications that in addition there may be true diurnal effects on secretion rates but they are relatively slight and the evidence is inconclusive.

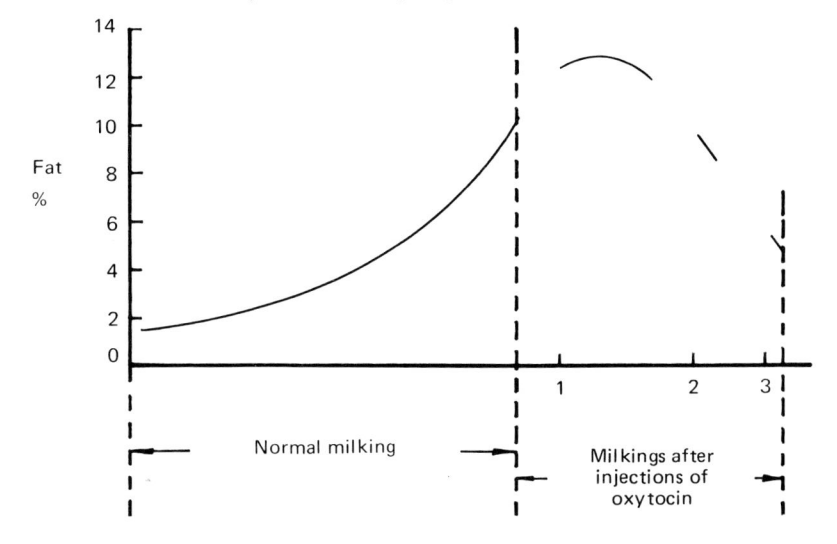

Fig VII 3. A typical curve showing the rise in fat percentage of the milk sampled during a normal milking and subsequently at further milkings after injections of oxytocin to remove the residual milk[4].

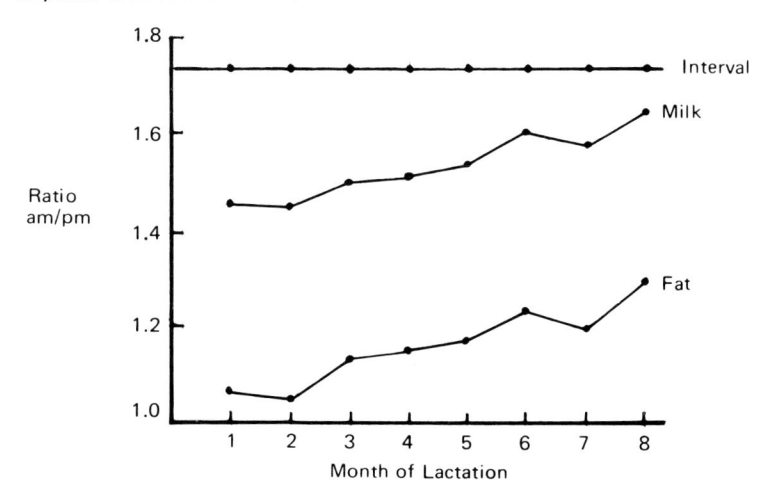

Fig VII 4. The mean changes in the ratios of the a.m. yields of milk and fat to the p.m. yields over the first 8 months of lactation of 97 cows milked at constant intervals of 15 and 9 hours[11].

184 *Milking Routines*

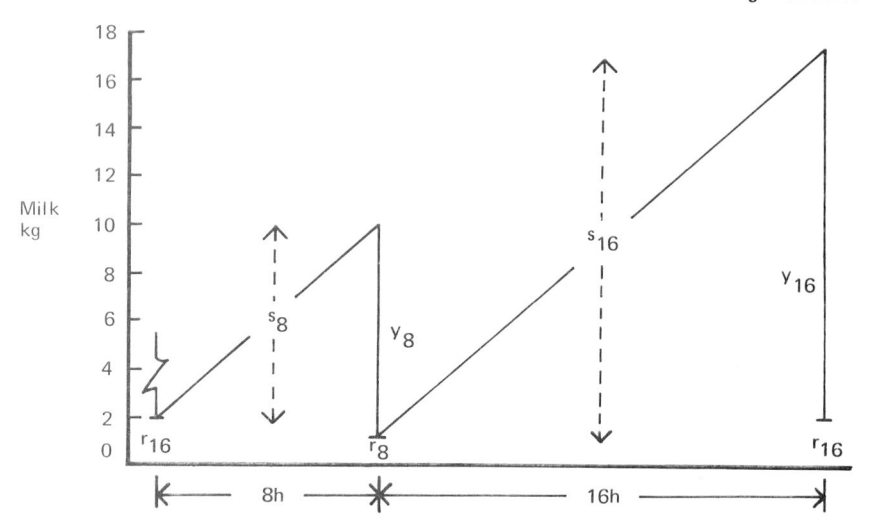

Fig. VII 5. The effect of unequal daily milking intervals on the yield and composition of milk, showing the result of the carryover of residual milk from one milking interval to the next.

Assumptions: milk secretion rate constant for up to 16h at 1 kg/h, quantity secreted (s) containing 4% fat; residual milk (r) is 11% of total milk present in the udder at milking (y + r); fat content of the residual milk is 12%. Then yield (y) of milk (or fat) at a.m. milking is $y = s_{16} + r_8 - r_{16}$

and at p.m. milking is $y = s_8 + r_{16} - r_8$

		Secreted	+	Net residual carryover	=	Yield	Milking intervals, h
Milk yield	a.m.	s_{16}	+	r_8 — r_{16}	=	y_{16}	16
		16	+	1 — 2	=	15kg	
	p.m.	s_8	+	r_{16} — r_8	=	y_8	8
		8	+	2 — 1	=	9kg	
	Ratio	2:1				1.7:1	2:1
		i.e. 16:8					
Fat yield	a.m.	$\frac{16 \times 4}{100}$	+	$\frac{(1-2) \times 12}{100}$		= 0.52kg	
	p.m.	$\frac{8 \times 4}{100}$	+	$\frac{(2-1) \times 12}{100}$		= 0.44kg	
	Ratio	2:1				1.2:1	
Fat %	a.m.	4.00				3.47	
	p.m.	4.00				4.89	

Note: After the short interval, yields of milk are greater than the amounts secreted in the interval and conversely with the longer interval. The effect of carryover on yield ratio is more marked for fat than milk with a corresponding large difference in fat % at the alternate milkings. The ratio of the yields of protein and lactose would be similar to that of the yields of milk.

INTRAMAMMARY MILK PRESSURE

The changes in intramammary pressure and secretion rates which occur whilst the udder is left unmilked are shown in Fig VII 6[12] using various sources of data[13, 14, 15]. At the completion of milking when all available milk has been removed the pressure within the teat sinus is atmospheric but within an hour it rises to 1.1 - 1.5 kPa (0.15 - 0.21 lbf/in^2) as milk residues drain and fill the collapsed teat and udder cisterns. Thereafter pressure increases slowly for 5 or 6 hours as the hydrostatic head of milk stored in the udder increases. Once the main ducts and cisterns are full the capacity is increased by the udder becoming distended with milk. Because of its structure and suspension the udder is well adapted to increase its capacity and this minimizes the increase in milk pressure. Nevertheless in the later phase of storage the milk pressure increases more rapidly until at the end of a normal milking interval pressures of 2 - 4 kPa (0.28 - 0.56 lbf/in^2) are reached (Fig VII 6). Even so most of the intramammary milk pressure measured in the teat is due to the hydrostatic head of milk and therefore the pressure in the secretory tissue in the upper udder will be much less.

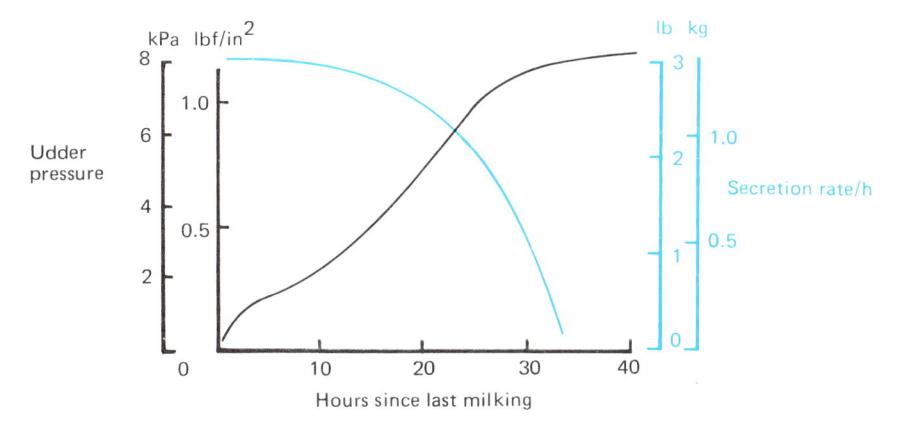

Fig VII 6. Increase in udder pressure with time after milking and change in milk secretion rate with the length of milking interval[12]. Data derived from various sources[13,14,15].

The milk pressure at the end of a milking interval depends upon the length of the interval, udder capacity and the yield of the cow, being greater in hind compared with front quarters and in early lactation. Characteristics other than yield are also important and the rate of increase in pressure per unit of milk secreted is lower in high yielding cows, old cows, in early lactation, in hind quarters and following the long night interval compared with the shorter day interval[13].

The trends in intramammary pressure during a milking interval are not closely related to the trends in the rate of milk secretion. Most of the change in intramammary pressure occurs when secretion rates are virtually constant. Nevertheless it is quite possible that intramammary pressure may have a marked effect on milk secretion rates but only after the milk pressure exceeds a certain threshold value (Fig VII 6).

MILKING INTERVALS WITH TWICE DAILY MILKING

The observation that the apparent rate of secretion (i.e. the milk yield per hour of milking interval) is much higher with the shorter day than with the longer night intervals has encouraged farmers to equalize the milking intervals of cows, at least for higher yielding animals and those in early lactation. The much higher fat concentrations of the milk after the shorter interval gave further support to the traditional view. However the results of the experiments that directly measured the true rates of secretion of milk and fat indicated that daily yields of milk and therefore lactation yields are unlikely to be affected by varying the lengths of milking intervals.

A series of experiments measured the direct effect of equal and unequal milking intervals on yield of milk and fat[16,20]. The milk yields were usually higher when the daily intervals were equal (i.e. 12 h and 12 h) compared with unequal intervals, but with intervals of 15 and 9, and 16 and 8 h the reduction in milk yields were less than 4% and were not statistically significant even with cows yielding over 5000 kg (11 000 lb) milk per lactation (Table VII 1). Markedly unequal milking intervals are widely adopted in the UK, probably more widely than in most other countries.

FREQUENCY OF MILKING

Changing the frequency of milking from twice to three or four times daily increases lactation yields of milk without materially changing milk composition. Neither the amount of the increase nor the physiological explanation are well established.

The effect of increasing the frequency of milking has been measured by analysing milk recording data and also by direct experimentation[21]. The estimates obtained from milk record data are obtained from large numbers of cows and generally indicate that the lactation yields of cows milked 3 times daily are 15 - 40% greater than those milked twice daily and that cows milked 4 times daily give 5% more than those milked 3 times daily. These figures probably overestimate the effect of more frequent milking. Commercial cows milked 3 times daily are not an unbiased sample but are those selected by farmers as likely to give more milk and to receive different feeding and management. These data do not necessarily indicate that if all cows are milked 3 or 4 times daily the average response will be a 20% increase in yield. Direct measurements of the effect of milking frequency have usually shown much smaller responses though the tests have been made with fewer cows. In such trials the comparisons are made between cows, sometimes within twins or between the half-udders of individual cows. These usually indicate increases of 5 - 15% for 3 times daily milking compared to twice daily although in one experiment it was as high as 32%[22,23,24].

Table VII.1 The yields of milk and fat produced by cows milked at equal and unequal intervals[12]

Intervals h	No. of cows	Length of record, days	Milk, kg	Fat, kg	Reference
12 - 12	7	232	2553	114	16
16 - 8	7	232	2477	112	
12 - 12	11	264	2920	143	16
16 - 8	11	264	2973	146	
12 - 12	7	Lactations	2514	96	17
14 - 10	7	Lactations	2562	97	
12 - 12	11	280	3203	129	18
15 - 9	11	280	3145	128	
12 - 12	17	280	3593	143	18
16 - 8	17	280	3470	142	
12 - 12	35	305	6242	236	
14 - 10	35	305	6222	243	19
16 - 8	35	305	6161	238	
12.5 - 11.5	82	266	4910	186	20
14.5 - 9.5	82	266	4800	181	

Limited experiments (i.e. for less than whole lactations) have considered the effects of reducing the frequency of milking from twice daily to once per day or of omitting one milking per week[25, 26]. The interest in the latter was in order to reduce the weekend labour required on farms operated by one man unable to obtain a relief milker. In both cases the yields were reduced. Milking once a day for complete lactations lowered yields by 40 - 50%[27] and omitting one milking per week reduced yields by 5 - 10%[28]. In these experiments the concentration of fat in the milk increased slightly.

Thus, although milk secretion rates appear to be constant at least up to 12 hours, milking more frequently than twice daily has been shown to increase milk production. Various explanations for the increases in yield that follow the change from twice to 3 and 4 times daily milking have been put foward but none is adequate[24, 29, 30]. A decline in the rate of secretion with time since the last milking is still given as the usual explanation but the rate of secretion experiments described earlier do not support this view. Other possibilities are:

that secretion is stimulated by an increase in the number of milkings per day; the so-called milking stimulus. That milking itself does maintain milk secretion has been demonstrated with laboratory animals and there is evidence that injections of oxytocin, the hormone released at milking time, stimulate milk secretion in cattle. However the increase in milk yields with 3 times daily milking has been demonstrated in half-udder experiments[22, 23], in which the two quarters milked 3 times daily gave more milk than the two quarters of the same udder milked twice daily. In such experiments similar levels of hormones transported in the blood must reach all four quarters and therefore seem an unlikely explanation for differences in secretion.

that with increased frequency of milking the yields at each milking are lower and therefore the residual milk will also be lower. It is possible that milk secretion rates are increased when residual milk is decreased but the evidence for this is inconclusive.

that lower lactation yields with less frequent milking are due to a delayed effect of the longer milking intervals – delayed in the sense that the main effect results from the milking intervals following a long interval rather than in the interval itself. This type of effect was demonstrated in experiments measuring the rate of milk secretion (Fig VII 2)[5,7]. If true, the experiments to determine the effect of milking frequency on milk yield would have failed because they were essentially short term.

In summary there is good evidence that milk yields are raised by increasing the frequency of milking. The extent of the rise in yield has not been established accurately, but it is probably within the range of 5 to 15% for an increase from 2 to 3 times daily milking, with the fat concentration falling slightly (i.e. less than 0.3 percentage units) and the concentration of other constituents being unaffected. The reasons for the increase in yield are not well established.

INCOMPLETE MILKING

The earlier references to the art of milking were usually found in books on domestic affairs. Markham in 1615 wrote in a book on the 'Complete

Housewife' that "the worst point of housewifery that can be is to leave a cow half milked"[30]. The belief that incomplete milking rapidly depresses milk yields and reduces the fat content of milk has been universally accepted for several hundred years.

There are two forms of incomplete milking. Some of the available milk may be left if *stripping* is incomplete, and milk is also left as *residual milk* which cannot be removed even by the most careful stripping. The strippings left in an udder by the modern milking machine, used without any manual assistance, are usually less than 0.5 kg and will average less than 0.25 kg (2%). Quantities of residual milk, however, lie within the range of 1 - 3 kg (10 - 20%) of milk per cow milking with a much higher fat content than strippings[4,12].

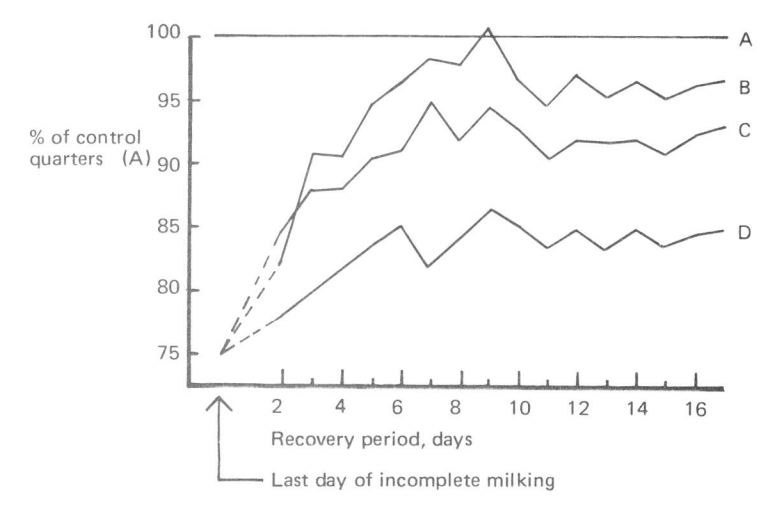

Fig VII 7. The quantities of milk obtained during 17 days following 4 days (B), 8 days (C) and 20 days (D) when only 75% of the expected milk yields were removed. Treatment comparisons were made between quarters within 6 cows and quantities of milk are expressed as proportions of the completely milked quarters (A). The first day's milkings after the incomplete milking period is not shown because it reflects the carryover of milk rather than secretion rate[32].

Contrary to popular belief the results of experiments on the effect of incomplete stripping indicate that its effect is small. The most extensive experiments were made many years ago and indicate that leaving about 0.5 kg of milk in an udder after milking reduced lactation yields by about 3% though the statistical significance of the results was not tested[31]. If much greater quantities of milk are left the effect on secretion is more marked. In a comparison between quarters of six cows, when only 75% of the normal yield was removed during four consecutive days, secretion fell to about 75% of normal and the recovery with standard milking took several days and was not complete. Longer periods of incomplete milking in other quarters of the same cows caused greater and more permanent reduction in secretion (Fig VII 7)[32]. The results of this type of experiment indicate that milking can temporarily affect the rate of milk secretion and also the rate of involution of active secretory tissue.

MILKING ROUTINES

The first section of this chapter has been concerned with the effect on milk secretion of the more direct ways of altering milk removal; for example by changing the frequency and lengths of the milking intervals. Milking can influence milk yield in less obvious ways through the routine operations carried out during the milking process. The study of the way that milking routines affect milk yield and composition is largely the investigation of the way in which management practices affect residual milk. The aim of a good milking routine is to obtain the highest proportion of available milk thus leaving the least residual milk. In so far as management can change the quantity of residual milk it does so by affecting milk ejection. If milk ejection is more effective then more milk is expressed from the alveoli and the fine ducts to the freer draining ducts and sinuses from which it can be removed by milking. This is important for unless milk ejection occurs less than one third of the milk in the udder can be removed by milking. Milking routines may also influence milk yield in other ways. If, after milk ejection, the available milk is not removed before the ejection reflex regresses then milk yield is reduced. In addition it is possible that the 'milking stimulus' may directly stimulate further milk secretion.

MILK EJECTION AND MILKING

From an agricultural standpoint milk ejection (dealt with fully in Chapter VI) has several important properties. It is a neuro-hormonal reflex initiated by various nervous stimuli which in turn cause the release of the hormone oxytocin from the posterior pituitary gland. The response to the nervous stimuli is 'unconditioned', that is it does not require any learning process[33]. A cow at first parturition immediately responds to the calf and milk ejection will occur. In modern milk production the calf is not present and cows have been selected that will come to accept other stimuli for milk ejection. The reaction to these stimuli is not inborn but develops and is therefore a 'conditioned' response, but nevertheless can be very effective. The release of the oxytocin causes a contraction of the alveoli and small ducts, expressing the milk towards the sinuses and teats giving a marked increase in intramammary milk pressure. The process is transitory and has its maximum effect for only a few minutes. The longer the interval between milk ejection and the end of milking the greater will be the quantity of residual milk. Additionally the whole process may be inhibited by adrenalin released in response to fright or pain between stimulation and the completion of milking. The need for adequate stimulation, the development of conditioned responses, the transitory nature of the whole process of ejection and the effects of adrenalin are all important considerations in building good milking systems[34,35].

MEASURING THE EFFECTIVENESS OF MILK EJECTION

With modern dairy cows milk ejection invariably takes place as milking begins though in a few cows it may be delayed for a period during which no milk is obtained and in others it may even occur in two or more stages[36]. Several methods are available to provide a quantitative estimate of ejection efficiency. The release of oxytocin has been measured directly by bioassay methods[37, 38]. Measurement of the effect of ejection is possible by determining the quantity of residual milk left in the udder at the completion of milking[39]. This can be done by milking again after injecting oxytocin preparations intravenously. The further yield of milk obtained cannot be the total residual milk but the evidence is that it constitutes by far the greater part[4]. A less direct estimate of ejection is to measure the rise in intramammary milk pressure after stimulation. The milk pressure in the teat sinus before stimulation will usually be in the range of 1.3 - 4.0 kPa (0.18 - 0.56 lbf/in²) above ambient atmospheric pressure. Within half a minute of stimulation the pressure rises rapidly to 4 - 8 kPa (0.56 - 1.12 lbf/in²) and occasionally even higher[40]. If milk is not removed directly after ejection the pressure declines slowly for an hour or more reaching a pressure that would have been expected if milk ejection had not occurred[40] (Fig VII 8).

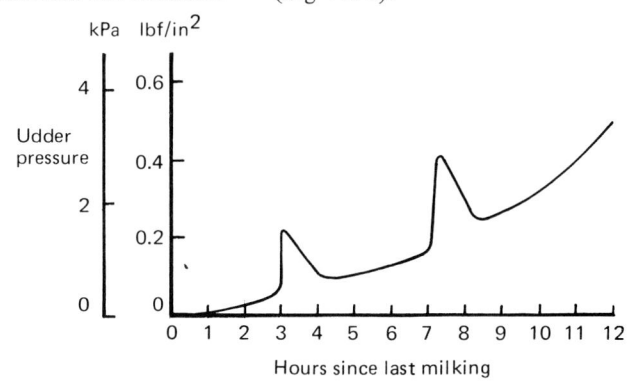

Fig VII 8. The effect of stimulation and milk ejection on the rise in udder pressure during a 12 hour period without milk removal. The peaks indicate the points when ejection occurred and the return to the expected pressure after about one hour[40].

STIMULATION OF MILK EJECTION AND MILKING ROUTINES

It was not possible to carry out useful research on milking until the significance of milk ejection as a separate physiological part of milk secretion had been demonstrated[34]. In commercial milk production the stimulation of milk ejection is complex. The absence of the calf and suckling eliminates the strongest natural stimulus for ejection though the stimulation provided by the manipulation of the teats during udder washing and foremilking may be as effective. In practice much of the stimulation comes from other management practices to which the cow becomes conditioned because they regularly precede milking by a short interval. Thus the cow moving into a milking stall, or receiving concentrates just before the start of milking is provided with a powerful stimulus for milk ejection. There are not adequate studies with cows of the methods of developing the most powerful stimuli for a conditioned response resulting in milk ejection[41]. From work with other species it might be expected that strong conditional reflexes leading to milk ejection will operate only if the milking routine fulfills certain conditions. The stimulus for a conditioned response (e.g. feeding concentrates) must always precede the stimulus for an inborn reaction (e.g. udder handling) and in this way the conditioned response is constantly reinforced. The time interval between the two types of stimuli must be reasonably constant and preferably short. Furthermore if several factors are developed to stimulate a conditioned response their effect may be additive. If there are several stimuli they should be applied in the same order. Finally conditioned responses are inhibited if the animal is frightened[34].

Once the importance of stimulating milk ejection was understood milking routines were proposed[35] that could be expected to give a more effective milk ejection and higher milk yields. These routines, concerned with the preparation of the cow for milking, emphasized that cows should be given a definite strong signal that milking was about to begin, that the signal should not cause pain or discomfort and that the routine preparation at each milking should be constant and precede milking by a regular, preferably short, interval. Milking routines are likely to be more variable in cowsheds than in parlours. In the former, the preparation for milking of each cow can be divorced from the application of the machine, and the order of events and timing may vary considerably from one milking to the next unless a strict routine is followed. In milking parlours the cow enters the parlour, is fed, the udder is prepared for milking and the machine is attached. It is difficult to vary the sequence of events and in practice there can be little delay between stimulation and milking if the milking parlour is to be used effectively.

The advice usually given on milking routines can be summarized as follows:

- provide an environment that gives the least stress to cows or stockmen
- adopt a routine such that the preparation for milking gives an adequate stimulation to milk ejection
- carry out the various operations (i.e. foremilking, udder washing, feeding concentrates, etc) in the same order at each milking immediately before the machine is attached.

Such milking routines are now accepted and are the basis of current teaching on milking. Nevertheless their value has not been adequately measured. The few experiments carried out usually compare the change from one milking routine to another, and not two *established* routines. This distinction is important because conditioned responses take time to develop and time to destroy. Thus if cows are conditioned to accept feeding as a stimulus to milk ejection by feeding one minute before milking, and then the routine is changed to provide an interval of 30 minutes between feeding and milking it is inevitable that milk yields will decline initially. Milk ejection will occur much too soon before milking *until* the old conditioned responses decay and new ones related to the changed routine develop, e.g. udder washing. Nevertheless there is some experimental evidence to confirm the importance of stimulation which indicates that levels of residual milk are lower with the improved milking routines[10, 39].

In recent years interesting New Zealand work has indicated that milk yields may be increased by massage of the udder for up to one minute during preparation for milking[42]. The very strong stimulation provided by the suckling calf has also been illustrated. A recent experiment[43] confirmed that calf suckling induced higher yields than machine milking in two groups of 12 cows. Further work is necessary to discover ways of improving the stimulation given in normal milking routines.

RATE OF MILKING AND LACTATION

Over a century ago farmers were aware that milk yields were related to the speed of milk withdrawal. In competitions at agricultural shows individual cows were milked more quickly by two men in order to maximize milk production, one milking on each side of the cow. Before 1900 experiments had demonstrated that if the four quarters of the cow are milked singly in rotation the first milked usually gives the highest yield with the greatest concentration of fat and the last milked the lowest yield with the lowest fat percentage[44]. At that time it was not known that the explanation for these trends was that milk ejection is transitory. Once the physiology of milk ejection had been described by Petersen and his colleagues interest in milking rate was increased, and farmers were encouraged to adopt 'quick milking' routines, to train their cows to milk quickly, or to take steps to prevent cows developing 'slow milking habits'. This was an over simplification. The important point was not really the encouragement of quicker milking, which with machine milking is governed largely by the properties of the machine, but to ensure that the interval between milk ejection and the end of milking is as short as possible. To do this the milking routine is often more important than the rate of milk withdrawal. Nevertheless research on milking rate has evolved in an interesting way.

MILKING MACHINE RATE AND THE DURATION OF MILKING

Early research using simple methods of measurement demonstrated a pattern in the rate of flow of milk from the udder during milking. Flow starts with a period, usually of less than one minute, when the rate of flow increases to a peak. This is followed by a maximum flow period when the flow is roughly constant until most of the milk has been obtained, and a final period of decelerating flow (see Fig V 1). This final period can be very variable in length. The rate of removal of the final portion can be accelerated by pulling downwards on the claw and massaging the quarters. The most striking observation of this early work was that there were considerable differences between the peak flow rates of cows[45], some milking at more than ten times the rates of others[46]. Furthermore, the patterns for individuals were relatively constant from day to day, during and in succeeding lactations. This suggested that milking rate was controlled by the anatomy of the teats rather than by the milk ejection mechanism and later it was demonstrated experimentally that nearly all the variation was removed if the diameters of the streak canals of the teats were made equal with uniform bore cannulae[47] (Fig VII 9). It was not surprising therefore that it proved to be impossible to train cows to milk more quickly (or slowly) either by changing the method of stimulating milk ejection or by regularly foreshortening the milking time of cows and heifers[48,49].

The duration of milking is determined not only by the rate of flow of milk from the udder but also by the quantity of milk to be removed. In addition there is a relationship between milking rate and milk yield, the higher the yield of a cow the faster its milking rate. However the increase in milking rate with higher yield is insufficient to balance the effect on milking time of increased yield, and therefore individual cows have their longest milking times at times of greatest milk yield. Thus cows take longer to milk at morning milkings, early in lactation and as milk yields increase with age[46].

Although the effectiveness of milk ejection has little influence on milking rate it may influence the duration of milking. Cows that react slowly to stimulation have a delayed milk ejection and take longer to milk. Milking routines that give inadequate stimulations may delay ejection but will not on average increase the duration of milking by more than 0.5 min per cow[39].

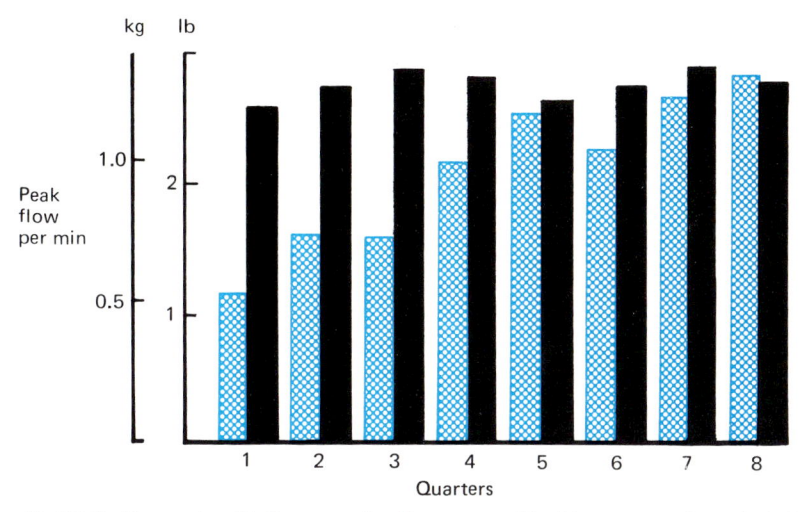

Fig VII 9. The peak milk flow rates for 8 quarters milked by teatcup through their normal teat orifices (blue) and through standard bore cannulae (black) inserted in the teat orifices at a vacuum of 54 kPa (15.9 inHg)[47].

BREEDING FOR FASTER MILKING

Since the main factor controlling milking rate is the anatomy of the streak canal of the teat it is not surprising that milking rate is highly heritable[50]. In some countries milking rate has been regarded as sufficiently important to be incorporated as a major selection factor in bull progeny testing schemes. With the high heritability good progress in breeding faster milking cows should be possible. However in farm practice it is not difficult to devise milking systems that overcome the main difficulties with slower milking cows (see Chapter VIII) but considerable inconvenience and delay is caused by the individual very slow milking cow and these are usually culled. Therefore the main requirement appears to be to identify and cull those bulls that produce very slow milking daughters. A final point is that susceptibility to udder infection is related to the patency of the streak canal, and therefore to milking rate. Fast milking cows tend to be more readily infected[51].

MILKING RATE AND MILK YIELD

Because milk ejection is transitory, data have been analysed to determine whether fast milking cows give higher lactation yields than slow milking cows.

Nearly all analyses demonstrate a direct relationship of this type. Typical data collected at the NIRD indicate that for each kg (2.2 lb)/min difference in peak milking rate of cows in early lactation a 400 kg (880 lb) difference in lactation yield can be expected[52]. This does not necessarily mean that cows give more milk because they are faster milking. Several workers believe that most of the relationship is a direct result of cows milking faster *because they are higher yielding*. So far this question has not been finally resolved. However an examination of the published data on milking rates shows that most of the variation is due to inherent anatomical differences between the cows. These are several times greater than the relatively minor variations in rate attributable to variations in milk yield. Furthermore multiple regression analyses indicate that the lactaction milk yield of cows are directly related to their milking rates measured in early lactaction even when correction is made for differences in milk yield at the time of recording the milking rates[53]. It would appear that milking rate does directly affect lactation milk yields.

SUMMARY

Since the recognition of the extent of residual milk and the effects of milking interval and completeness of milking upon the quantity of milk left in the gland, many studies have been made of the influences of milking routines on the rates of secretion of milk and fat. In general these investigations have sought to determine whether equalizing milking intervals or milking more frequently will result in greater milk yield.

Although cows vary in their ability to adapt to longer milking intervals the various experiments to measure the relationship between secretion rate and milking interval show little effect below 16 hours. However, these short term studies to measure secretion rates are limited in two ways. Firstly, to measure secretion the quantity of residual milk has to be determined, and its removal could itself affect the subsequent secretion rate in proportion to the amount removed. This may mean, for example, that for most cows secretion rates are unaffected for 12 to 13 hours, rather than for 16 hours. Secondly, in short term studies with intervening or control intervals of 12 hours, the shorter interval effects may be limited by these previous intervals, for we know that after changes in secretion rates more than one milking interval is required before the rates of secretion recover. Longer term studies of uneven milking intervals and frequency of milking have produced confusing results and many hypotheses have been advanced to explain them.

In practice, the benefits in extra yield from either shortening the night interval below 16 hours or of milking more than twice in 24 hours will rarely justify the extra labour costs involved.

The evidence of benefits to justify increasing the stimuli for a conditioned response to aid milk ejection is not convincing, although it has been proved that less complete milking can result from inadequate milk ejection associated with irregular milking routines which include long periods between stimulation of the cow and milk removal. The efficient use of modern milking parlours demands a routine at milking time. This makes it impossible to depart from a regular sequence of events which provides the stimuli for a conditioned response for milk ejection. Greater care is necessary to maintain a consistent routine in smaller milking parlours and especially in cowsheds, where the opportunities for variation are much greater. Wherever cows are milked the need to avoid any disturbance which may inhibit milk ejection remains an underlying principle.

There is no doubt that the work we have described marks an advance in understanding the non-nutritional factors that affect milk yield and composition. Most of the information has been obtained in the last 25 years and follows the demonstration that milk secretion and milk ejection are separate but essential processes in lactation and that there is considerable carryover of milk in the udder from one milking interval to the next. Nevertheless most of the new data are concerned with describing and understanding the processes of lactation in cattle and too little with the exploitation of this information in dairy farming. The evidence on the effect on lactation milk yields of varying the frequency of milking, incomplete milking and different milking routines is not good. Furthermore the extent to which the considerable differences between herd milk yields is due to these factors has still to be determined. Traditionally these non-nutritional aspects of management are held to be important; it is the conventional wisdom and routinely taught to agricultural students. Nevertheless the evidence for the importance of 'cowmanship' is slim. There is now an increasing interest in behaviour studies but these also seem to be rather academic. They measure the way in which cows react to one another rather than the way their milk production is affected by the way that men and production systems impose upon them. 'Stress' in cattle is said to be important for milk production and disease yet there is little evidence that our systems of milk production produce measureable stress. The reason that we have pointed to these deficiencies in this branch of animal husbandry is to stimulate more research. We believe that such work would be important for dairy farming. Observation and indirect evidence suggest that cowmanship is economically valuable but to be useful the research must determine those specific aspects of stockmanship that are important so that others can be taught the skills[54].

REFERENCES

(1) Gaines, W.L. & Sanmann, F.P. (1927) *American Journal of Physiology*, 80, 691.

(2) Ragsdale, A.C., Turner, C.W. & Brody, S. (1924) *Journal of Dairy Science*, 7, 249.

(3) Edwards, J. (1936) *Journal of Dairy Research*, 7, 211.

(4) Johansson, I. (1952) *Acta Agriculturae Scandinavica*, 2, 82.

(5) Bailey, G.L. Clough, P.A. & Dodd, F.H. (1955) *Journal of Dairy Research*, 22, 22.

(6) Turner, H.G. (1955) *Australian Journal of Agricultural Research*, 6, 514.

(7) Elliot, G.M., Dodd, F.H. & Brumby, P.J. (1960) *Journal of Dairy Research*, 27, 293.

(8) Marx, G.D., Linnerud, A.C., Miller, G.E., Caruolo, E.V. & Donker, J.D. (1963) *Journal of Dairy Science*, 46, 626.

(9) Wheelock, J.V., Rook, J.A.F., Dodd, F.H. & Griffin, T.K. (1966) *Journal of Dairy Research*, 33, 161.

(10) Elliott, G.M. (1961) *Journal of Dairy Research*, 28, 123.

(11) Bartlett, S. (1929) *Journal of Agricultural Science*, 19, 36.

(12) Schmidt, G.H. (1971) In: *Biology of Lactation*. San Francisco: W.H. Freeman & Co., p.149.

(13) Korkman, N. (1953) *Kungliga Lantbrukshögskolans Annaler*, 20, 303.

(14) Schmidt, G.H. (1960) *Journal of Dairy Science*, 43, 213.

(15) Tucker, H.A., Reece, R.P. & Mather, R.E. (1961) *Journal of Dairy Science*, 44, 1725.

(16) McMeekan, C.P. & Brumby, P.J. (1956) *Nature*, 178, 799.

(17) Koshi, J.H. & Petersen, W.E. (1954) *Journal of Dairy Science*, 37, 673.

(18) Hansson, A., Claesson, O., Brännäng, E. & Gustafsson, N. (1958) *Acta Agriculturae Scandinavica*, 8, 296.

(19) Schmidt, G.H. & Trimberger, G.W. (1963) *Journal of Dairy Science*, 46, 19.

(20) Ormiston, E.E., Spahr, S.L. Touchberry, R.W. & Albright, J.L. (1967) *Journal of Dairy Science*, 50, 1597.

(21) Elliott, G.M. (1959) *Dairy Science Abstracts*, 21, 481.

(22) Ludwick, L.M., Speilman, A. & Petersen, W.E. (1941) *Journal of Dairy Science*, 24, 505.

(23) Cash, J.G. & Yapp, W.W. (1950) *Journal of Dairy Science,* **33**, 382.

(24) Elliott, G.M. (1961) *Journal of Dairy Research,* **28**, 209.

(25) Hesseltine, W.R., Mochrie, R.D., Eaton, H.D., Elliott, F.I. & Beall, G. (1953) *Bulletin, Storrs Agricultural Experimental Station,* No. 304.

(26) Parker, O.F. (1965) *Proceedings of the Ruakura Farmers' Conference Week,* p.236.

(27) Claesson, O., Hansson, A., Gustafsson, N. & Brännäng, E. (1959) *Acta Agriculturae Scandinavica,* **9**, 38.

(28) Claesson, O. (1962) *Dairy Farmer,* **9**, 36.

(29) Morag, M. (1968) *Annales de Biologie Animale, Biochimie, Biophysique,* **8**, 27.

(30) Markham, G. (1615) In: *The English Housewife,* London: Brewster & Sawbridge, p.143.

(31) Woodward, T.E., Hotis, R.P. & Graves, R.R. (1936) *Technical Bulletin, United States Department of Agriculture,* No. 522.

(32) Dodd, F.H. & Clough, P.A. (1962) *International Dairy Congress 16, Copenhagen 1,* p.89.

(33) Baryshnikov, I.A. (1959) *Dairy Science Abstracts,* **21**, 47.

(34) Ely, F. & Petersen, W.E. (1941) *Journal of Dairy Science,* **24**, 211.

(35) Miller, K. & Petersen, W.E. (1941) *Journal of Dairy Science,* **24**, 225.

(36) Whittlestone, W.G. (1946) *New Zealand Journal of Science and Technology,* **32A**, 1.

(37) Cleverley, J.D. (1968) *Journal of Endocrinology,* **40**.

(38) Van Dongen, C.G. & Hays, R.L. (1966) *Endocrinology,* **78**, 1.

(39) Dodd, F.H., Foot, A.S. & Henriques, E. (1949) *Journal of Dairy Research,* **16**, 301.

(40) Tgetgel, B. (1926) *Schweizer Archiv für Tierheilkunde,* **68**, 335.

(41) Brandsma, S. (1969) *Proceedings of the Symposium on Machine Milking 1968.* Reading: National Institute for Research in Dairying, p.119.

(42) Phillips, D.S.M. (1960) *Proceedings of the New Zealand Society of Animal Production,* **20**, 93.

(43) Walsh, J.P. (1974) *Dairy Science Abstracts,* **36**, 3791.

(44) Babcock, S.M. (1889) *Report, Wisconsin Agricultural Experimental Station,* p.42.

(45) Foot, A.S. (1935) *Journal of Dairy Research,* **6**, 313.

(46) Dodd, F.H. (1953) *Journal of Dairy Research,* **20**, 301.

(47) Baxter, E.S., Clarke, P.M., Dodd, F.H. & Foot, A.S. (1950) *Journal of Dairy Research,* **17**, 117.

(48) Dodd, F.H. & Foot, A.S. (1947) *Jounral of Dairy Research,* **15**, 1.

(49) Dodd, F.H., Foot, A.S., Henriques, E. & Neave, F.K. (1950) *Journal of Dairy Research,* **17**, 107.

(50) Johannson, I. (1961) In: *Genetic Aspects of Dairy Cattle Breeding,* London: Oliver and Boyd, p.149.

(51) Dodd, F.H. & Neave, F.K. (1951) *Journal of Dairy Research,* **18**, 240.

(52) Dodd, F.H. & Foot, A.S. (1953) *Journal of Dairy Research,* **20**, 138.

(53) Clough, P.A. & Dodd, F.H. (1957) *Journal of Dairy Research,* **24**, 152.

(54) Seabrook, M.F. (1972) *Journal of Agricultural Labour Science,* **1**, 45.

Chapter VIII

MACHINE MILKING IN COWSHEDS AND MILKING PARLOURS

P A Clough

Whatever assessment is made of the detailed knowledge of the milking machine and the dairy cow presented in the previous chapters, the fact remains that the basic milking machine is substantially the same as when designed over seventy years ago, and the lactating dairy cow, herself working twenty-four hours a day, imposes the age-old rigid discipline on the lives of dairy farmers and their employees. Every day of the year begins and ends with milking. The task occurs with monotonous regularity and has a dominant influence on daily work schedules.

Progress in breeding and feeding has been responsible for the highly productive modern dairy cow typified by the docile co-operative animals of the Friesian breed. Loose housing, particularly cubicle housing, has given the deserved freedom of movement and activity, denied to animals fastened by the neck and confined to stalls in traditional cowsheds, and facilitated mechanization of the movement of food to and effluent from buildings.

The development of machine milking installations, milking techniques and properly planned work routines, especially in milking parlours, has made it possible for the larger herds of loose housed cows to be milked with less effort by fewer people. This chapter includes a brief account of machine milking in traditional cowsheds, but is principally concerned with milking parlours and factors which affect milking performance. Detailed information of buildings and fixed or movable equipment which is not part of the milking machine installation is not included.

COWSHEDS

Traditional cowsheds provide suitable accommodation for cows under conditions which permit individual rationing of all foodstuffs, but considerable manual labour is required to distribute foodstuffs and litter, and to remove dung at least twice daily. Tractors and trailers can be used in some cowsheds with wide doorways and passages, but it has not been possible to eliminate, or even reduce the frequency of any of the daily tasks. Simple feeding systems such as self-feeding or easy-feeding of silage and hay cannot be practised in cowsheds.

MILKING EQUIPMENT FOR COWSHEDS

Between 1940 and 1953 developments in machine milking in cowsheds were hindered by the interpretation of the Milk & Dairies Regulations which forbade the tipping or cooling of milk in cowsheds. This encouraged milking machine manufacturers to design mobile milking machines complete with one or more milk cans on wheeled trolleys or suspended from overhead tracks. These machines were cumbersome and expensive and did not solve the problem of transporting milk to the milk room; which was the main cause of poor milking performance in cowsheds.

The majority of farmers accepted the deficiencies of cowshed milking until they were in a position to change to loose housing with a milking parlour. A small number of farmers using bucket milking units made use of simple wheeled trolleys, which enabled all the milking equipment to be moved along the cowshed and reduced the number of visits to the milk room to one or two at a milking (Fig VIII 1).

The introduction of refrigerated bulk milk tanks on dairy farms and the success of circulation cleaning for pipeline milking machines in milking parlours encouraged the installation of pipeline machines in cowsheds from 1958 onwards. A fixed milking pipeline above all the stalls in a cowshed directly connected to a releaser or milk pump at the bulk tank overcame the problem of milk transport, but the installation was costly and far from satisfactory. Poor design of circuits led to high vacuum fluctuations in the milking pipelines resulting in slow and incomplete milking. Many pipelines were also difficult to clean. The general dissatisfaction expressed by farmers through the National Farmers' Union was largely responsible for the production of the British Standard Institute Code of Practice on Pipeline Milking Machine Installations, CP 3007, which established criteria for the design and installation of pipeline milking machines. A simple pipeline system in which a mobile container in the cowshed was linked via a flexible plastics tube to a releaser or pump in the milk room was offered by one manufacturer in 1960 as a means of reducing milk transport time associated with bucket milking. Very few of these dump stations were sold and they are no longer available.

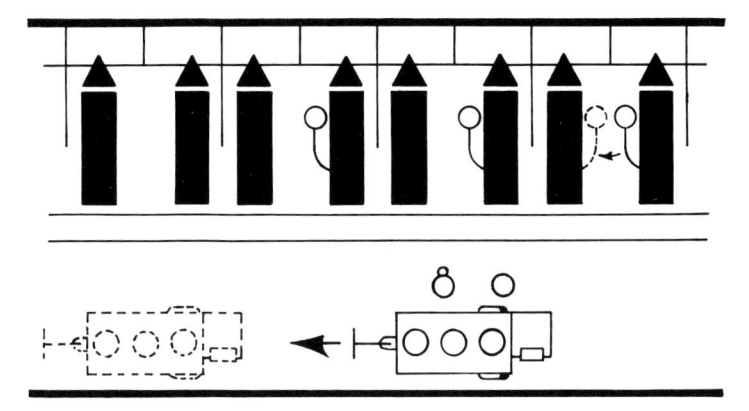

Fig VIII 1. Bucket milking in a cowsned using a wheeled trolley to carry three 45 ℓ milk cans and recording and udder washing equipment. The trolley is moved along the cowshed as milking proceeds and few visits are made to the milk room. Milk is cooled in the cans.

THE ORGANIZATION OF MACHINE MILKING IN COWSHEDS

The work done during milking should be confined to the jobs that are essential to machine milking: preparing cows for milking and moving milking units from cow to cow. Where bucket milking units are used some time must be directed to transporting milk to the milk room where it will be cooled and stored. At least two transport cans full of milk should be taken on each trip to the milk room.

The feeding of concentrates and washing of udders need not be included in the milking work routine. Both of these tasks should be completed for all the cows before the start of milking. Preparation for milking would then be confined to the withdrawal of foremilk before putting on the teatcups. On the completion of the milking of each cow the teatcups should be removed and the teats dipped in a disinfectant solution as soon as the milking unit is in use on the next cow.

The distance walked by the milker(s) during milking will be reduced if the cows are milked in succession from one end of the shed and the milking units kept close together. In high yielding herds equalization of milking intervals may be accomplished by reversing the order of milking to allow cows in early lactation to be milked first in the morning and last in the evening. When bucket milking units are used the milk yield of individual cows should be measured by weighing at the trolley before the milk is tipped into the transport cans. Milk meters have been developed for use in cowsheds equipped with fixed milking pipeline machines. The yield of a cow is indicated by the proportional sample collected from the milk flowing through the meter so that little or no delay occurs when the sample is cleared between cows.

The number of cows milked per man hour is controlled by the work routine of the milker, the milk yields of the cows, and the number and type of milking units in use. When bucket units are used a reasonable minimum performance would be 30 cows milked per man hour. A fixed milking pipeline machine discharging directly into a refrigerated bulk milk tank provides conditions very similar to those in a two stalls per unit abreast parlour. The main difference is that in a cowshed the units are moved instead of the cows. A farmer should not be satisfied with less than 40 cows milked per man hour and there is no reason why this should not be increased up to 60 cows per man hour in large cowsheds housing 60 or more cows. Cowshed milking work routines are shown in Table VIII 1, and potential performance at different milk yields in Table VIII 2 for an assumed unit idle time per cow of 0.7 min.

Table VIII 1 Cowshed milking work routine; the times taken by the stockman to complete the various elements in the work routine[1]

Operation	Bucket, min/cow	Pipeline, min/cow
Foremilk	0.20	0.15
Change unit / Change bucket	0.40	0.25
Disinfect teats	0.15	0.15
Tip milk and move trolley	0.20	
Walking to and from cow	0.20	0.15
Transport milk	0.25	
Miscellaneous	0.05	0.05
Work routine time, min/cow	1.45	0.75
Maximum no. of cows milked per man hour	41	80

Type of milking installation

MILKING PARLOURS

A milking parlour is a building equipped with a limited number of stalls and milking units to which cows are brought to be milked. The equipment in a parlour is arranged for easy movement of the animals and to enable the milker to manage several milking units without undue effort. In Europe and America it is common practice to provide a manger with each stall to allow each cow to be fed concentrate food whilst in the parlour.

Table VIII 2 Potential performance of milking installations* in relation to the number of units used and level of milk production[4]

TYPE OF INSTALLATION				NUMBER OF MILKING UNITS											
1 Stall/unit yield	2 Stalls/unit yield	Cowshed yield	Unit time, min	3		4		5		6		7		8	
				P	AWT	P	AWT	P	AWT	P	AWT	P	AWT	P	AWT
	3.2		3.7	49	1.2	65	0.9	81	0.7	97	0.6	114	0.5	130	0.5
	4.5	3.2	4.2	43	1.4	57	1.1	71	0.8	86	0.7	100	0.6	114	0.5
3.2	5.9	4.5	4.7	38	1.6	51	1.2	64	0.9	77	0.8	89	0.7	102	0.6
4.5	7.3	5.9	5.2	35	1.7	46	1.3	58	1.0	69	0.9	81	0.7	92	0.7
5.9	8.6	7.3	5.7	32	1.9	42	1.4	53	1.1	63	1.0	74	0.8	84	0.7
7.3	10.0	8.6	6.2	29	2.1	39	1.5	48	1.2	58	1.0	68	0.9	77	0.8
8.6	11.3	10.0	6.7	27	2.3	35	1.7	45	1.3	54	1.1	63	1.0	72	0.8
10.0	12.7	11.3	7.2	25	2.4	33	1.8	42	1.4	50	1.2	58	1.0	67	0.9
11.3		12.7	7.7	23	2.6	31	1.9	39	1.5	47	1.3	55	1.1	62	1.0
12.7			8.2	22	2.7	29	2.1	37	1.6	44	1.4	51	1.2	59	1.0
			9.0	20	3.0	27	2.3	33	1.8	40	1.5	47	1.3	53	1.1
			10.0	18	3.3	24	2.5	30	2.0	36	1.7	42	1.4	48	1.2

* P = cows milked/man hour; yield = average yield of the cows at a milking, kg;
AWT = time available to stockman to complete the routine work on each cow, min.

PARLOUR TYPE AND LAYOUT

Milking parlours may be divided into two basic types: those having a milking unit at each stall are known as 'one stall per unit' and those with a unit shared between two stalls are known as 'two stalls per unit' parlours. Both basic types have the stalls in several different arrangements in relation to the area in which the milker works (Figs VIII 2 and VIII 3).

BASIC TYPES

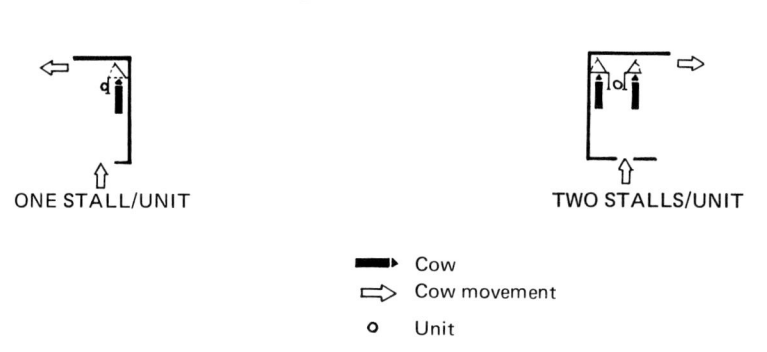

ONE STALL/UNIT   TWO STALLS/UNIT

■▶ Cow
⇨ Cow movement
o Unit

EFFECT OF BASIC TYPE IN ALL LAYOUTS

Most routine work done while unit is idle	Most routine work done while unit is milking
Fewer cows milked/unit h	More cows milked/unit h

IN PARLOURS WITH INDIVIDUAL COW ENTRY

Less feeding time/cow (milking time + 1 work routine time)	More feeding time (approx twice milking time)

IN STATIC HERRINGBONE PARLOURS WITH:

- same work routine and equal numbers of stalls

More available milking time/cow Same feeding time/cow	Less available milking time/cow Same feeding time/cow

- same work routine and same available milking time

Fewer stalls Less feeding time/cow	More stalls More feeding time/cow

Fig VIII 2. The basic types of milking parlour.

P.A. Clough

ONE STALL/UNIT — abreast

TWO STALLS/UNIT — abreast

tandem

chute

double tandem

double tandem

herringbone

herringbone

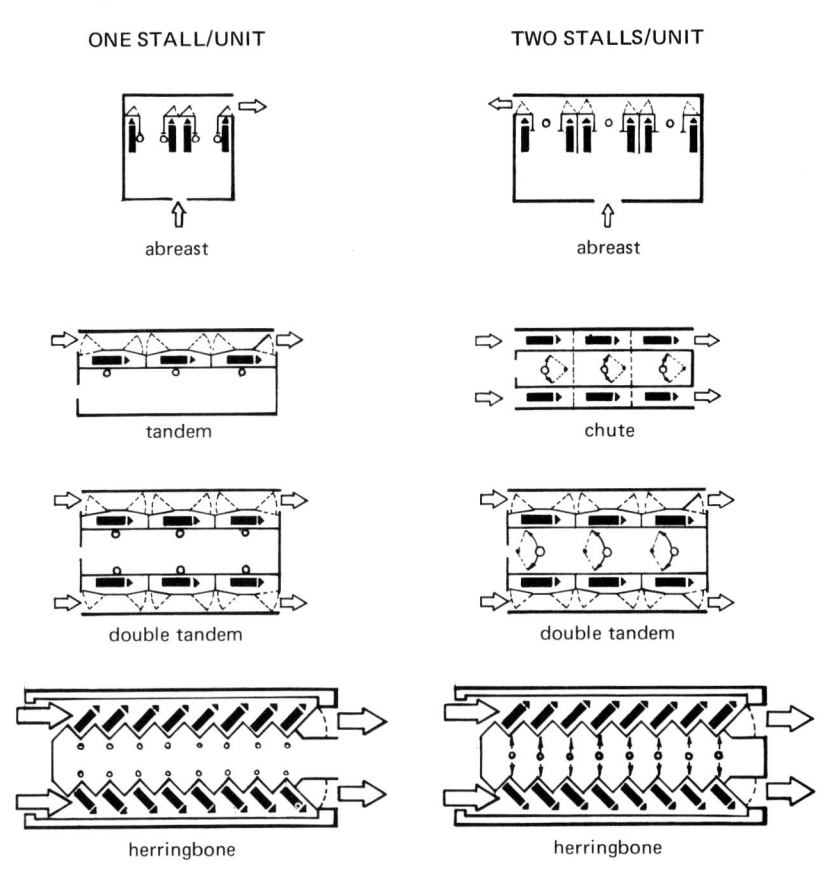

Fig VIII 3. Diagrams of static milking parlours.

In the *abreast layout* cows stand side by side in stalls which have a gate at the front and a chain at the rear. To enter a stall a cow walks across the work space of the milker. After milking the front gate is opened and the cow leaves by a special exit passage. It is useful for the floor of the stalls to be raised some 250 to 400 mm above the level of the floor area of the milker by a single step to make the udder more accessible to the milker and discourage the cow from moving during milking. More than one step breaks up the work area and is a serious hazard to the milker. The main objection to the abreast parlour is that it is not practical to have sufficient difference between floor levels to allow the milker to work standing erect.

The abreast parlour was developed in the form of a movable bail in 1925 and is still popular as an inexpensive prefabricated fixed milking installation.

In *tandem layouts* the stalls are arranged in line on one or both sides of a pit 1.3 to 1.8 m wide in which the milker works. The floor of the pit is 0.8 to 0.9 m below the level of the stalls to allow the milker to work without stooping. Each stall is a rectangular crate 2.4 m by 0.8 m and has an entry and exit opening onto an access passage on the side away from the milker. The cows do not cross the work space of the milker when entering or leaving a tandem parlour, but the layout should not be used with more than four stalls in a line because the distance between units is 2.4 m.

The *chute parlour,* which was developed in the USA in 1952, is a double tandem parlour without the separate access passages for the cows. The cows are moved as a batch through a row of stalls when the cross division in each stall is withdrawn. This layout can be installed in a building 3.5 m wide compared with 4.7 m width required for a conventional double tandem parlour. The cost of the stalls is less than for a conventional tandem and this layout is suitable for up to 4 stalls in line. The chute parlour was introduced to Great Britain in 1955 and was becoming popular when the herringbone parlour was introduced in 1957.

The *herringbone layout* is a modified double tandem without formal stalls in which the cows stand at an angle of about 30° on both sides of the pit in which the milker works. Cows are moved in and out in batches when the entry and exit gates are opened, and in the parlour the cows overlap each other so that the distance between milking units is only 0.9 m. This most compact parlour layout was designed in Australia in 1910 but did not become popular until 1954 when it was introduced into New Zealand to accommodate the relatively small cows of the Jersey breed. The early models installed in Great Britain were less effective than expected because there was no feeding equipment and the dimensions of the cow standings and rump rails were unsatisfactory for cows of the British Friesian breed. By 1961 milking machine manufacturers were supplying feeding equipment for herringbone parlours and during the following six or seven years considerable improvement was made to the design and layout of all the equipment used in herringbone parlours.

The majority of herringbone parlours in Great Britain are two stalls per unit layouts with a single row of receiver jars along the centre of the pit at eye level to the milker, or above the head of the milker. Since 1967 one stall per unit herringbones have been available with a row of receiver jars along each side of the parlour under the cow standings, and more recently with jars at eye level on each side of the parlour. The majority of new milking parlours erected during the past six or seven years have been herringbone or abreast layouts with the popularity of the herringbone increasing as the design and performance has improved.

Rotary milking parlours are those in which the cows, stalls and milking equipment are carried on a rotating platform. All have one stall per milking unit. The first rotary milking parlour was constructed in the USA in 1930 and named the *Rotolactor*. This was an abreast parlour with 40 stalls on a circular platform on which the cows stood side by side facing a wall internal to the platform. The operators were stationed in a work area 0.8 m below and outside the perimeter of the platform. Cows crossed a bridge over the work area to enter the stalls and after milking left the parlour through a gap in the internal wall, down a spiral ramp in the central area and through a tunnel below the stalls to the outside of the building (Fig VIII 4). This was a large and very expensive installation. One similar parlour was constructed in Australia and a few installations for 40 stalls and milking units were made in Russia, East and West Europe between 1960 and 1968.

A simple inexpensive *rotary abreast parlour* was designed by a New Zealand farmer in 1969 (Fig VIII 5) in which cows back off the platform at the exit position. This is the most popular rotary milking parlour in New Zealand, Britain and Europe and is available in sizes from 15 to 40 stalls and milking units. Operation may be by one to three persons depending on the number of stalls and whether the milking clusters are removed automatically at the end of milking. The prime operator positioned alongside the entry race in the work area surrounding the raised platform is able to encourage hesitant cows to enter the stalls. This parlour is compact, has simple stallwork and is the least expensive design of rotary parlour.

In the *rotary tandem* layout, Fig VIII 6, the formal tandem stalls, with cow nose to tail, are mounted on a narrow platform rotating within a perimeter wall and surrounding a circular area in which the milker works at a level of 0.8 m to 0.9 m below the platform. This arrangement has relatively expensive stall work and the highest building costs because each stall is approximately 2.5 m long and this gives a large diameter circle. Rotary tandem parlours were first installed commercially in Italy in 1965 and a few have been installed in other European countries with 13 to 22 stalls and milking units. In these large parlours the platform is rotated continuously and operation may be by one or two persons depending on whether the teatcup clusters are removed automatically at the end of milking or not.

Smaller versions of the rotary tandem layout are available with 5 - 10 stalls and milking units. In these parlours the platform is rotated through the length of one stall when the operator is ready to change cows. When the platform stops the cow which has been milked leaves, moving forward through an exit doorway in the perimeter wall and is replaced by a cow entering through an adjacent doorway from the collection yard. The platform is rotated once the teatcups are on a cow ready for milking. The 8 stall 8 unit version of this type of rotary tandem layout has been the most successful commercially, but sales have been restricted because the equipment and building costs are much higher than for fixed herringbone milking parlours of similar capabilities.

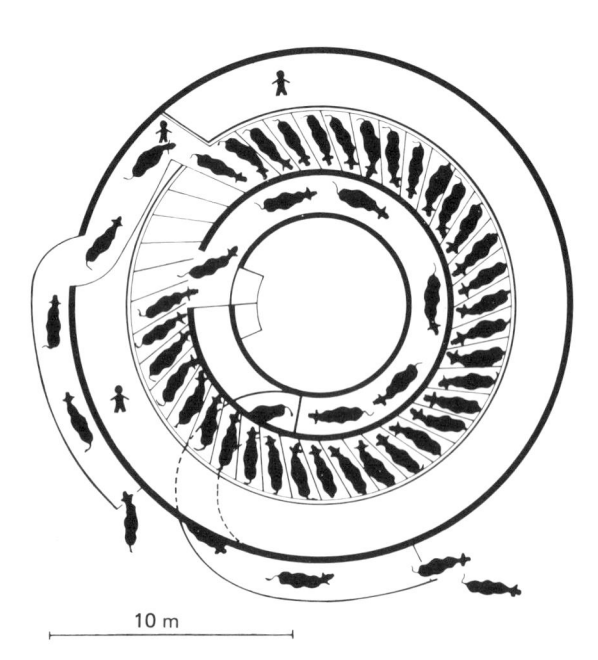

Fig VIII 4. Rotolactor milking parlour and layout diagram for 40 stalls and milking units.

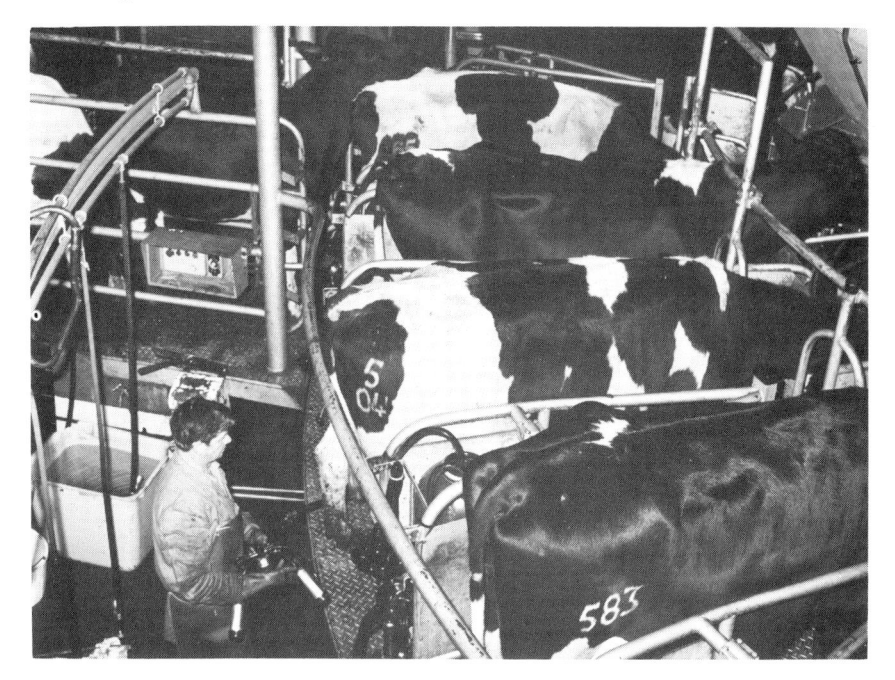

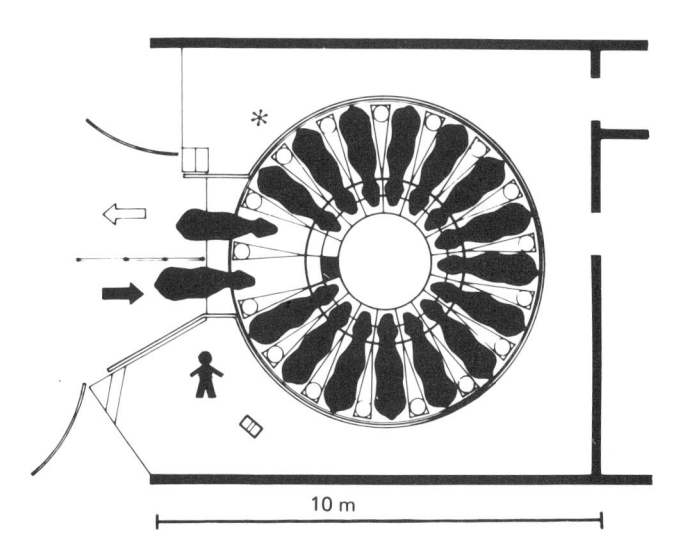

Fig VIII 5. Rotary abreast parlour in which cows back off the platform (17 stalls and units with recorder jars). The asterisk indicates the position of a second operator in the absence of automatic cluster removal.

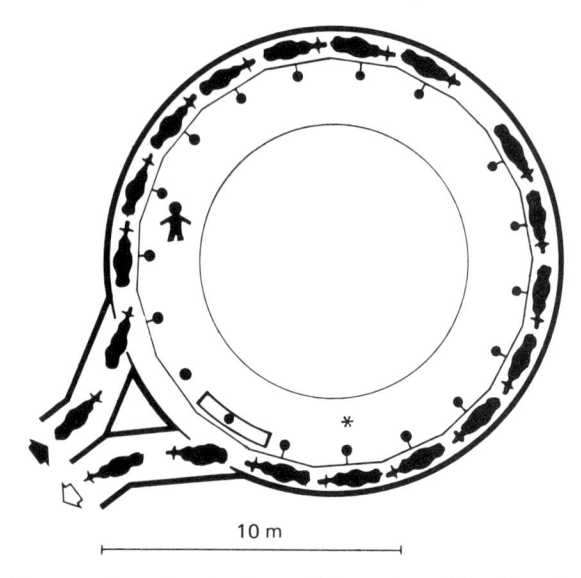

Fig VIII 6. Diagram of a rotary tandem milking parlour (18 stalls and units). The asterisk indicates the position of a second operator in the absence of automatic cluster removal.

The *Unilactor* is a form of tandem milking parlour in which separate stalls are on wheels attached to a track around a rectangular work area some 0.8 to 0.9 m below the level of the platform. This parlour was designed for one man operation with automatic udder washing and teatcup cluster removal to allow one milker with 14 - 22 stalls and units to milk 80 to 120 cows per hour (Fig VIII 8).

Rotary herringbone milking parlours have been in use in Russia and Eastern Europe for about ten years. They are known as rotary herringbone parlours because the cows, on a raised platform round the perimeter of a circular work area, overlap each other as in a static herringbone parlour (Fig VIII 7). This arrangement allows twice the number of cows and units to be accommodated in a given circle than in the rotary tandem layout.

The first rotary herringbone installed in the UK, in 1971, was developed from a Russian design in which a gate at the rear of each stall was operated automatically to control entry and exit and position cows during milking. Parlours of this and other commercial designs incorporating dividing gates between stalls are now available in several sizes with 12 to 28 stalls. Simple stallwork without dividing gates was featured in the rotary herringbone designed by an Australian farmer in 1968. Control of cows at the exit was inadequate and an automatic neck yoke at each manger was included in the commercial version which is marketed in Europe and the UK.

Fig VIII 7. Rotary herringbone parlour and layout diagram (18 stalls and units). The asterisk indicates the position of a second operator in the absence of automatic cluster removal.

Most of the rotary milking parlour installations are the compact abreast herringbone layouts with twelve or more stalls and milking units to allow a single operator to milk 100 or more cows per hour when automatic equipment is used for teatcup cluster removal and other routine operations.

MILKING EQUIPMENT IN MILKING PARLOURS

The movable abreast milking parlour or bail first used in 1925 was equipped with a simple pipeline milking machine cleaned by drawing a hot detergent solution through each milking cluster to a container in the milk room and then forcing steam in the reverse direction.

Subsequently some fixed milking parlours were equipped with pipeline milking machines, and glass receiver jars of 4 or 5 gal (18 to 23 ℓ) capacity to measure the milk yields of individual cows. Neither type of pipeline milking machine was completely satisfactory and abreast, tandem and chute parlours were equipped with direct-to-can milking units from 1950 onwards. The transport cans of 8 to 12 gal (36 to 54 ℓ) capacity were fitted with a special lid with connectors to a milking cluster and the main vacuum line. Milk yields of individual cows were measured by suspending the cans from spring balances attached to the milking stalls. A method of cleaning the stainless steel can lids. teatcups and rubber tubes was devised in which the milking equipment was immersed in a solution of 3% caustic soda between milkings. The combination of direct-to-can milking, immersion cleaning and milk cooling in the cans by means of turbine or sparge ring coolers was the cheapest effective system of milking available for several years.

The introduction of refrigerated bulk milk tanks, and the herringbone milking parlour in 1957 revived interest in pipeline milking machines. At that time milking pipeline machines gave high vacuum fluctuations during milking, there was no provision for the measurement of milk yield, and it was not possible to inspect the milk prior to discharge into cans or a bulk tank. Some pipeline milking machines were equipped with 4 or 5 gal glass jars suspended from a spring balance at each milking unit; milking was satisfactory with these machines but the rate at which the jars were emptied when milk was rejected or discharged to the milk room was very slow, caused delays in the milking procedure and reduced the number of cows milked per man hour.

The method of circulation cleaning for simple pipeline milking machines, devised in New Zealand in 1954, would not clean the existing machines with weigh jars and the associated valves. A pipeline milking machine with glass jars specifically designed for in-place circulation cleaning using hot chemical detergent and disinfectant solutions was eventually developed. Where possible, without materially affecting the cost of components, joints and crevices were eliminated or simplified to facilitate cleaning. The usual components were re-arranged to minimize the time spent preparing for milking or cleaning, flow through the units was equalized automatically and the rate of transfer of milk from the jars increased by the use of rubber tubing of 18 - 25 mm internal diameter.

The pulsators were connected to a galvanized iron pipeline from a self draining auxillary trap in the milk room or milking parlour. A borosilicate glass pipeline of 25 mm bore from the self draining trap served as the vacuum line during milking and the liquid supply line during washing. A second glass line was used to transfer milk and return cleaning fluids from the jars in the parlour to a receiver jar which was connected to the self draining trap and to a milk releaser. Two pinch valves operated by a single lever controlled vacuum supply to the top and bottom spigots of each jar for milking and milk transfer and during milking the rubber tube to a jetter assembly was pinched off by a metal clip. Each unit was ready for washing when the lever was in neutral position; with both rubber tubes open and the cluster fitted to the jetter assembly. The cleaning circuit was completed by the rotation of a three way valve and the removal of the pump discharge tube from the bulk tank or D-pan. The pulsators were in operation throughout the cleaning process. Cleaning and disinfection problems associated with the lids and gaskets of the jars and the vacuum operated releasers were resolved by the use of rigidly mounted one piece glass jars and a stainless steel centrifugal milk pump. Further details of this type of recorder plant are given in Chapter III.

The cleaning of pipeline milking installations was simplified and heat disinfection made more effective by the acidified boiling water process introduced in 1964 (Chapter X). One manufacturer substituted an air compressor for the milk pump to increase the rate of milk transfer and raise the temperature in the milking system above 78°C during acidified boiling water cleaning.

In static herringbone milking parlours having two stalls per milking unit (Fig VIII 9) the obvious position for the glass jars is down the centre of the milker's pit. When installed near floor level the inlet from the teatcup clusters is below the level of the udders of the cows and this arrangement will provide the

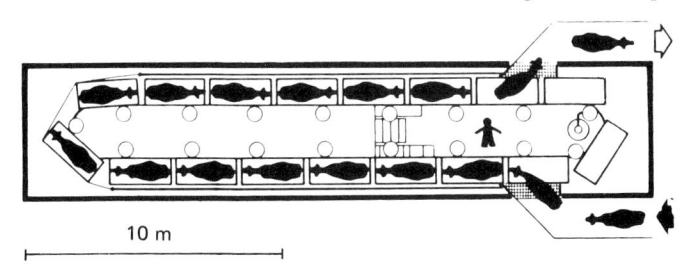

Fig VIII 8. Diagram of a Unilactor milking parlour (17 stalls and milking units).

most constant vacuum conditions at the teats during milking. At this level, however, it is difficult to measure milk yield, inspect and sample or reject milk into a container, and the milker is also faced with the problem of avoiding the long milk tubes from the clusters to the jars when moving along the pit.

Fig VIII 9. Static herringbone parlour with two stalls per unit and jars along the centre line of the pit at eye level to the milker. The use of simple hooks to support the long milk tubes allows free access for the milker along both sides of the pit throughout milking.

Jars set at eye level, the inlet tubes to the jars being 1.8 to 2.0 m above the floor of the pit, are convenient for visual inspection, sampling, rejection and transfer. When the long milk tubes are supported on hooks along the rump rails there is no obstruction to the milker. Objections to having equipment along the centre of a herringbone pit have been met in several ways. A single line of jars along the centre line of the pit may be mounted 2 m above the floor of the pit so that the milker moves below all the milking equipment. With this arrangement relatively high vacuum fluctuations occur during milking because milk is lifted almost 2 m from the cows to the jars and it is difficult to measure milk yields. Rejection of milk is possible, though inconvenient, but milk transfer rates are high. Another alternative has been to fit a milking unit *at each stall* with the line of jars along each side of the pit below the cow standings. The work area of the milker is clear of equipment, vacuum fluctuations at the teats during milking are minimized, and the time for milk transfer from the jars extended. This layout is claimed to be of particular value for high yielding herds because a milking unit is always available when a cow is in position for milking. Herringbone parlours are also available with jars at eye level along each side of the pit to facilitate observation of milk yield, sampling and rejection.

In rotary milking parlours the milking installations and cleaning procedures are similar to those in static parlours. Recorder jars are mounted at eye level or below the level of the rotating platform.

MACHINE MILKING PERFORMANCE

The serious exploitation of the milking machine as a means of making a substantial increase in the number of cows milked per man hour and the number of cows managed per man has occurred during the past twenty-five years. Previously only a few pioneering farmers had shown interest in milking machine installations in milking parlours. The majority of farmers accepted hand milking with the inevitable limitation of six to eight cows milked per man hour and ten to twenty cows managed per man. During the Second World War there was a gradual adoption of milking machines to replace hand milking in cowsheds, but there was rarely any positive effort towards increasing the number of cows milked per man hour. As recently as 1959 the results of a Milk Marketing Board survey of dairy farms in England and Wales[2] showed that the number of cows milked per man hour was fourteen in cowsheds and sixteen in milking parlours.

Many people believe that milking performance should be measured in terms of gallons of milk produced per hour, either for each man or milking unit. In fact the number of gallons of milk produced per man hour will fluctuate at morning and evening milkings throughout the year according to the milk yields of the cows in a herd, but the work done by the milker remains virtually the same for every cow at every milking. Thus the number of cows milked per man hour is the best measure of the effectiveness of the man and the milking installation.

Many efforts were made to establish planning data for milking parlours, particularly in the USA, by measuring and comparing the number of cows milked per man hour in parlours of different size and layout. Various combinations of milkers and assistants with different numbers of milking units and stalls were subjected to time study measurement, but the published results were of value only as case studies.

The underlying principles of machine milking performance were not apparent until 1956 when Chetwynd[3] produced work study data which showed that when a milker was fully occupied with routine work when milking each cow (changing units, changing cows, foremilking, udder washing, etc), the maximum number of cows which could be milked per hour would be found by dividing the *work routine time* per cow into 60 minutes. In practice performance would be affected by the number of milking units used by a milker and the milking times of the cows in a herd because any interruption of the regular repetition of routine work, which would occur should the milker have to wait for a cow to complete milking, deal with an emergency or attend to milk transport or cooling, would result in fewer cows being milked per hour.

In the interval from the time the teatcups of a milking unit were attached to the teats of a cow until the milker returned to that milking unit, a full work routine would be carried out at each of the other milking units in the milking parlour. Maximum performance would only be possible if this interval was longer than the milking time of every cow in the herd, but many cows would be subjected to overmilking. A relationship between the average milk yield in a herd at a milking and average milking time per cow was established from data collected in 20 commercial herds which could be expressed by the equation:

$y = 0.164x + 2.33$
where y = herd milking time, min/cow,
and x = herd milk yield, lb/cow[4]

This information was used in a method of calculating the potential performance of milking installations up to the limit imposed by work routine time. The authors defined *unit time* (UT) as the total time a milking unit was associated with a cow; this was the milking time of the cow plus the time the milking unit was idle between successive cows milked with that milking unit. The maximum number of cows which could be milked per milking unit per hour would be calculated by dividing the unit time/cow into 60 minutes, and the potential performance of the installation (P) found by multiplying the number of cows milked/unit/hour by the number of milking units (N) used. The potential performance of the milking installation would be achieved if the work routine time was less than the *available work time* (AWT) which is calculated by dividing the potential performance into 60 minutes. Data from different milk yields and numbers of milking units in milking parlours with one stall/unit and two stalls/unit shown in Table VIII 2 were calculated from

$$P = \frac{60}{UT} \times N \text{ and } AWT = \frac{60}{P} = \frac{UT}{N}$$

The unit idle time allowed for 1 stall/unit parlours was 1.2 min/cow and for two stalls/unit was 0.2 min/cow.

This table of milking performance and method of calculation has been used extensively in advisory work to indicate the size of installation required for particular farms and as a guide in the assessment of milking performance of existing milking parlours.

Scott[5] pointed out that the concept of *available work time* was only applicable to milking parlours with two stalls/unit where unit idle time is about 0.2 minutes while the unit is transferred from one cow to another. In one stall/unit milking parlours most of the routine work is done whilst a milking unit is idle and any change in work routine time would have a direct effect on the number of cows milked per man hour. It should be realized that the milking performance of a milking parlour on a farm rarely coincides with that predicted from the general data used in the compilation of Table VIII 2, and it is misleading to rate actual performance as a percentage of that predicted. Occasionally extravagant claims have been made for a particular type and size of milking parlour in terms of cows milked per man hour without reference to the work routine time and the milk yields of the cows at the observed milking. In many autumn calved herds a shorter work routine is employed during the summer months when the cows are grazing and milk yields low. The number of cows milked per man hour in such circumstances cannot be used as an indication of the performance to be expected on that or other farms with a comparable milking parlour, at other times of the year unless the average milk yield of the cows and work routine time are recorded at the observed milking.

More recent investigations in 28 herds with a total of 1934 cows milked in static herringbone parlours[5] has shown a reduction in the mean milking time of cows relative to their milk yields. This was to be expected after a period of 14 years during which time measurable improvements were made to the design and effectiveness of teatcup liners, pulsators and other milking machine components supplied to dairy farmers. The change to the British Friesian breed improved milk yield and an increase in the inherant milking rates of cows in the national herd would result. The regression lines for data collected in 1972 are shown in Fig VIII 10 with the time limits within which 95% of herds would occur for mean milk yields of 11.3 kg/cow at a milking.

The information on the herd mean milk yields and milking time per cow can be used in combination with work routine times to establish the potential performance of various types of milking installations. Once the teatcups of a milking unit have been attached to the teats of a cow the milker carries out routine operations at each of the other milking units before returning to remove the teatcups. The duration of the interval between successive visits to a particular unit is determined by the work routine time and the number of milking units in the installation. This period of time is the *available milking time* and must exceed the actual milking time per cow to allow the maximum number of cows to be milked per man hour. The multiple activity chart (Fig VIII 11) illustrates the effect of the pattern and duration of routine manual operations on the available milking time per cow in a static herringbone parlour with eight stalls with eight milking units.

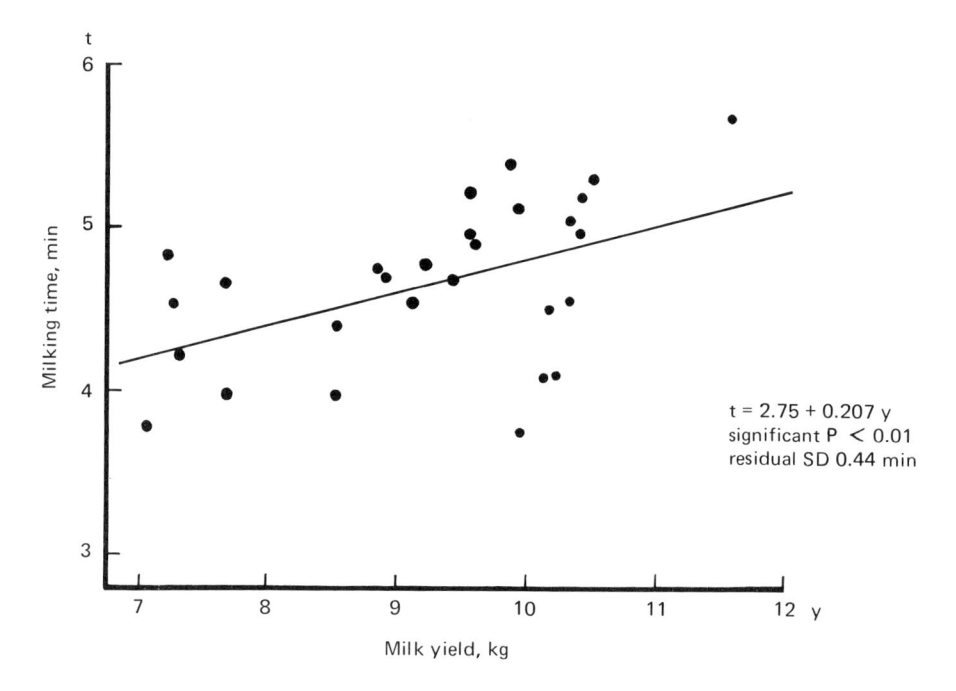

$t = 2.75 + 0.207 y$
significant P < 0.01
residual SD 0.44 min

Fig VIII 10. The relationship between mean milking time, min/cow, for 28 herds at a morning milking and mean milk yield, kg/cow. Mean milking time can be estimated from the milk yields, for example:

Milk yield/cow	Milking time, min	
kg	Estimate	95% limits
11.3	5.10	4.22 − 5.98

For a particular milking installation the upper limit of the number of cows that can be milked per hour by a stockman is set by the number of units he uses and the milking times of the cows. He will achieve this performance only if he can complete the routine work in the required time, and in practice this is the most important factor. While it is possible for a milker, properly organized, to speed up certain of the tasks, major improvements can be made only by mechanization or the task is found non essential and so can be discarded. In the further consideration of milking performance the times taken to complete a number of basic routines are used (i.e. Tables VIII 3 and 5, routines A - H).

The manual operations in work routine A of Table VIII 3 cover all the requirements of machine milking, management and hygiene discussed in Chapters VII, IX and X and can be carried out in the times listed without any particular manual dexterity[7]. Ergonomic assessment in a static herringbone parlour when 50 cows were milked/hour did not reveal any signs of mental or physical stress in the milker[8]. Adjustments to the equipment and in the technique of working reduced the time spent on some operations and the work routine time to 1.00 min/cow (routine B). The omission of foremilking and recording of milk yield resulted in a work routine time of 0.80 min/cow (routine C).

The lactating cows in nearly all herds are milked twice daily, more milk being produced at morning milking because this follows an interval of 13 - 16 hours in the UK. There is usually a time of year when herd milk yield reaches a peak becaust it is common practice to calve most of the cows in a herd in a short season of 2 - 4 months. Whatever the type of milking installation the number of milking units should be sufficient to ensure that the available milking time per cow, with the selected work routine, is not less than the average milking time per cow at morning milking when herd milk yield is at the peak level.

In herds of Friesian cows it is unlikely that average morning milk yield per cow at the time of peak herd milk yield will be less than 11.3 kg (25 lb) and the data in Fig VIII 10 indicate that it would be unwise to have a milking installation that allowed less than 5.5 min available milking time per cow.

Table VIII 3   Examples of practical work routines applicable to static herringbone and stop-start tandem parlours

Operation	A min/cow	B min/cow	C min/cow
Disinfect teats	0.10	0.07	0.07
Change and feed cows	0.25	0.20	0.18
Foremilk	0.10	0.08	
Wash and dry udder	0.25	0.25	0.25
*Change unit	0.30	0.25	0.25
Record Yield	0.15	0.10	
Miscellaneous	0.05	0.05	0.05
Total, min/cow	1.20	1.00	0.80
Maximum no. cows milked/man hour	50	60	75

In one stall/unit parlours this is in two elements, take off and put on cluster

222   *Milking in Cowsheds and Parlours*

	8 stall 4 units Left side	Milking units 1 2 3 4	Right side	Time in min
0	Disinfect teats of 4 cows			0
	Let out 4 cows			
1	Let in and feed 4 cows		4 units milking cows on right of parlour	1
	Foremilk 4 cows			
2	Wash teats of 4 cows			2
	Dry teats of 4 cows			
3	Put on cluster 1		Remove cluster 1 Record yield 1	3
	Put on cluster 2		Remove cluster 2 Record yield 2	
4	Put on cluster 3		Remove cluster 3 Record yield 3	4
	Put on cluster 4		Remove cluster 4 Record yield 4	
	Miscellaneous		Miscellaneous	4.8
5			Disinfect teats of 4 cows	5
			Let out 4 cows	
6	4 units milking cows on left of parlour		Let in and feed 4 cows	6
			Foremilk 4 cows	
7			Wash teats of 4 cows	7
	Remove cluster 1 Record yield 1		Dry teats of 4 cows	
8	Remove cluster 2 Record yield 2		Put on cluster 1	8
	Remove cluster 3 Record yield 3		Put on cluster 2	
9	Remove cluster 4 Record yield 4		Put on cluster 3	9
			Put on cluster 4	
	Miscellaneous		Miscellaneous	9.6
10	Disinfect..........			10

Available milking time   ■ 4.55
Unit idle time   ☐ 0.25
Work cycle time, 8 stall, 4 unit 4.8 min/cow

Available feeding time   ▤ 8.6 min

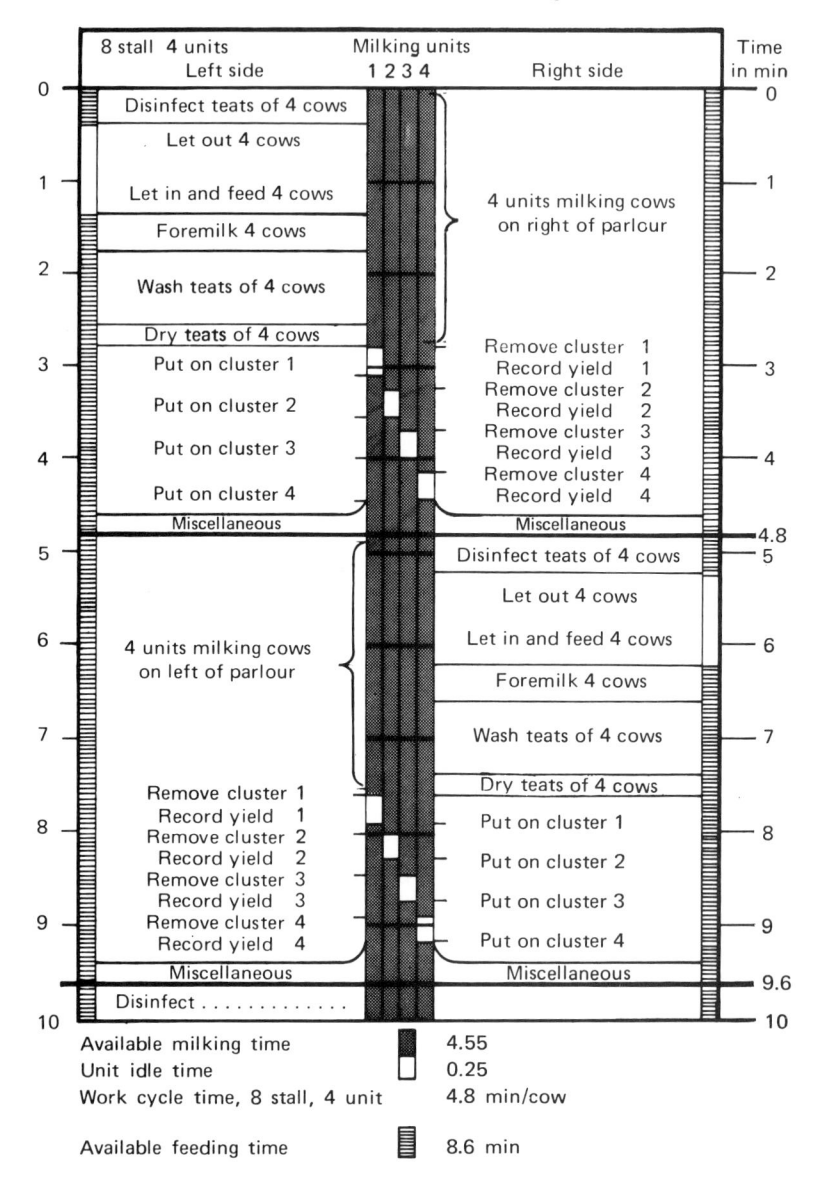

Fig VIII 11. Multiple activity chart for a work routine of 1.2 min/cow in a static herringbone parlour, with 8 stalls and 4 units and 8 stalls and 8 units. Each unit is on a cow for a longer period in the 8 unit layout than in the 4 unit layout because units

P A Clough

8 stall 8 unit Left side	Milking units 1 2 3 4 5 6 7 8	Right side

Disinfect teats of 4 cows
Let out 4 cows
Let in and feed 4 cows
Foremilk 4 cows
Wash teats of 4 cows
Dry teats of 4 cows
Put on cluster 1
Put on cluster 2
Put on cluster 3
Put on cluster 4
Miscellaneous

Miscellaneous
Remove and record 5
Remove and record 6
Remove and record 7
Remove and record 8

Disinfect teats of 4 cows
Let out 4 cows
Let in and feed 4 cows
Foremilk 4 cows
Wash teats of 4 cows
Dry teats of 4 cows
Put on cluster 5
Put on cluster 6
Put on cluster 7
Put on cluster 8

Miscellaneous
Remove and record 1
Remove and record 2
Remove and record 3
Remove and record 4
Disinfect . . . .

Available milking time   5.8
Unit idle time   3.8
Work cycle time, 8 stall 8 unit   9.6 min/cow

Available feeding time   8.6 min

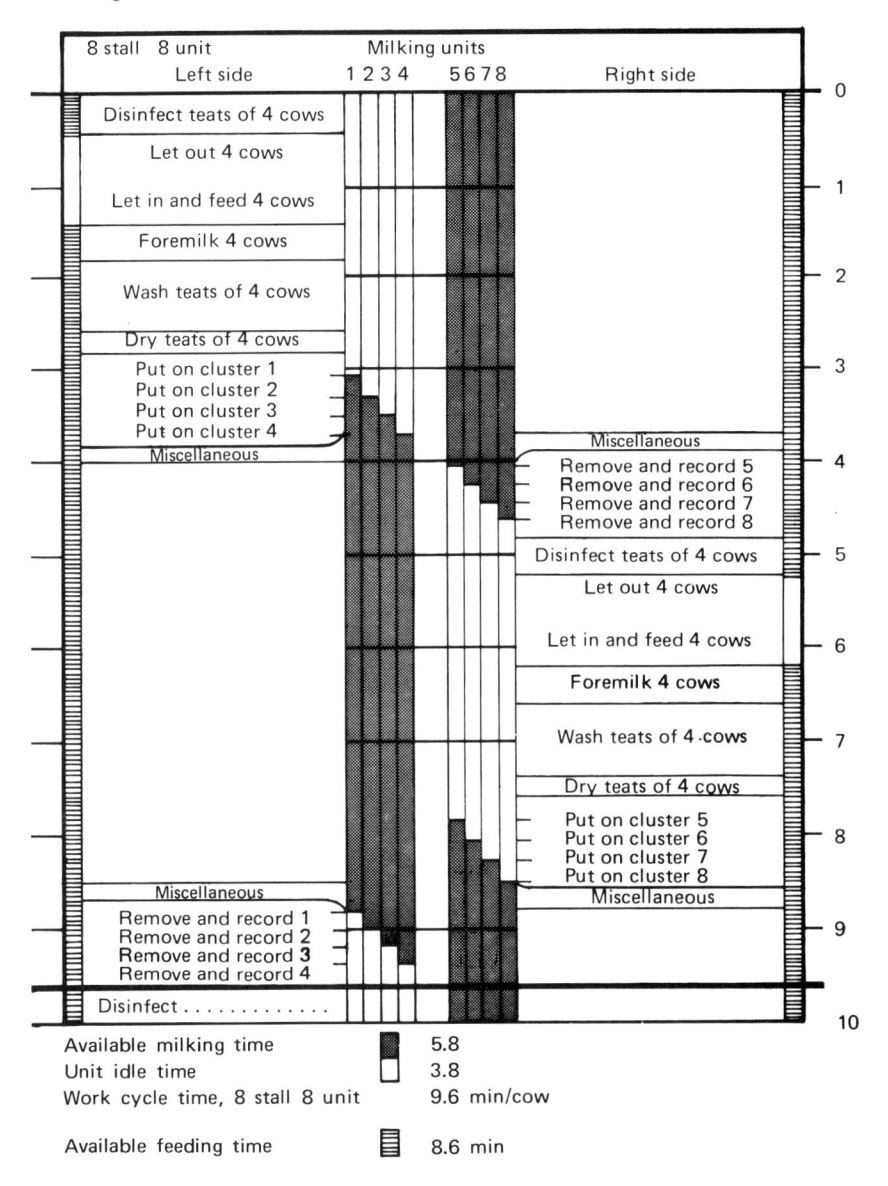

on one side of the parlour remain on the cows until the units on the other side are in use after batches of cows have been changed.

Table VIII 4 *The effect of work routine time on the available milking time per cow in static herringbone milking parlours*

Work routine time, min/cow		A 1.2	B 1.0	C 0.8
Size of parlour				
No. of stalls	No. of milking units	Available milking time, min/cow		
8	8	5.8		
10	5	5.8		
10	10	7.1	6.0	
12	6	7.0	5.8	
12	12	8.5	7.2	5.7
14	7	8.2	6.8	5.4
14	14		8.4	6.5
16	8		7.8	6.2
16	16			7.3
20	10			7.8
Maximum no. cows milked/man hour		50	60	75

Table VIII 4 is based on multiple activity charts drawn up for different sizes of static herringbone milking parlours with the work routines A, B and C, described in Table VIII 3. This shows that at least 8 stalls and 8 units or 10 stalls and 5 units are necessary for the available milking time per cow to exceed 5.5 min with work routine A and even more units for the shorter routines.

In rotary tandem parlours with 8 stalls and 8 milking units the available milking time per cow for work routines A, B and C would be 8.4, 7.0 and 5.6 min/cow respectively (Fig VIII 12).

SELECTION OF A MILKING PARLOUR

Within the framework of a farm management programme a farmer must decide on herd size, concentrates feeding and calving policy, the times of day when milking will begin and how long it will last, and the work routine to be followed during milking. *From this it will be possible to specify the number of cows to be milked per hour.* Should this be more than the work routine will allow, a larger parlour for two operators or a shorter work routine time will be necessary. The operations of changing batches of cows and changing milking units from cow to cow have to be included in the milking work routine, but the other operations listed in Table VIII 3 are not directly essential to machine

P.A. Clough

WORK ROUTINE TIME/COW

WORK ROUTINE

		MAN MIN/COW
	REMOVE CLUSTER	0·0
	DISINFECT TEATS	0·0
	IDENTIFY TO FEED	0·05
	WASH AND DRY TEATS	0·20
	PUT ON TEAT CUPS	0·20

DIRECTION OF ROTATION

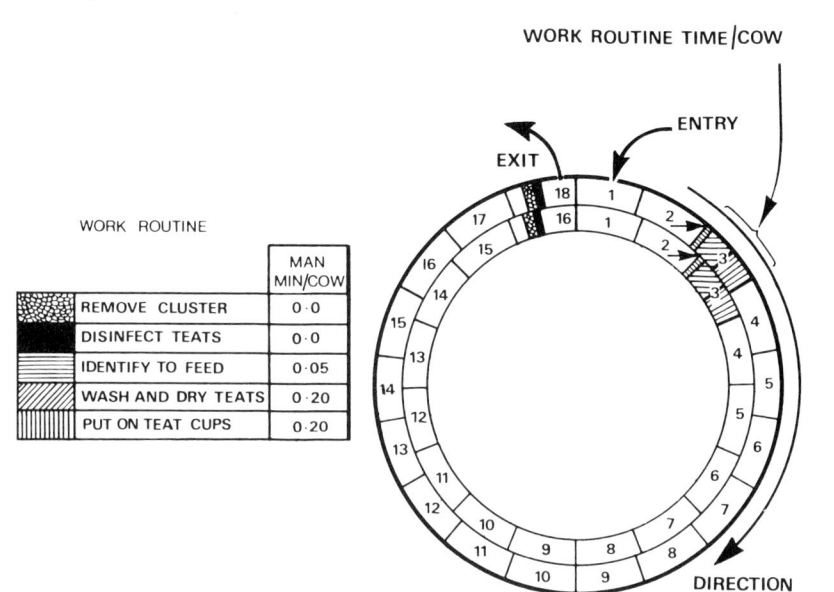

Fig VIII 12. The relationship between the maximum number of cows milked/man hour, the work routine time and the rotation time of rotary milking parlours.

a Rotary abreast, tandem and herringbone parlours with *continuous* platform rotation. Diagram for 16 and 18 unit parlours.

Assuming one milker, a cow exit time of 0.3 min, automatic cluster removal and automatic teat disinfection.

Rotation time (min) = No. of stalls x work routine time/cow (min)

Available milking time (min) = No. of stalls x work routine time/cow −
(3 x work routine time/cow + 0.3)

= (No. of stalls −3) x work routine time/cow −0.3)

Max. cows milked/man hour = 60 ÷ work routine time/cow

Available feeding time (min) = Rotation time −(2 x work routine time/cow + 0.4)

b Rotary tandem parlours with *intermittent* platform rotation and all routine work done whilst platform is stationery

Rotation time (min) = No. of stalls x work routine time/cow

Available milking time (min) = No. of stalls x work routine time/cow − work routine time/cow

= Rotation time − routine time/cow

Max. cows milked/man hour = 60 ÷ work routine time/cow

Available feeding time (min) = Rotation time − time to change cow and feed milking. The order of priority for inclusion in the milking work routine will depend on the likely contribution to herd health or management and the production of uncontaminated milk. The mechanization of some routine operation should also be considered (see Table VIII 5) as a means of reducing work routine time per cow.

The considerable capital investment involved in the establishment of a new milking parlour merits a detailed forecast of the likely changes in herd size and management over the following ten years. The parlour building and fixed equipment will have an expected life in excess of ten years and must be sited correctly in relation to other buildings and yards. Within this period changes in the milking installation necessary to raise the number of cows milked per man hour should not require alterations to the building and fixed equipment.

The most direct approach to future expansion would be the installation of a 16 stall 8 unit or 20 stall 10 unit static herringbone layout equipped with automatic teatcup removal units in a building designed to accommodate a one stall/unit layout, should the conversion become necessary to increase milking performance. This means of limiting initial capital investment could be even more beneficial for herds where there would never be justification for a herringbone parlour with more than 12 stalls, because the introduction of automatic cluster removal units could be deferred until the time of conversion from 6 to 12 milking units.

There are, in fact, many existing 16 stall 8 unit and 20 stall 10 unit static herringbone milking parlours, where two men are under-employed milking 80 - 150 cows, which could be operated by one person with the aid of automatic teatcup cluster removal units to prevent overmilking of individual cows. Further mechanization to cut down work routine time could be introduced as equipment and capital become available. This would be more sensible than conversion to a one stall per unit installation and the retention of two operators.

Recently many farmers have been advised by the representatives of milking machine manufacturers and others to install a one stall per unit layout in a new herringbone parlour on the grounds that the additional cost of milking units, compared with a two stall per unit layout with two more stalls and a longer building, is offset by the more flexible operating arrangements in the shorter building. Whilst true at the time of sale, this advice should be given only when it is known that herd size, management and work routine time will not be altered throughout the expected life span of the milking parlour.

At present interest is centred on milking parlours with 10 to 16 stalls for farms where one person milks a herd at each end of the working day. Farmers with limited capital, who are content to accept 50 cows milked per hour now, should be encouraged to plan a programme of investment over the next few years which would lead to a milking performance of 75 or more cows per man hour. This is the only way in which it will be possible to cut down the number

Table VIII 5 The times of the various routine operations (min) for work routines in high performance milking parlours

Routine operation	Static herringbone D	Rotary parlours with continuous or automatic intermittent rotation			
		E	F	G	H
Disinfect teats	automatic	automatic	automatic	automatic	automatic
Change cows	) 0.15	automatic	automatic	automatic	automatic
Identify cows	)	0.07	0.05	0.05	automatic
Feed	)	automatic	automatic	automatic	automatic
Foremilk	omit	0.08	omit	omit	omit
Wash teats, dry teats	) 0.20	0.20	0.20	automatic 0.10	automatic
Remove teatcups	automatic 0.20	automatic 0.20	automatic 0.20	automatic 0.20	automatic 0.20
Replace teatcups	omit	automatic	automatic	automatic	automatic
Record milk yield	0.05	0.05	0.05	0.05	0.05
Miscellaneous					
Total min/cow	0.60	0.60	0.50	0.40	0.25
Maximum no. cows milked/man hour	100	100	120	150	240

of hours worked each day to a socially acceptable level. The most suitable size of two-stall/unit parlour should be installed, automatic teatcup cluster removal units and other equipment for reducing work routine time introduced as soon as convenient and, if necessary, the conversion made to a one-stall/unit layout.

Farmers who decide to install a one stall/unit layout, but cannot meet the capital requirement for the final size of parlour equipped with automatic teatcup cluster removal units etc., should install the number of milking units which will meet current needs in a building which will accommodate the final number of stalls and milking units.

Although capital investment is lower and the organization of labour for regular and relief milking easier when a herd is milked by one person each day, a farmer may prefer to have a larger parlour with two operators. The previous remarks about one stall and two stall per unit layouts apply in these circumstances, but a new herringbone parlour for two operators should have 20 or 24 stalls. During milking each operator should be responsible for a particular number of milking units and work independently except when batches of cows are changed. At these times the person at the cow entry end of the parlour may be responsible for the identification and feed allocation to all the cows in the batch and the second person may prepare more than half the number of cows for milking.

Many variations in labour organization are possible in the management of herds of 250 or more cows, particularly where the herd is divided into groups of 60 - 120 cows and the size of the milking installation such that each group can be milked in 1 to 1½ hours. Depending on herd size one or more specialist milkers may be fully employed milking all the cows in the herd once, twice or three times each day. In such circumstances the interest and efficiency of the milker or milkers is maintained by the inclusion of adequate rest periods between milking sessions.

HIGH PERFORMANCE MILKING PARLOURS

The information in Table VIII 4 covers work routines in static herringbone parlours in which there is little dependence on automatic equipment other than automatic teatcup cluster removal units used principally to avoid overmilking of individual cows. In seeking ways of cutting down the time devoted to herd milking and the number of hours worked each day, whilst maintaining or improving the standard of herd management, dairy farmers have become interested in high performance milking parlours in which automatic devices are used to replace manual operations. Milking work routines D, E, F, G and H of Table VIII 5 in which manual work is progressively reduced from 0.6 to 0.25 minutes per cow would make it possible for one person to milk 100 - 240 cows per hour.

It is relatively easy to mechanize and automate all the routine operations listed in routine H, except for attaching teatcups to the teats of cows, and make use of electronic data collection, storage and analysis equipment to control herd management from the milking parlour. In practice the main problems have been in the cost and unreliability of sensitive equipment in the adverse conditions of a milking parlour. The resources now committed to research and development work in this area of livestock engineering by state and commercial organizations in many countries should guarantee steady progress in the design and production of automatic equipment for milking parlours.

Where less emphasis is placed on management activities during milking, work routine D would be applicable in static herringbone layouts with 16 stalls and 16 units or 20 stalls and 10 units in which the available milking times/cow would be 6.0 and 5.8 min respectively. For high yielding herds a static herringbone parlour with 20 stalls and 20 units would provide 7.2 min of available milking time/cow.

Although the basic cost per stall and milking unit is higher for rotary milking parlours than static herringbone layouts, the cost of automatic equipment is often less for these parlours because a single stationary device will serve all the stalls or milking units and the movement of the platform can be employed to operate various valves and switches. The movement of cows in and out of the stalls one at a time, and the fact that the operator is stationed at one position with all the necessary controls and equipment to hand makes it easier for the operator to identify each cow, prevent contaminated milk being transferred to the bulk storage tank, and direct any particular cow into an inspection pen adjacent to the exit race from the parlour.

Table VIII 6 *The effect of work routine time on available milking time/cow in rotary milking parlours*

Routine*	Operator's work routine time, min/cow	Maximum no. cows milked/ man hour	Available milking time, min/cow			
			14	16	18	20 stalls
			14	16	18	20 milking units
E	0.60	100	6.3	7.5	8.7	9.9
F	0.50	120		6.2	7.2	8.2
G	0.40	150			5.7	6.5

*see Table VIII 5

The available milking times per cow in different sizes of rotary milking parlours have been calculated in respect of work routines E, F and G, and are listed in Table VIII 6. The method of calculation is shown in Fig VIII 12a. It must be realized that since milking units are attached in order of rotation the available milking time cannot be extended for any individual cow by a change in the order of attachment and removal of the teatcup clusters. In practice rotary milking parlours should be constructed with 16 or more stalls and milking units because there is no way in which the size of the milking installation can be increased, other than the complete renewal of the milking parlour.

Dairy farmers who have recognized the potential of rotary milking parlours have anticipated the introduction of automatic equipment and installed parlours with up to 40 stalls and milking units in which two or three operators co-operate in the manual operations during milking to milk 60 - 80 cows/man hour. There is no doubt that once reliable automatic equipment is available that will provide the detailed information on each cow, which in smaller herds of cows has been dependent on the powers of observation and memory of the experienced cowman, it will be used by dairy farmers to obtain milking performances of 150 or more cows per man hour.

There will always be a need for dedicated stockmen with respect and sympathy for the cows they manage and the aim of all concerned with dairy farming should be directed towards making the work associated with cows sufficiently interesting and rewarding to encourage intelligent young people to make a career in dairy herd management. Shorter working hours, better working conditions and higher salaries will have to be accompanied by a change in the role of the operator in a milking parlour. There must be less physical and mental effort involved in respect of each cow milked, and any change in output, health or behaviour of a cow indicated to the milker by sensitive instruments and equipment. The efforts of the milker should be directed towards the anomalous cow rather than dissipated on equal attention to every animal during milking.

REFERENCES

(1) Belshaw, D.G.R. & Scott, A.H. (1962) *Technical Report (Work Study) No. 1.* Farm Economics Branch, School of Agriculture, Cambridge University.

(2) Hodges, J. & O'Connor, L.K. (1961) *Personal Communication,* Milk Marketing Board, Thames Ditton, Surrey.

(3) Chetwynd, K.J. (1956) *Agricultural Review,* **12**, 35.

(4) Clough, P.A. & Dodd, F.H. (1959) *NAAS Quarterly Review,* **43**, 1.

(5) Scott, A.H., (1963) *Technical Report (Work Study) No. 3* Farm Economics Branch, School of Agriculture, Cambridge University.

(6) Clough, P.A., Westgarth, D.R. & Williams, Dilys F. (1973) *Proceedings of the British Society of Animal Production,* **2**, 73.

(7) Quick, A.J. (1968) *Esso Farmer,* **20**, 18.

(8) Tomlinson, R.W. (1970) *Journal of the Institution of Agricultural Engineers,* **25**, 18.

Chapter IX

MACHINE MILKING AND MASTITIS

R G Kingwill, F H Dodd and F K Neave

Udder disease occurs in all cattle, but it is particularly important in herds of cows kept for commercial milk production. In most of these mastitis is a continuing and costly scourge contained only by the frequent use of antibiotics. Over 10 000 000 tubes of antibiotic are used annually in Britain to treat clinical mastitis. The disease is not new; the agricultural texts of the 18th and 19th centuries gave rules for avoiding 'garget' and teat damage, and mastitis must have been a major difficulty facing the inventors of milking machines. Pulsation and the double chambered teatcup were introduced to avoid teat and udder damage, and cannula milking was finally abandoned because it was impossible with this type of milker to avoid mastitis. Most of the basic properties of the milking machine (e.g. vacuum level and pulsation rate) were arrived at by experience, including observations on mastitis, and have changed little in the past 70 years (see Chapter I).

It would be wrong to attribute present levels of mastitis to the introduction of machine milking; mastitis was often a serious problem in hand milked herds, and occurs frequently in suckled beef herds[1] and dry cows. Probably current levels of infection are similar to those when most herds were hand milked, but there are few comparable data. Much of the early literature on milking machines and mastitis was concerned with the effects of the *change* from hand milking and most surveys and experiments indicated higher levels of infection and more clinical mastitis in machine milked herds. How much this was due to the types of machine then available, the way they were used, or resulted from other contemporary changes in therapy and management, or even from changes in methods of diagnosis, is unknown.

In this chapter we have attempted to put into perspective the existing information on the relationship of udder disease to milking machine design and properties and machine milking practice. For those unfamiliar with the definition, diagnosis and epidemiology of mastitis, we have included brief introductory sections describing udder diseases and the mode of infection. Because most transfer of pathogens between cows occurs at milking time, hygiene in milking is dealt with at some length. Finally, the development of control systems based on husbandry and the problems of their application on farms are described.

This chapter does not review the very numerous publications on milking machines and mastitis. It should be read in conjunction with other reviews[2,3,4].

UDDER INFECTION AND MASTITIS

Mastitis is inflammation of the udder caused by infection, injury, secretory malfunction or physiological change. The disease of economic importance is nearly always associated with microbial infection. The major pathogens, responsible for over 90% of udder disease, are *Staphylococcus aureus*, *Streptococcus agalactiae, Str. dysgalactiae* and *Str. uberis* though in some herds other bacteria such as *Escherichia coli* and *Corynebacterium ulcerans* can cause serious occasional problems. *C.pyogenes* is an additional cause of sporadic but severe mastitis particularly in dry animals.

Nearly all cows calving for the first time are uninfected, and pathogens cannot be detected in aseptically taken milk samples, the milk yields of corresponding fore or hind quarters of the udder are nearly equal, the milk composition of all four quarters are almost identical and the somatic cell counts are normally less than 100 000/ml of milk. Lactating and dry cattle of all ages are exposed to contamination by pathogens widespread in the environment but once lactation begins, the udder is also exposed externally to larger numbers of pathogenic bacteria transferred from infected cows during milking. Normally the streak canal of the teat prevents the entry of the pathogens into the teat sinus but penetration can occur and may be followed by intramammary infection. The cow's udder may be infected in one or more quarters with the same or different pathogens, and from about 10% of infected quarters two, or rarely more, types of pathogens can be isolated from the milk. Once a quarter is infected pathogens can nearly always be isolated from milk samples and characteristic changes in the milk detected.

The course of an infection varies but the general pattern is described by Fig IX 1[5]. Often the initial stage is subclinical, that is, the changes in milk composition and the inflammation are so slight that they cannot be detected by any visual examination of the milk or palpation of the udder. Many infections are never detected by the milker and recovery occurs spontaneously but more often the infection results in clinical mastitis, days or months after the initial infection, with obvious inflammation of the quarter and changes in milk appearance. Most common of these are clots formed from a mixture of leucocytes, epithelial cells and precipitated milk proteins. The milk may be discoloured and have marked changes in chemical composition. Occasionally severe clinical symptoms develop when the udder becomes hard, tender or swollen and milk secretion is much reduced or may cease. Infection can be systemic with a rise in body temperature, leading in the most severe cases to death. When signs of mastitis are detected by the stockmen it is normal to give a course of intramammary infusions, usually with antibiotic preparations. As a result the clinical signs always disappear quite quickly and often the infection is eliminated.

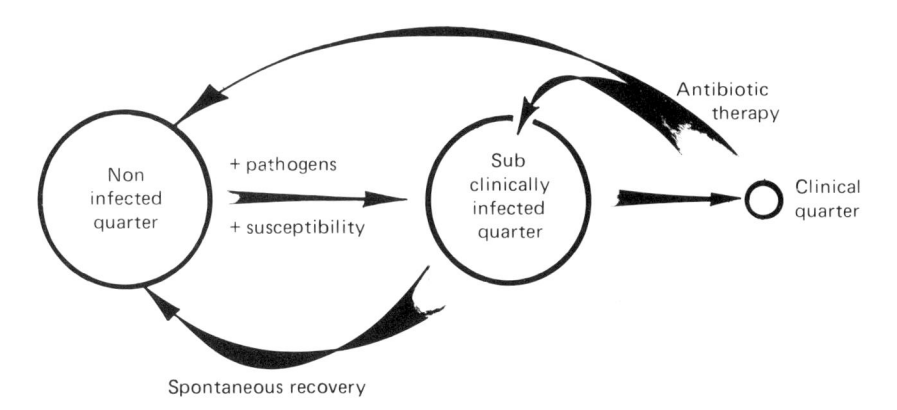

Fig IX 1. The pattern of intramammary infection and clinical mastitis. Note that at any time only a small proportion of infections show clinical signs, antibiotic therapy does not always eliminate infection and some subclinical infections recover spontaneously[5].

In this chapter the description of udder disease is restricted to infections by the major pathogens. In addition infections with minor pathogens such as micrococci and *Corynebacterium bovis* are common occurrences but they are excluded from consideration because normally they result in little inflammation, rarely cause clinical mastitis and have little effect on milk yield and composition[6].

A small proportion of cases of mastitis may be 'non-specific' that is without any infectious agent. These cases may result from physiological changes such as those that occur at calving or drying off, with persistent incomplete milking or from hormonal influences.

Many factors affect the probability of an udder quarter becoming infected. Some of these are mainly genetic, for example the greater susceptibility of fast milking cows[7], others such as the higher new infection rates at drying off, calving, and in older cows shown in Fig IX 2, are partly physiological[9] and many are environmental such as differences in methods of housing, milking and possibly feeding[4].

Most cows surviving for several lactations become infected once or more often. The proportion of infections caused by the different pathogens varies, and herd levels of infection range from less than 10% to more than 90% of cows infected (see Fig IX 3). On average, unless infection is controlled, about half the cows in a herd are likely to be infected in about two udder quarters per cow. Nevertheless at any one time only about 2% of cows have clinical mastitis. These large differences between herds in infection level stem from many factors, and certainly cannot be attributed entirely to machine milking. Other factors include

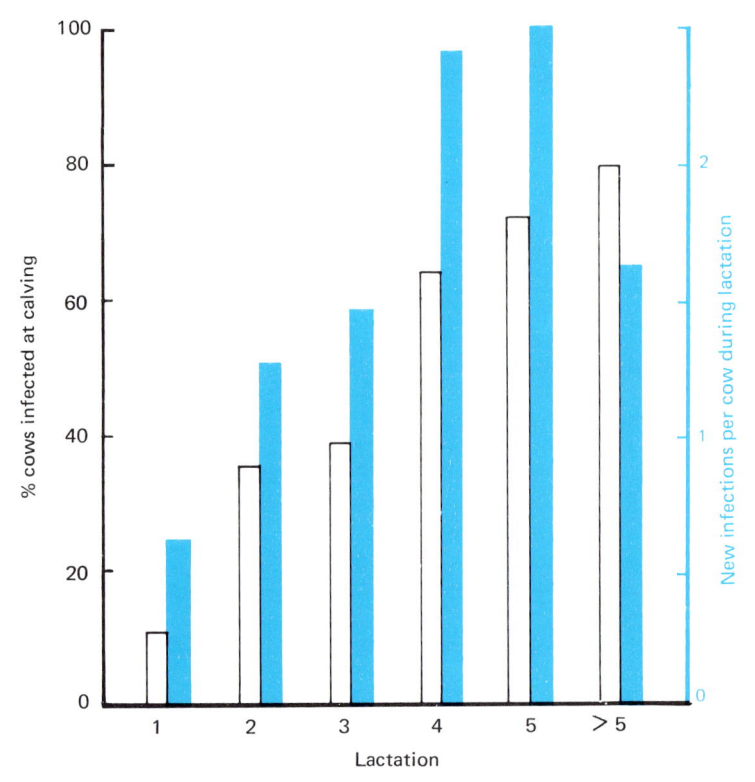

Fig IX 2. Both the prevalence of infection at calving and the occurrence of new infection during lactation is greater in older cows[8].

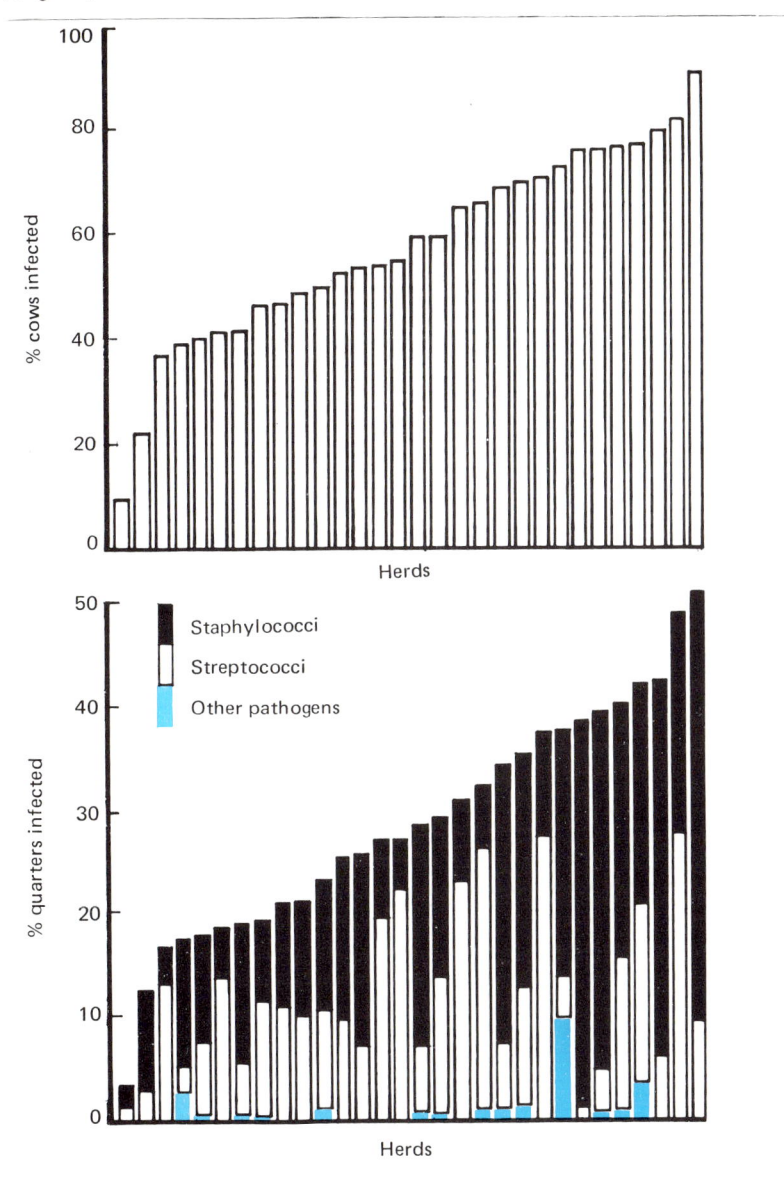

Fig IX 3. Levels of infection in cows and quarters in 30 herds not using teat dips or drying off therapy[10].

the genetic differences between the herds and between the strains of pathogens in the herds, the occurrence of teat lesions, herd culling policies, and differences in the care in detecting clinical mastitis and in the treatment given.

DIAGNOSIS OF INFECTION

Although inflammation can be detected from the changes that occur in the composition of the milk, infection can be diagnosed only by the recovery of pathogens from milk samples taken aseptically from individual quarters. The initial bacteriological test usually includes surface plating of fresh or refrigerated milk samples on aesculin blood agar and an examination of the appearance of the colonies that grow during 48 h incubation at $37°C$[11]. Unfortunately it cannot be concluded with certainty that the presence of pathogens in a single milk sample indicates infection, or the converse. The milk sample may be contaminated because of poor sampling technique or because the streak canal is colonized or lesions are present at or near the tip of the teat. Alternatively, if low numbers of pathogens are present in the milk of an infected quarter none may be detected in the 0.01 or 0.1 ml of milk examined[12]. Because microbiological tests for infection are complicated and involve delay, and do not assess the degree of inflammation present, other tests for udder disease have been developed. Most common of these are measurements of the somatic cell counts of milk, or tests based upon changes in cell count or the chemical composition of milk such as the California Mastitis Test, the Whiteside Test and electrical conductivity measurements[4]. Such indirect tests require careful interpretation because similar changes also occur at calving and drying off and often persist after infection has been eliminated.

EFFECT OF MASTITIS ON MILK YIELD AND COMPOSITION

Milk yield is invariably depressed by udder infection but to a widely varying degree. Severe clinical mastitis may completely halt milk production, but the cumulative losses from subclinical disease, which by definition passes undetected, constitute much the greatest costs of the disease.

Infection of an udder quarter usually causes a change in milk composition followed by depression in yield, due to impairment of the secretory tissues. Permeability of this tissue is much increased allowing freer exchange between secreted milk and the blood. The lactose and potassium diffuse from the milk to the blood and therefore their concentration in the milk declines. To maintain the osmotic status, sodium and chloride ions enter the gland and there is also an increase in the serum protein in the milk. Changes also occur in other constituents[13]. The effects on the concentration of milk fat are less regular.

Table IX.1 Relative depression in yield of milk, fat and solids not fat from 24 quarters during and after infection with **Staph. aureus**, compared to corresponding non-infected quarters of the same udders[14].

	No. of quarters	Relative depression: milk yield (%)	fat (%)	solids not fat (%)
Quarters showing spontaneous recovery:				
Infected quarter	9	24.4 ± 5.73**	0.28 ± 0.113*	0.23 ± 0.057**
In *same* lactation after recovery		27.3 ± 6.03**	0.19 ± 0.072*	0.01 ± 0.042
In *next* lactation after recovery		10.3 ± 5.82	−0.07 ± 0.092	−0.04 ± 0.048
Quarters recovering after treatment:				
Infected quarter	15	34.5 ± 5.65**	0.30 ± 0.063**	0.29 ± 0.077**
In *same* lactation after recovery		39.1 ± 5.71**	0.10 ± 0.036*	0.12 ± 0.049*
In *next* lactation after recovery		11.1 ± 5.39*	−0.11 ± 0.059	−0.01 ± 0.021
Total number of quarters:				
A. Infected quarters	24	30.7 ± 4.17**	0.29 ± 0.056**	0.27 ± 0.055**
B. In *same* lactation after recovery		34.7 ± 4.31**	0.13 ± 0.061*	0.08 ± 0.035*
C. In *next* lactation after recovery		10.8 ± 3.91*	−0.09 ± 0.059	−0.02 ± 0.022

* $p < 0.05$ and > 0.01; ** $p < 0.01$

Significance of differences between A and B	Not significant	$p > 0.05, < 0.10$	$p < 0.01$
Significance of differences between A and C	$p < 0.01$	$p > 0.01$	$p < 0.01$
Significance of differences between B and C	$p < 0.01$	$p < 0.05$	Not significant

The comparison of the yield of corresponding infected and healthy quarters within the same udder is probably the best method of assessing losses, although such a method will slightly exaggerate the total losses since some compensatory secretion by the healthy quarters is known to occur. A study of 72 lactations including 92 quarters infected with *Staph. aureus* showed wide variation but the *relative* yield of quarters during the period of infection was depressed by a mean of 15.3 ± 2.5%, the fat percentage by 0.19 ± 0.03, and the solids not fat percentage by 0.20 ± 0.02[14]. The mean relative depression in yield of the quarters already infected at calving was 24.7 ± 3.2%, compared with 8.0 ± 3.5% for the quarters becoming infected during lactation. The difference between these means indicates the serious effects of infection in the late dry period and early lactation (see also[15]).

The data included 24 cows that had been recorded while infected, again in the same lactation after recovery, and in the following lactation. After elimination of infection there was a rapid but incomplete recovery in both the fat and solids not fat percentages of the milk, and after a dry period this recovery was usually complete. There was no recovery in milk yield after elimination of infection until the following lactation, and even then the recovery was incomplete[14] (Table IX 1).

Estimates of the effect of udder disease on lactation yield are complicated by the many other factors affecting milk yield, such as variation in age and stage of lactation, and plane of nutrition. A typical study with a herd with prompt treatment of clinical mastitis showed lacation yields of milk, solids not fat and fat to be depressed by 10, 11 and 12% respectively in the presence of infection[16].

MODE OF INFECTION

The mode of infection is not fully understood but it is accepted that infection follows the penetration of the streak canal of the teat by a pathogen. New infections are established comparatively rarely. The chance of a single uninfected lactating quarter becoming infected in a 24 h period is less than one in five hundred even in herds with more than 50% of the cows infected. Therefore, either pathogens penetrate the streak canal infrequently or penetration is often not followed by infection. The length of the streak canal is usually within the range 7 - 15 mm (0.3 - 0.6 in) and a pathogen may pass through to the teat sinus as a result of multiplication within the canal, by direct or indirect propulsion or perhaps partly by both methods. The relative importance of these methods is not known but there is experimental evidence that growth and propulsion occur, though complete penetration by growth is unlikely during a single normal milking interval, but can occur within 24 h in an unmilked cow. Probably the relative importance of growth and propulsion varies at different times. During lactation, particularly when disinfectant teat dips are used, direct penetration may be more important whereas in the dry period bacterial growth could be the main way in which pathogens penetrate the streak canal. When bacteria enter the teat sinus they must grow before an infection is established and this does not always occur.

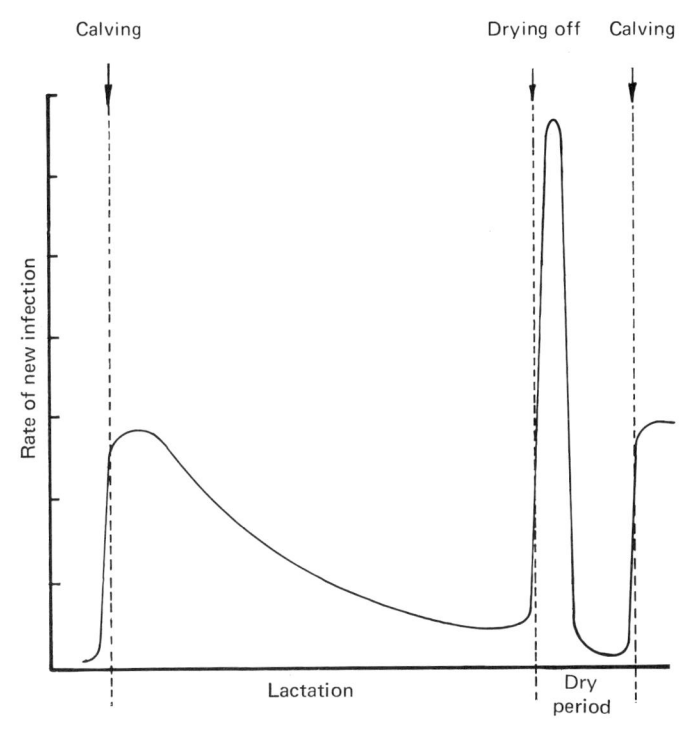

Fig IX 4. The relative frequency of new infection during lactation and the dry period. The diagram summarizes the ideas expressed in the text, but is based on estimates of new infection with time.

A number of studies have demonstrated the importance of milk removal in preventing new infection. Infection developed in 8% of quarters inoculated in the teat duct with *Staph. aureus* when the animals were milked twice daily, and in 23% of quarters when the first post inoculation milking was omitted[17]. When small numbers of bacteria were infused into quarters immediately *before* milking there were very few new infections but infection was common when the same number of bacteria were infused *after* milking[18]. When only the external surface of the teat was exposed to pathogens the new infection rate was 10 times greater from unmilked quarters compared with those milked twice daily[19]. Finally, the careful removal of foremilk at the start of milking may reduce new infections[20, 21]. These results suggest that bacterial penetration may be much more frequent than is indicated by the new infection rate but that the pathogens are quickly expelled from the udder before an infection is established. Milking, or at least milk removal, is an important mechanism for preventing new infection.

The changing frequency of new infection during lactation and the dry period is shown diagrammatically in Fig IX 4. In summary, new infection occurs infrequently even when exposure to pathogens is high. Therefore the natural defence mechanisms are normally effective and in this respect the patency of the streak canal and the regular removal of milk appear to be most important. Other factors which help to prevent infection are the destruction of bacteria by leucocytes and also the naturally occurring inhibitors in milk, in the secretion of the dry udder, and in the streak canals of the teats[22].

EXPOSURE OF THE UDDER TO PATHOGENS

The main reservoirs of staphylococci, *Str. agalactiae,* and *Str. dysgalactiae,* are infected quarters. In addition these pathogens, and also *Str. uberis,* may be isolated from the skin of teats and udders of cows with intramammary infections and from the tonsils, vaginas and the coats of cows. However, with the possible exception of *Str. uberis* these organisms will not normally persist long on *healthy* teat skin and their presence indicates recent contamination. Infected teat lesions and colonized teat canals can be important secondary sites and on occasions even the most abundant source of pathogens (e.g. when teat chaps are common). Normally most other pathogens causing severe mastitis do so on few occasions and usually their sources are environmental rather than from infected quarters. Thus pseudomonads probably emanate from contaminated water, soil or dirty milking equipment, and coliforms from dung and bedding. Little is known of the sites of *C.ulcerans.*

In average herds with about half the cows infected, it is not surprising that the teat skin of all animals, and the surfaces of most of the equipment used during milking, will be contaminated with pathogens. In the absence of hygienic precautions most teat lesions will also be infected and many teat ducts colonized. Although the common pathogens are well distributed among cows and on equipment during milking, the number that can be recovered from healthy teat skin is normally small except when the previous cow milked is shedding large numbers of pathogens. Furthermore, during milking most of the pathogens at or near the teat end may be washed away by the milk. Those that remain may colonize teat orifices and teat lesions thereby greatly increasing the exposure of the teat to infection[23].

HYGIENE

The object of using hygiene in the control of mastitis is to prevent the exposure of the cows' teats to pathogenic bacteria. In practice hygiene can reduce the transfer of bacteria from cow to cow, but because none of the current methods can prevent all spread, hygiene must also aim to destroy those pathogens that contaminate the teat skin.

Hygiene has no direct effect on the main source of pathogens because it does not cure any intramammary infections, but good hygiene reduces both the prevalence of teat lesions, and also the number of lesions and teat ducts that are colonized by staphylococci and *Str. dysgalactiae*. In some herds this is an important reduction in the exposure to pathogens. Furthermore, efficient methods of cleaning milking equipment and housing cows will reduce exposure to coliforms.

PREVENTION OF TRANSFER OF PATHOGENS DURING MILKING

A mastitic quarter may excrete over 10^8 bacteria/ml of milk. Pathogens in the milk from an infected quarter may be transferred to non-infected quarters of the same or other cows as shown in Fig IX 5. The milking cluster transfers pathogens between cows, and may also be an important cause of cross contamination between quarters of the same udder due to the reverse flow of milk within the cluster[24]. In addition measurements at various stages of milking show that even when disinfectants are used some bacterial transfer occurs on hands and equipment whenever the milker handles the udder during foremilking, udder washing and when machine stripping[8,25,26] (Table IX 2).

Foremilking and disinfection of hands

In Britain there is a statutory requirement to remove foremilk before milking. The detection of abnormal secretion enables the milker to exclude mastitic milk from the herd bulk milk, and to treat clinical mastitis promptly (see Fig IX 6). Handling the teats during foremilking spreads pathogens from cow to cow, even when smooth gloves are worn and the hands are frequently dipped in disinfectant during milking[8] although if foremilking precedes udder washing contamination of the hands will be reduced. This contamination can be avoided by use of in-line filters which allow monitoring of all the milk and require dismantling only when clots are detected[27]. However, there is evidence that careful expulsion of the contents of the teat sinus can reduce new udder infection by removing any pathogens in the sinus[20, 21] although the practical significance and economic benefit of this requires further examination under varied conditions, particularly when teat disinfection is practised.

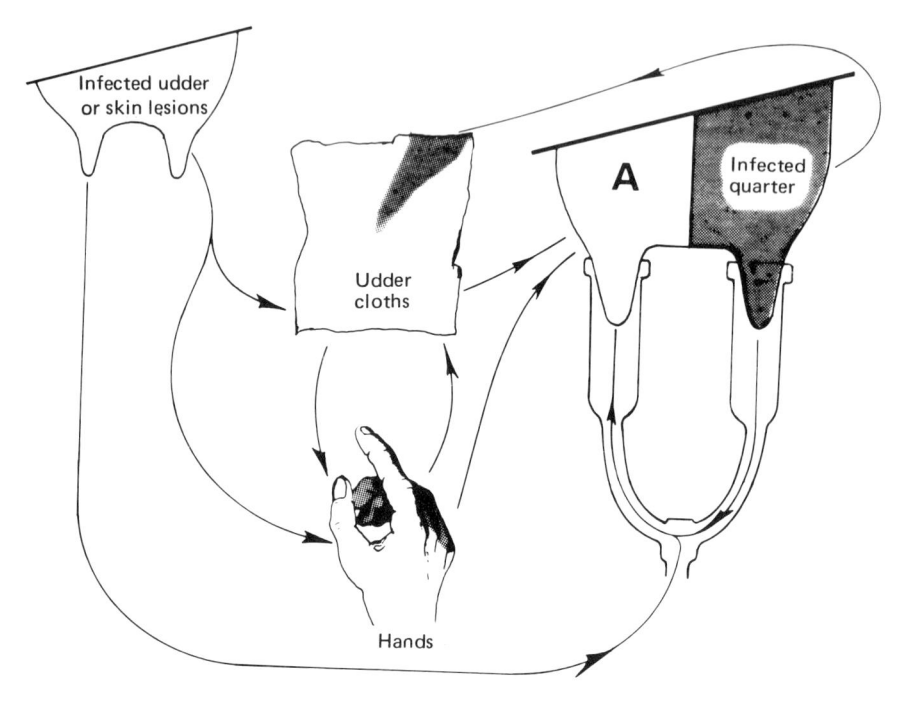

Fig IX 5. Exposure of uninfected quarter A to pathogens from an infected quarter of the same udder and to pathogens from another udder.

Udder washing

Udder washing is included in a milking routine as a stimulus for milk ejection and because the production of milk with minimum contamination requires the cow's teats to be clean at milking time. Clean housing and no mud underfoot are the first requirements to reduce soiling of udders and teats both for clean milk production and to prevent mastitis. Clean teats are less likely to be affected by chaps and sores, and lesions will heal more rapidly. The cleaning of dirty teats and particularly teat ends at milking is difficult, and in the time available udder wash disinfectant solutions cannot destroy all the bacteria on teat skin or on udder cloths. The use of a cloth on successive cows always transfers pathogens between animals[23, 25], although this may be reduced by using several cloths in rotation to increase disinfection time. Udders are best washed with running water containing disinfectant (Fig IX 7) (Table IX 3), and dried with a single service paper towel. Drying is desirable because even if cow to cow transfer of bacteria is avoided the risk remains that washing udders may concentrate bacteria at the teat ends.

Table IX.2 *Transfer of* **Staph. aureus** *at stages during milking from infected to uninfected udders*[23].

Hygiene	Swabs at each stage	Teats before taking foremilk	Teats after taking foremilk	Teats after udder washing	Teats after cow milked	Teatcups before milking test cow
Water only	24 - 32	0	29	63	97	100
Disinfectant Paper towels Gloves worn	18 - 24	0	16	39	79	100

% swabs contaminated with **Staph. aureus** from:

Tests carried out on cows known to have teat skin uncontaminated with **Staph.** *aureus and milked immediately after cow infected with* **Staph.** *aureus.*

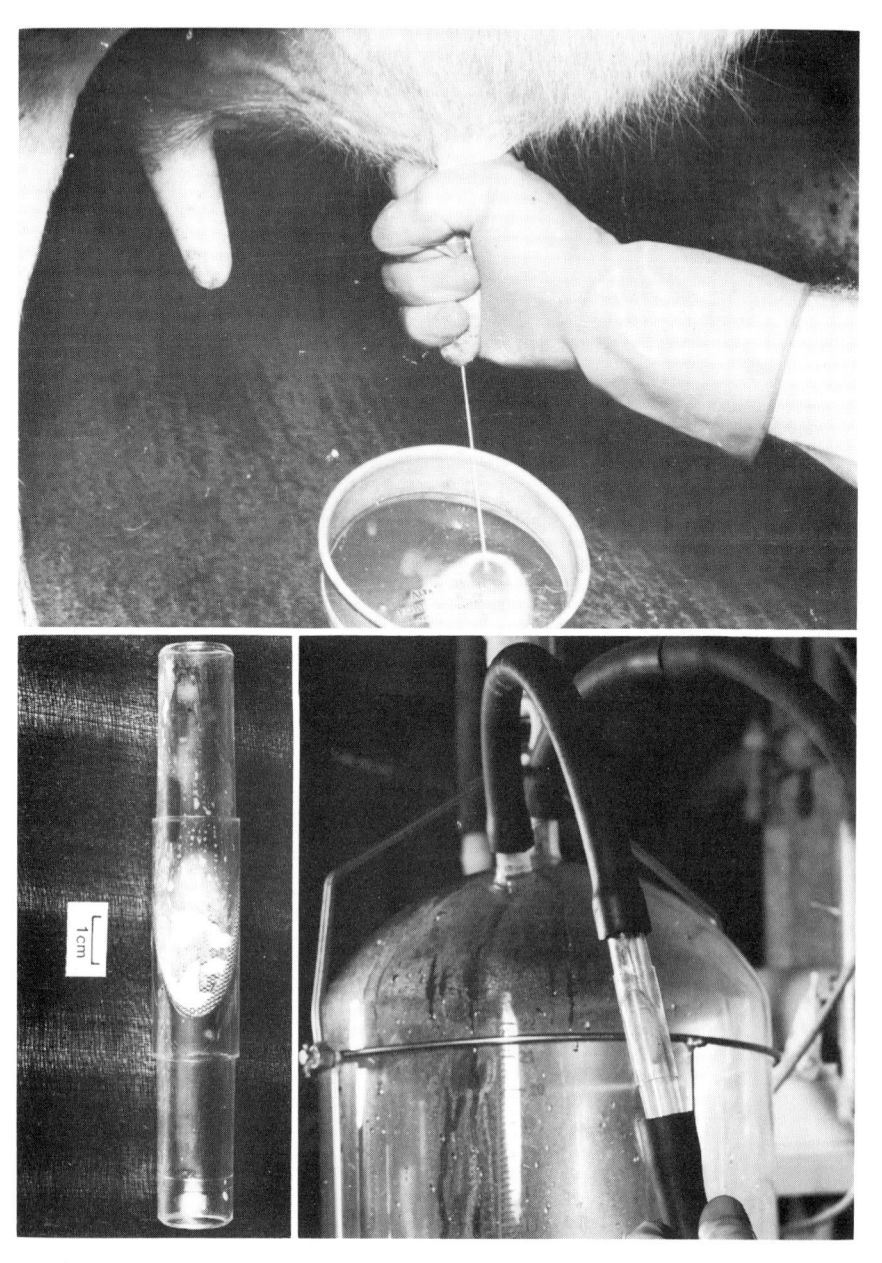

Fig IX 6. Examination of the foremilk for abnormal appearance (above). The use of an in-line filter (below) is an alternative method of detecting clots in the milk[27].

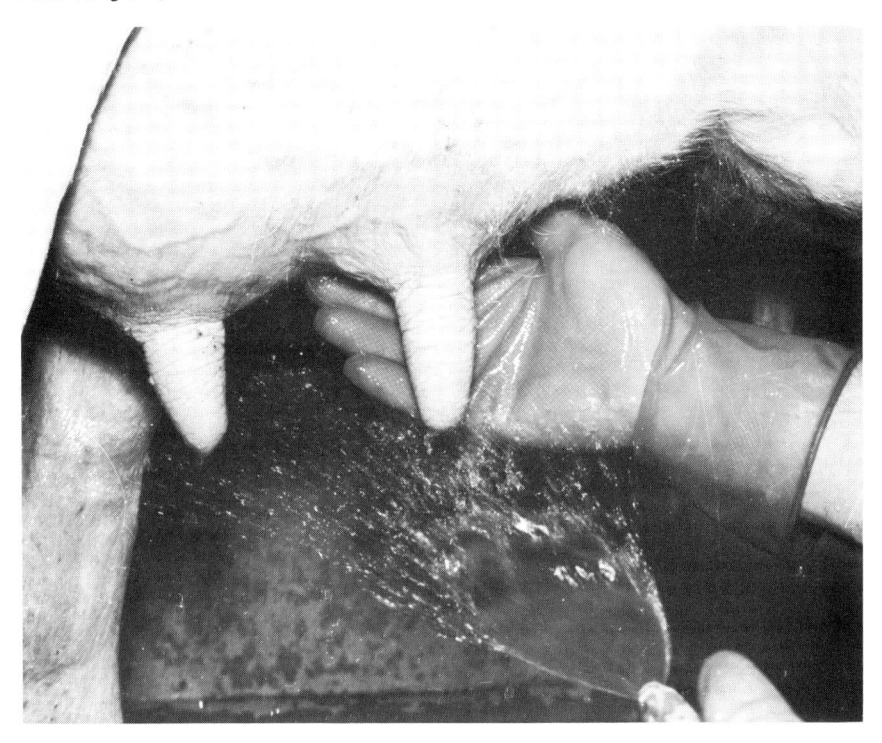

Fig IX 7. Udder washing with gloved hands and running water containing disinfectant minimizes the transfer of bacteria at this time.

For mastitis control, if cows could be kept clean and dry at all times udder washing would be best avoided. Unfortunately it is rarely possible to prevent gross contamination of at least a proportion of cows. Teat cleaning is therefore necessary for clean milk production, and should be done in a manner least likely to increase contamination of the teats with mastitis pathogens.

Teatcup liner disinfection

Bacterial transfer on teatcup liners can be much reduced by dipping the milking machine cluster in disinfectant or flushing with running water before each cow is milked, but it is not eliminated. To prevent transfer completely requires heat treatment of the cluster before each cow is milked (see Fig IX 8). Pasteurization for five seconds at a temperature of 85°C is almost completely effective but expensive and normally impractical (Table IX 4).

Teat disinfection

In practice it is impossible to eliminate all hand contact with the udder during milking, and difficult to make practical completely effective teatcup

Table IX.3 *The efficiency of washing methods in removing* **Staph. aureus** *from teats artifically contaminated*[23]. *Mean washing time/udder = 20 seconds. Teat swabs taken 1 min after washing teats*

Treatment of teats[a]	Geometric mean colony count of *Staph. aureus* recovered from teats [b,f]
Foremilk taken, no washing	23 000
Mains water spray for 15 s.	7 940
Water wash, bucket, paper	1 790
Water wash, bucket, cloth	1 630
Chlorine wash, bucket, paper[c]	937
Water wash, hose, hand	630
Water wash, hose, hand[d]	193
Chlorine wash, bucket, cloth[e]	137
Chlorine wash, bucket, cloth	128
Chlorine wash, hose, hand	69

a Foremilk taken twice using finger and thumb *before* washing teats, (except d and e).
b Each treatment is mean of 18 teat swabs.
c Chlorine in all cases 0.06%.
d Foremilk taken while hose in use.
e Foremilk taken after udder washing.
f Treatments joined by same vertical line not significantly different.

Table IX.4 *Disinfection of teatcup clusters after removal from cows with mastitis. Each cluster was sterilized by steam before milking and was tested after milking only one infected cow or a cow with artificially contaminated teats*[23].

Teatcup clusters

Disinfection of cluster	Time	No. tested	% positive after disinfection	No. *Staph. aureus* recovered per cluster
Cold water flush	5 s	19	100	100 000 − 800 000
Cold hypochlorite circulation (0.03%)	3 min	19	100	50 − 2000
Circulation of water at 66°C	3 min	18	22	0 − 80
Circulation of water at 74°C	3 min	85	0	0
Circulation of water at 85°C	5 s	530	3	0 − 15

disinfection. Therefore it is inevitable that some bacterial transfer to the teats occurs. To reduce this contamination, disinfectant teat dipping first suggested by Moak in 1916[28], has been re-examined and developed (Fig IX 9).

Fig IX 8. Two methods of cluster disinfection. Flushing with cold water will reduce contamination but only the use of hot water (as on the right) or steam achieves rapid sterilization of rubber components. See also Table IX 4.

An ideal disinfectant teat dip would destroy all pathogens on the teat skin at the end of milking, give a residual protection to deal with further contamination during the milking interval, avoid damaging healthy teat skin and assist in healing teat lesions. So far none of the many disinfectants tested is ideal by this demanding standard, but the number of pathogens surviving on teat skin can be substantially reduced by teat dips, as illustrated in Fig IX 10. Of the disinfectants tested the hypochlorites, iodophors and chlorhexidine appear to be the best for teat dipping (Table IX 5). Iodophor teat dips (0.5% available iodine) proved successful under a variety of farm conditions, though they are less effective and more expensive than sodium hypochlorite (4% available chlorine) in destroying skin contamination.

Hypochlorite solutions are commonly believed to cause damage to the teat skin of cows. The evidence from field experiments is that the introduction of

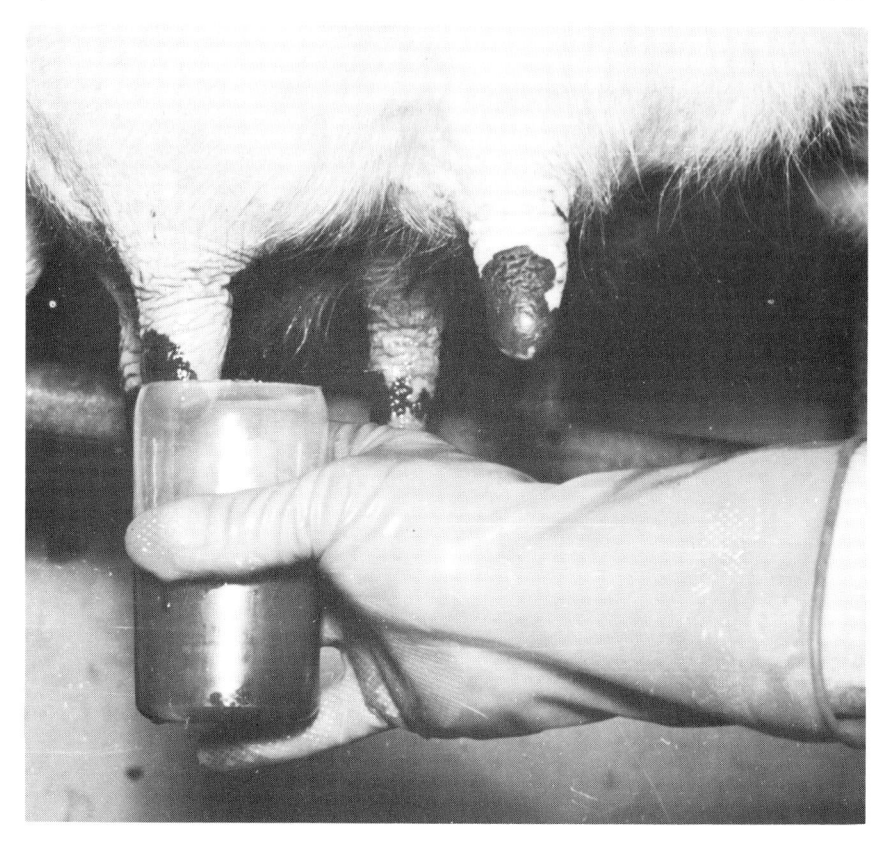

Fig IX 9. Dipping the teats in a disinfectant immediately after milking reduces the chances of pathogens surviving on the skin.

hypochlorite teat dips (and sometimes also of iodophors) may cause temporary skin irritation for 1 - 2 weeks on some cows in some herds. If their use is continued the skin condition improves and the prevalence of teat lesions declines to a level lower than when dips are not used. Teat disinfectants are particularly effective in promoting the healing of infectious teat lesions and minor scratches. The prevention of teat chaps and promotion of their healing is aided by the addition of emollients such as glycerol with teat dips. It is a disadvantage of hypochlorite solutions that they are unstable with emollients. Conversely the popularity of iodophor teat disinfectants in spite of higher cost is explained partly by their stability with emollients and their capacity to stain the skin.

Teat dipping adds only 6 seconds per cow to a milking routine, but the convenience of the task depends on the milking system. The mechanization of teat disinfection is complicated by the difficulty of accurately delivering disinfectant to the whole surface of all four teats[29]. Teat spraying often fails to reach the standard of disinfection that can be achieved by immersing teats in disinfectant. Hand held teat sprays carefully used require more time and more disinfectant solution than teat dipping, and satisfactory automatic sprays have still to be developed. Nevertheless in large milking parlours, especially with automatic cluster removal, automatic teat disinfection is necessary. The practical difference in prevention of infection between teat end disinfection and whole teat immersion, obviously of importance to mechanization, remains to be determined.

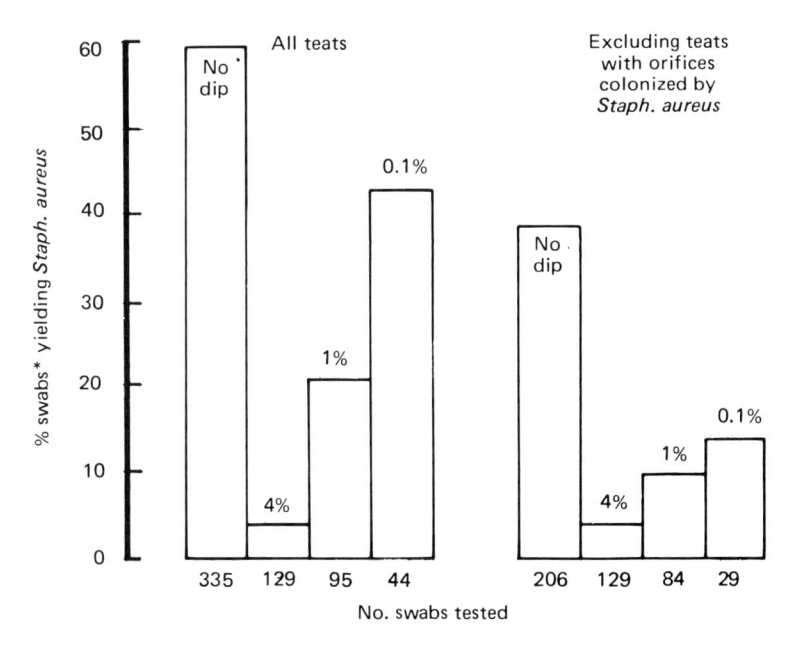

Fig IX 10. The effect of teat dipping with three concentrations of sodium hypochlorite on the recovery of *Staph. aureus* from teats free from blemishes[5].
* Swabs were taken before milking from uninfected cows. At the preceding milking each of the cows had been milked immediately after a cow infected in all four quarters with *Staph. aureus;* no hygiene precautions were taken except the teat dip after milking.

THE PREVENTION OF NEW INFECTION BY HYGIENE

The purpose of hygiene in the control of mastitis is to prevent intramammary infection. To do this the procedures recommended must be shown to be effective, easily incorporated into practical routines, and acceptable to farmers. There has been ample evidence for many years that, under the degree of control

Table IX.5 The skin disinfecting properties of some teat dips*[5]

Teat dip	Concentration	Other additive	pH	Geom. mean count of Staph. aureus from teat swabs[†]
Iodophor	0.5% available iodine	dichlorophenol and 5% glycerol	4.4	538
Hypochlorite (Na)	0.1% available chlorine		8.8	416
Hexachlorophane	0.5%	alcohol	3.9	234
Iodophor	0.5% available iodine	33% glycerol	4.6	206
PVP Iodine	0.5% available iodine	33% glycerol	5.2	139
Iodophor	0.5% available iodine	15% glycerol	4.7	107
Chlorhexidine	1.0%	pyrollidine	6.2	40
Iodophor	0.5% available iodine	2.5% lanolin	2.2	23
Iodophor	0.5% available iodine		4.9	17
Hypochlorite (Na)	1.0% available chlorine		10.3	14
Hypochlorite (Na)	4.0% available chlorine		10.9	6

† Figures differing by approximately 3.3 fold are significant at 5% level.

* Each dip was tested on 18 to 24 teats using 55 cows free of teat lesions, udder infections and teat contamination. An hour before using the teat dips the teats were immersed once in milk containing 5 x 10^7 Staph. aureus/ml. Swabs of the teats were taken one hour after using the disinfectant dips.

that can be achieved in a research herd, new infections can be markedly reduced by various hygiene measures[20, 23] (Table IX 6). In spite of this, until recently effective hygiene methods were not generally used by farmers. They found the recommendations complicated and time consuming, and, probably more important, hygiene did not reduce mastitis quickly. It was evident that more research was necessary to devise simple practical systems that were effective under farm conditions. The NIRD has carried out such a programme using extensive field experiments in order to develop a control system.

NIRD hygiene field experiments

The object of the first experiment was to compare the effect on new infection of the fullest hygiene system practical under the conditions of a supervised field experiment ('full hygiene') with a system in which the only hygiene was udder washing with plain water and a common udder cloth ('no hygiene'). Seven hundred and twenty cows in fourteen herds were used, seven herds applying each hygiene treatment for one year[8]. The full hygiene routine consisted of udder washing with individual boiled cloths or paper towels using a 0.01% solution of chlorhexidine. The milkers wore rubber gloves which were rinsed in the udder wash disinfectant between operations. The milking machine clusters were pasteurized for 5 seconds using water at 85°C before each cow was milked and the teats dipped after milking in a 0.5% solution of chlorhexidine (see Fig IX 11). To detect infections milk samples for bacteriological examination were taken at regular intervals, at calving and drying off, and whenever antibiotic therapy was given. Antibiotic therapy was limited to cases of clinical mastitis detected by the cowmen and to alternate cows at drying off in each herd. At the beginning of the experiment 51% of the cows were infected; 5% were infected with *Str. agalactiae*, 13% with other streptococci, 40% with staphylococci and only 1% with other pathogens. The levels of infection varied between herds from 16 to 76% of cows infected with staphylococci and 5 to 42% with streptococci. The effects of the different hygiene methods on new infection rates can be seen in Table IX 7. There were large differences between the results from the herds, but those using the 'full hygiene' system contracted about half the new infections that occurred in the 'no hygiene' herds. The reduction in the new infections was much greater with streptococcal than with staphylococcal infections.

In a second experiment three hygiene systems were compared using about 1500 cows in 15 herds[25]. One of the systems was based on the 'full hygiene' of the first experiment, except that 0.01% iodophor solution was used for udder washing and a 0.5% solution for teat dipping; rubber gloves were worn. The second, or 'partial hygiene' system was identical with the 'full hygiene' except that the most expensive operation, teatcup pasteurization, was omitted and the teatcup liners were not disinfected in any way. The experimental control was similar to the 'no hygiene' of the first experiment. Each of the three hygiene

252   *Mastitis*

Fig IX 11. A milking parlour equipped with experimental automatic cluster pasteurizing units for research purposes (see inset for detail). Foremilk cup A, paper towel dispenser B, disinfectant bucket C and teat dip container D, are all placed conveniently for use.

systems was applied in every herd for 6 months at a time. The main result of this experiment is shown in Fig IX 12. The 'full hygiene' reduced the new infections by 58%, but even the simple system reduced the infections by 44%.

Table IX.6 The effect of six hygiene routines on the rate of new infection. The table summarizes the results of three experiments in which exposure to pathogens was artificially increased[5,25]

Treatment of teatcup clusters	None	None		Disinfected* Disinfectant†		Pasteurized at 85°C Disinfectant†	
Udder wash		Water					
Cows on experiments		54		38		16	
Teat dip	None	None	Yes††	None	Yes††	None	Yes††
Cows infected	23	19		7	0	0	0
Quarters infected	31	27		10	0	0	0
Cases clinical mastitis	30	25		10	0	0	0

* Clusters rinsed in either 0.02% chlorhexidine solution or an iodophor solution (0.02% available iodine).

† Udders washed with a sodium hypochlorite solution (0.05% available chlorine).

†† Two teats of each cow dipped in either 0.5% chlorhexidine or in iodophor (0.2% available iodine).

Table IX.7 *The new subclinical infections occurring during lactation in herds using either no hygiene other than plain water udder washing or a 'full hygiene' routine. Results from 351 cows in seven herds not using hygiene and 370 cows in seven herds using the hygiene routine*[8].

Infections	All infections		Excluding possible cross infections within cows*	
	No hygiene	'Full hygiene'	No hygiene	'Full hygiene'
Staph. aureus	255	189	161	91
All streptococcal	310	115	249	71
Others	65	59	64	57
Total	630	363	474	219

* A new infection occurring in a cow already infected in another quarter with the same pathogen.

Staphylococcal and streptococcal infections were reduced equally but other types of infection, though very low in number, were not affected. It would appear that, providing the teats and particularly the ends of the teats are disinfected immediately after the removal of the machine, many infections can be prevented.

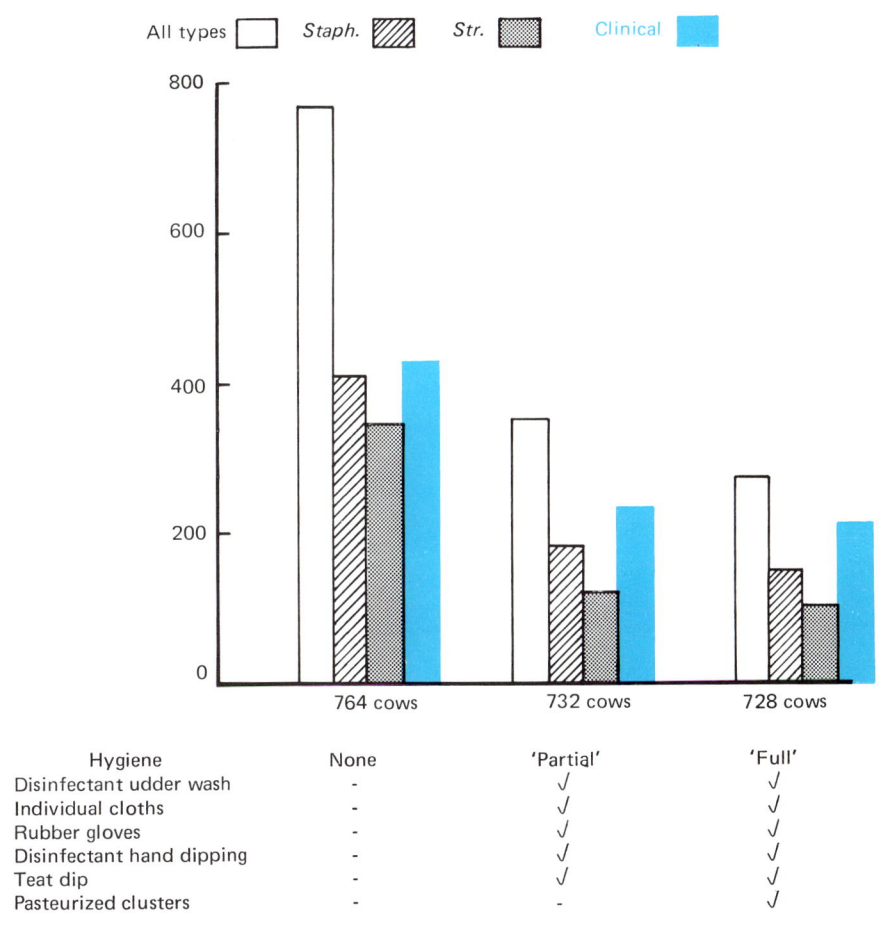

Fig IX 12. The effect of hygiene methods on new infections and clinical mastitis in lactating cows (MFE 2). Three hygiene routines were compared in 15 herds[25].

It is tempting to conclude that the relative success of the hygiene routines, particularly the value of teat dipping after milking, indicates that a high proportion of infections originate from bacteria that penetrate the streak canal of teats and enter the udder during the intervals between milkings. This is not necessarily so. The main results of teat dipping may be to prevent bacterial colonization of the streak canals of the teats but it also reduces general skin contamination (see Fig IX 10). This will reduce the exposure during the milking interval *and also* during milking. The timing and mechanism of bacterial penetration of the streak canal is far from clear.

TEAT LESIONS

Teat lesions are important in mastitis control because they are frequently colonized by staphylococci and *Str. dysgalactiae* thus becoming major sources of these pathogens. High new infection rates and frequent clinical mastitis caused by these pathogens often occur at a time when teat lesions are most common and this may account for the generally much higher level of clinical mastitis during winter months.

Bovine teat skin is devoid of sweat glands and hair, forming a relatively smooth covering interrupted only at the apex where the outer layer invaginates to form the lining of the streak canal. Teat skin is usually free from lesions until first calving and during the dry periods of older cows, but during lactation they are common. Milking and some of the associated management practices are therefore important, directly or indirectly, in determining the condition of teat skin.

The act of milking aggravates all types of teat lesions, but machine milking is a cause of teat canal erosion (orifice erosion), haemorrhagic blisters near the teat end, and of much teat chapping (see Fig IX 13). Orifice erosion and teat chaps are the most common teat lesions which in some herds can affect most teats in the winter months[30]. Teat chaps are fissures due to the loss of elasticity of the skin and their occurrence is closely related to drying weather conditions, increasing when the temperature falls and the dew point is low. Dirty bedding, poor udder washing, and flies all increase the severity of chapping. Neither of these types of lesions is caused by mastitis pathogens, but both are rapidly colonized by them.

Infectious lesions are caused by paravaccinia virus and much less frequently by bovine herpes mammilitis virus and staphylococci[31]. Paravaccinia (pseudocowpox) sores occur in most herds; they rarely affect many cows. All infectious lesions tend to be seasonal, being most common in winter and spring. Mammilitis is very occasionally a serious herd problem which can be controlled by vaccination.

In general teat lesions are most common on the teats of older cows and in early lactation, and less common in herds washing udders with iodophor or hypochlorite, rather than plain water or udder washes containing surface active agents.

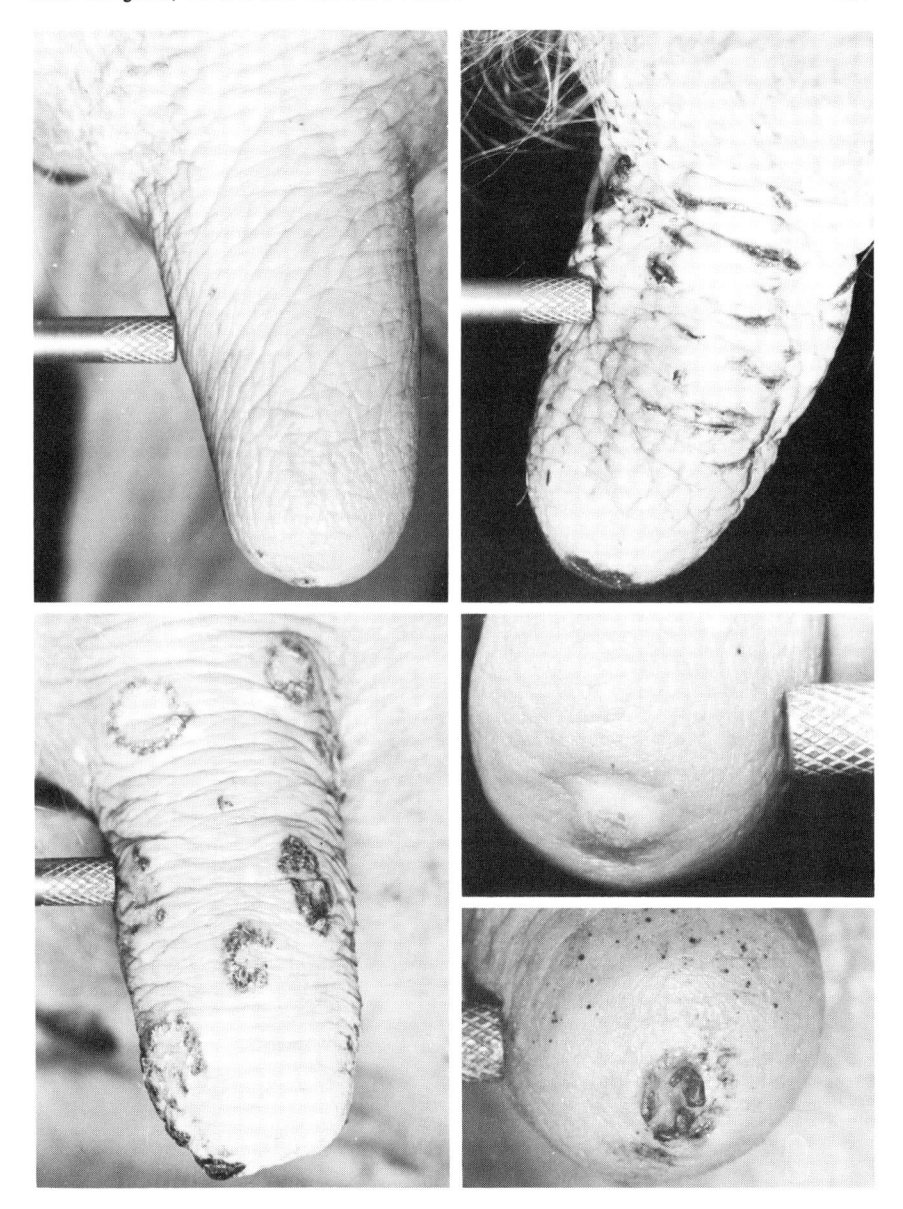

Fig IX 13. Teat lesions. Normal teat (above left) and teats affected with chaps (above right) and 'ring' sores (probably caused by *Paravaccinia* virus) (bottom left). A normal orifice and an eroded orifice surrounded by small haemorrhagic blisters are also shown (bottom right). (Photographs by M.F.H. Shearn.)

While it is possible to reduce all types of teat lesion by good husbandry, there is no completely effective way of preventing teat chaps. Usually teat lesions are treated with ointments or salves but these make good hygiene difficult. Disinfectant teat dips including emollients such as iodophor solution containing up to 9% glycerol produce better results[32].

SUMMARY OF THE VALUE OF HYGIENE IN CONTROLLING MASTITIS

The published data on the effect of hygiene show that hygiene techniques correctly used greatly reduce skin contamination, the colonization of streak canals and teat lesions, improve healing of some skin lesions and wounds, and can halve the rate of new infections by the major pathogens. Infections caused by coliforms, pseudomonads and other pathogens emanating from dung, soil and dirty water are much less influenced by conventional hygiene than are infections by staphylococci, *Str. agalactiae* and *Str. dysgalactiae*. Although the most complete hygiene routines give the greatest reductions in the infection rate, it appears that with present levels of infection post milking teat disinfection is the most important component. Foremilking and udder washing are the two machine milking operations that make really effective hygiene difficult. Even when hand dipping is practised and disinfectants used for udder washing, bacteria are transferred from cow to cow. With the development of effective in-line filters foremilking is not necessary for detecting mastitis. Udder washing is required as part of the routine for clean milk production but it should be carried out so that cross contamination between cows is minimal relying on the housing system to keep the cows clean.

The effects of practical hygiene routines on new infection are sufficient to make them the basis of a mastitis control system. Nevertheless even when a hygiene system is used and the levels of infection fall in most, if not in all herds, much variation remains between herds in the rates of new infections (see Fig IX 19). Herds differ more than tenfold in this respect. This means that other environmental factors have in total a greater influence on the new infection rate than conventional hygiene methods.

The field experiments also show that even when the new infection rates in herds are halved, the levels of infection (i.e. proportion of cows infected) decline by only about 10 - 15% in a year. This slow decline makes a control system *based on hygiene alone* inadequate. In addition it is probably the main reason why the various hygiene routines that have been introduced in the past have appeared to be so ineffective.

MILKING MACHINE FACTORS

The machine milked cow is much less likely to contract a new intramammary infection during lactation than during the unmilked dry period, assuming equal exposure to pathogenic bacteria[19]. Nevertheless, the techniques employed during milkings can have an important influence on new infection rate for two reasons. As described above most transfer of pathogens occurs at the time of milking and the milking machine is an important agent in increasing teat contamination and teat lesions. In addition, it appears that milking machines, either intrinsically because of their basic physical properties, or in the way that they are used, can assist pathogenic bacteria to penetrate the streak canal of the teat. It is possible, but less likely, that they also help in the establishment of infection after penetration. The precise factors have not been determined, but in total they could be a major reason contributing to the wide differences between herds in the rates of new infection (see Fig IX 19).

Considerable literature exists on the relationship between milking machine factors and mastitis, and the earlier work has been fully reviewed[2,3]. These reviews revealed the lack of well established relationships because so many of the papers report field observations that are difficult to interpret or experiments using too few cows, carried out for too short a length of time and with inadequate diagnostic techniques.

The milking machine factors most likely to affect the penetration of bacteria through the streak canal are the vacuum, the pulsation and the way in which these are applied to the teat in the teatcup liner. These factors will also be influenced by the performance of the vacuum pump, the arrangement and dimensions of the milking machine pipelines, the milk collecting vessels and air admission. The effect of these on the physical properties of the milking machine has been discussed in Chapter V.

VACUUM LEVEL

Constant exposure of the teats to the vacuum levels normally used in milking machines (40 - 60 kPa; 12 - 17 inHg) causes damage to the teat skin particularly when the teatcup liners do not collapse completely during each pulsation cycle. This damage is greatest at high vacuum levels. During milking the intramammary milk pressure exceeds ambient air pressure until the moment when milk flow ceases when fluctuations in vacuum occur within the teat sinus. These are synchronous with the pulsation cycles and result in levels of vacuum in the sinus reaching about half or three quarters of the milking machine vacuum[33].

Because greater operating vacuum levels will result in increased teat apex damage and higher vacuum levels in the teat sinus at the end of milk flow it is reasonable to expect that this will also lead to a rise in new infection rates. This is generally accepted and there are a number of field observations which attribute increased levels of clinical mastitis in herds to high vacuum levels. The results of the controlled experiments are equivocal and although the usual outcome is a somewhat greater infection with high vacuum, the increase is relatively small and then only occurs when a very high vacuum is used[34]. Vacuum, within the limits of 40 - 55 kPa (12 - 16 inHg) does not appear to influence infection to any marked degree.

VACUUM FLUCTUATIONS

There are two main causes of fluctuation in the vacuum level in the teatcup liner. Firstly, cyclic fluctuations are induced by pulsation (Chapter V). These occur in all types of milking machines when milk is flowing and particularly if the milk tubes are filled with milk and there is no air admission at the claw. When the milk ways are restricted, these vacuum fluctuations may range from 30 - 60 kPa (9 - 17 inHg) in each pulsation cycle and the liner may flood with milk preventing its complete collapse. Secondly, irregular vacuum fluctuations occur randomly when air is allowed into the vacuum system, for example if units fall off or through careless changing of milking units from cow to cow. The extent of such fluctuations will depend upon the basic design of the milking equipment, vacuum pump reserve capacity and the skill of the milker in avoiding excess air admission during manipulation. They are much less frequent than the cyclic fluctuations (Fig IX 14) but they cause a decrease in milking rate and also modify liner wall movement.

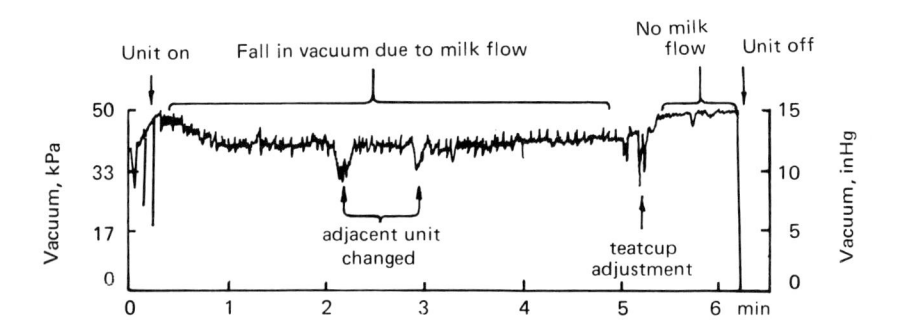

Fig IX 14. Variations in vacuum level within the claw during the milking of a single cow, illustrating continuous cyclic vacuum fluctuation during milk flow, and occasional irregular fluctuations. This record was obtained using a bellows recorder which was insufficiently sensitive to record the full amplitude of the cyclic fluctuations. See also Chapter V.

The demonstration by Nyhan and Cowhig in Ireland[35] of a relationship between irregular vacuum fluctuations and udder disease stimulated research into mechanisms by which the milking machine could lead to increased infection.

Experiments using half udder milking machines applying minimal and exaggerated vacuum fluctuations to cows' teats heavily contaminated with mastitis pathogens showed high rates of new infection with the larger vacuum fluctuations[36]. Bacterial penetration of the teat canal deep enough to increase the risk of infection was shown to result from the impact of contaminated milk on the end of the teat[37]. For the highest rates of new infection both cyclic and irregular fluctuations in vacuum are needed simultaneously, and the later in milking such impacts occur the greater is the chance of infection[38]. Impacts can also be generated when the cluster is being removed from the udder[39].

Shields or deflector plates fitted within the teatcup liner above the opening of the short milk tube to intercept milk travelling back to the teat prevented bacterial penetration under experimental conditions. Shields of this type were also fitted in two diagonal teatcups of each cluster in 15 herds, but resulted in a benefit in terms of preventing new udder infections in only 3 of the 15 herds[39].

This result, indicating only a modest benefit in practice from the avoidance of the impact effect, suggests that the vacuum fluctuations that occur on farms are not a major factor influencing new infection or that shields give inadequate protection. All the evidence is that shields prevent impacts but the conditions leading to fluctuating vacuum may have other adverse effects not yet discovered.

The general advice to ensure adequate vacuum pump reserve capacity and free movement of air in the cluster seems to be justified. The absence of a clearly demonstrated benefit from minimizing all vacuum fluctuations indicates that further knowledge is required before stronger recommendations are made.

PULSATION

Pulsation was introduced at an early stage of milking machine development in order to overcome the effects of constant vacuum. In the conventional double chambered teatcup the rhythmical movements of the liner do not prevent the teat apex from being continuously exposed to vacuum during milking but do reduce blood congestion by squeezing the teat.

The main aspects of pulsation that may influence the occurrence of udder disease are: pulsation rate and ratio, speed of movement and completeness of collapse of the liner walls, and whether the four liners in a cluster pulsate alternately or simultaneously. No comprehensive study of the effects of pulsation on mastitis has yet been made, but field observations indicate that a complete breakdown in pulsation can result in a serious mastitis problem. Although experimental results are inconsistent the indication is that more infections occur when pulsation ratios are very wide and rates are very high, i.e. under pulsation conditions when consistent opening and closing of liners is likely to be impaired.

Since pulsation is the cause of cyclic vacuum fluctuations within the liner, contributing to the impact mechanism described above, and is a factor in the reverse flow of milk leading to the contamination of teats with pathogens from another or previously milked teat[24], there have been renewed attempts to dispense with conventional pulsation. Continuous suction machines such as the Murchland were used for many years at the beginning of this century (see Chapter I). Recent developments confirm that single chambered teatcups are well tolerated by cows providing measures are taken to avoid teat congestion, and also that satisfactory rates of milking can be obtained[40,41]. Machines which avoid reverse flow of milk also offer potential advantages in hygiene, but so far there is insufficient evidence to determine whether they confer benefits in the control of udder disease.

TEATCUP LINER DESIGN

The forces causing milk flow during machine milking are applied through and modified by the teatcup liner and it is generally believed that the design of liners influences levels of mastitis. There is little direct evidence indicating that liner design affects new infection rates but certain liners increase teat orifice erosion and other teat lesions particularly around the orifice and where the liner mouth contacts the teat. There is also evidence that some liner designs increase the frequency of clinical mastitis even when the new infection rate is unaltered. The particular design factors responsible for these effects have not been identified.

CLAW DESIGN AND AIR ADMISSION HOLES

Claws are usually fitted with air admission holes to aid the removal of milk from the liner into the long milk tubes and thereby to reduce 'flooding' and cyclic vacuum fluctuations. The possibility that blocked air admission holes could contribute to increased levels of udder disease was indicated in experiments carried out in New Zealand over 30 years ago[42] but even now no conclusive evidence is available. Enlarging the bore of claw nipples and of the short milk tubes from about 6 mm to 12 mm greatly reduces cyclic vacuum fluctuations under the teat.

In recent years there has been a trend to increase the volume of claws, but air admission and the diameter of the short milk tubes are more important in maintaining uninterrupted milk flow.

MILKING MACHINE PIPELINES AND LAYOUT

With the increase in herd size there has been a change from cowshed to parlour milking and to the use of pipeline milking machines. The properties of these new designs of milking machine, particularly with regard to the vacuum fluctuations which occur at the teat end are often very different from those of bucket or recorder jar machines. However, the effects of the many variations in design of milking machines are complex and as yet no clear relationship between basic plant layout and udder disease has been established. Nevertheless the bore of pipelines when milking pipeline machines are installed in cowsheds has been increased to 38 mm (1.5 in) and risers are avoided if possible. In parlours the milking pipeline may have even larger bore tubes and there is a trend to fit them below udder level. Most recently, when milking is into a recorder jar, the fashion has been to mount jars below the level of the udders.

MACHINE MILKING PRACTICE

Field experiments show that herds can have different levels of udder disease even when milked by machines of the same type and manufacture. To the extent that the machine is responsible for such differences this may be explained by poor machine maintenance or the way in which the milking equipment is used. It is widely believed that the skill of the milker is of primary importance. Skilled, observant stockmen may reduce mastitis by following good hygiene practices, endeavouring to provide clean cow housing to minimize teat lesions and exposure to pathogens, by promptly detecting and treating clinical mastitis, culling susceptible cows and by good maintenance of milking equipment. In addition, husbandry methods such as the preparation of the cow for milking, the attachment of the units, the length of time that the teatcups are in contact with the teats and the techniques employed at the end of milking may influence udder disease.

PREPARATION FOR MILKING

When milking is preceded by udder washing this may influence both the exposure of the udder to pathogens and their penetration through the streak canal. If udder stimulation is inadequate, ill-timed or painful the milking machine may be attached to the udder before milk ejection occurs. Little milk will be removed before ejecting and until then the situation may be similar to that occurring during overmilking. In the absence of increased intramammary pressure it is possible that reduced pressures will occur within the teat sinus, although this has not been demonstrated. As dairy cows rapidly become conditioned to any reasonably constant preparation routine, the delay between attaching the teatcup and milk ejection will rarely exceed 30 seconds, though this may be considerably exceeded with poor routines or with some cows in late lactation.

OVERMILKING

Traditionally, leaving the milking machine attached to the udder after milk flow has ceased has been regarded as one of the main factors pre-disposing to udder infection. Evidence exists that persistent overmilking increases teat erosion and causes damage to the mucous lining of the teat sinus[43] and this could be expected to increase infection. Nevertheless, controlled experiments have failed to demonstrate a major adverse effect of overmilking on infection. Furthermore, although overmilking occurs more frequently in fore quarters, because of their tendency to be milked out more quickly than hind quarters, the levels of infection are usually lower in fore than in hind quarters. Some overmilking occurs even in the best managed herds and although it seems desirable to avoid more than the minimum to keep the milker fully occupied, it does not appear to be a factor of major importance in udder disease.

METHODS OF STRIPPING AND CLUSTER REMOVAL

The oldest rule of good dairy cow management is the necessity for careful stripping at the end of milking, which is supposed to be essential for the maintenance of secretion and udder health. Nevertheless, in the past 25 years, farmers have abandoned hand stripping after machine milking relying instead on machine stripping, and more recently have given up all forms of stripping for most cows. There is no indication that changes in stripping methods affect new infection rates but infections may be less likely to show clinical signs if all the available milk is carefully removed at the end of milking. Finally, to minimize irregular vacuum fluctuations within the machine and to reduce the risk of back flow of milk within the cluster (see above) it is good practice to shut off the vacuum *before* removing the cluster at the end of milking.

HOUSING AND OTHER HUSBANDRY FACTORS

The incidence of clinical mastitis is often highest during the winter period, and this is usually attributed to environmental factors. In Britain, however, for economic reasons most cows calve during this period and the higher incidence of clinical disease associated with early lactation must account for part of the increase found at this time. Nevertheless poorly designed and badly managed winter housing leads to more teat lesions, can cause teat damage and increases clinical mastitis[44]. This is particularly so when cow stalls or cubicles are too short and narrow or if parlour milked cows are held in exposed collecting yards. The latter probably accounts for the higher level of chapped teats found in parlour milked cows. Hard, brittle straw and abrasive wood chippings cause many small scratches on the udders and teats which, following colonization, may become important sources of pathogens.

Coliform mastitis can be a serious problem usually associated with housing conditions that give high exposure to coliforms. Most frequently, outbreaks occur in herds using sawdust bedding, which under certain conditions can permit rapid growth of coliform bacteria, even when the cows appear quite clean. The outbreaks usually affect freshly calved cows, but can often be overcome by changing the type of bedding, as for example from sawdust to washed sand[45,46].

There are a number of published reports as well as field observations comparing the effects of different housing systems on mastitis, all of which appear to incriminate specific systems. While these reports are useful in pointing out particular design features to be avoided, they do not enable a valid comparison to be made between the systems (i.e. cowsheds, yards and cubicles). The effects on mastitis of the different management methods found within each of the main housing systems will be much greater than the differences between the systems themselves.

CONCLUSIONS AND RECOMMENDATIONS

Most farmers, advisers and research workers believe that machine milking, either intrinsically or because of the way it is done, is an important factor influencing udder disease and can account for much of the very large differences in herd levels of infection. As this brief review indicates, there is no doubt of the evidence that many individual milking machine factors predispose to infection or can cause clinical mastitis. Nevertheless such evidence as exists on the value of the currently recommended designs and the methods of use of milking machines on mastitis is poor, and none is available to show that their general adoption would result in marked reductions in the level of udder disease. It is quite possible that the main reason why a clear picture does not emerge from this research is that the factors are complex. Alternatively it may be because most of the studies were small and utilized inadequate techniques or even that the major factors in milking machine design as they relate to mastitis are still undiscovered. This lack of clear evidence has not been helpful to farmers and advisers. As economic conditions change and new equipment becomes available farmers must modify their systems of milk production, and veterinary and other advisers faced with mastitis outbreaks must take immediate action; neither can wait for the results of further research.

The following is a brief guide for good milking and housing practice. Some of this is distilled from experimental evidence but, where this is not available, the conventional recommendations have been accepted.

Vacuum

A pipeline vacuum of 40 - 55 kPa (12 - 16 inHg) is recommended, avoiding excessive irregular fluctuations by providing adequate vacuum pump reserve capacity, and changing the cluster with minimum air admission. Worn liners should be discarded. Ensure that air bleeds in the claws are not blocked and that the dimensions of the tubes with a dual function of conveying milk and connecting liners to the vacuum system are large enough to avoid excessive vacuum fluctuations and 'flooding' of the liners[47].

Pulsation

Pulsation rate should be within the range 50 - 60 cycles/minute and the milking ratio such that the liners are more than half open for 50 - 65% of the cycle with at least a perceptible dwell time at atmospheric pressure.

Teatcup liners

Use liners with a barrel diameter not greater than the average diameter of the teats after milk ejection (18 - 22 mm; 0.7 - 0.85 in) and a mouthpiece that does not cause obvious damage to the teats or leave excessive strippings.

Pipeline layout

Recommendations for layout are found in the British Code of Practice[47]. Length of pipeline should be kept to a minimum, it should form a ring with the ends having a separate full bore connection to the receiver vessel, the bore of the pipeline should be adequate for the number of milking units served, milk entry ports must not be more than 2 m above cow standing level and there must be a continuous fall from this point to the receiver vessel. Risers in milking pipelines should be avoided. All fittings must be made of recommended materials, capable of withstanding internal pressure bewteen 70 kPa (20 inHg) vacuum and 200 kPa (28 lb/in^2) gauge and capable of being cleaned satisfactorily. Recommendations are also made for the use of recorder jars.

Maintenance

The milking equipment should be maintained in good working order. Minimum requirements are described in the British Code of Practice[47] and the equipment should be checked and if necessary returned to these standards regularly. The milker should be responsible for seeing that his machine is in working order day by day by simple checks of the vacuum regulator, air admission holes and pulsators.

Use of equipment

Adopt a regular milking routine, avoid excessive overmilking by using the correct number of units in relation to the work routine (see Chapter VIII), avoid machine stripping, and remove units only after shutting off the vacuum. Look for any indication of milking machine damage to the teats and investigate the cause.

Housing

Care must be taken in the design and management of cow housing, attention being paid in particular to siting and protection in collecting yards if herds are parlour milked, to the dimensions of cubicles and cow stalls, and to the material used for bedding, keeping the latter dry and as clean as possible. If severe clinical mastitis occurs where sawdust is used, the sawdust should be changed promptly or replaced by sand.

MASTITIS CONTROL

The high levels of udder infection in most herds indicate that in general good control methods have not been used. For many years advice was inconsistent, farmers were not convinced of the importance of the problem and had little confidence in the changing advice they received.

An effective control system must relate to the pattern of udder disease (see Fig IX 1) and therefore must be based on the prevention of new subclinical infection rather than the elimination of infections by treatment with antibiotics. This is because the main problem is to reduce subclinical infections which are at least twenty times more numerous than cases of clinical mastitis and are the main source of mastitis pathogens. This does not conflict with farmers' and veterinary surgeons' concern with clinical mastitis because clinical signs are preceded by infection.

There is no prospect of preventing all new infections by the major pathogens or controlling mastitis by breeding more resistant cows or by immunization. Improved husbandry is the only practical basis of a control system but many management practices have, at one time or another, been associated with mastitis. The problem has been to discover one or two simple methods that, on their own, would substantially reduce new infection in most herds. At present control cannot be achieved through improved milking machine design or use. The hygiene techniques described above provide a basis for control, but alone they do not constitute a control system because their effect in reducing levels of infection is too slow.

THE STRATEGY OF THE CONTROL OF MASTITIS[48]

Str. agalactiae can be eradicated from herds because the only important source of this pathogen is infected quarters which respond well to penicillin therapy. Once these infections are eliminated *Str. agalactiae* disappears from the environment within a few days. In the future eradication of *Staph. aureus* and *Str. dysgalactiae* may become possible, but the problem is much more difficult, particularly with *Staph. aureus* which is not easily eliminated by antibiotic therapy, and both these organisms can persist in secondary sites (i.e. teat sores, colonized teat canals). The eradication of *Str. uberis, E. coli, Pseudomonas spp.* and some other pathogens would appear to be impossible because they are widespread in the cow's environment. Therefore the aim of a control system is to reduce levels of infection rather than complete eradication. In practice there are important economic losses due to mastitis in *most herds*. Therefore it is impractical for control to be based on investigating herd problems, discovering the specific causes of high levels of infection and providing a control scheme tailored to the needs of each individual herd. A better approach seems to be to have a control system which is simple and effective under nearly all herd conditions and does not require detailed laboratory investigation on farms. It must also be economic to the farmer and reduce staphylococcal and streptococcal infection and clinical mastitis to low levels reasonably quickly.

For practical reasons mastitis control is measured by its effect on the *prevalence* or *level* of infection (i.e. the proportion of cows or quarters infected or the cell count of the milk) rather than the reduction in the *rate* of new infection. The prevalence of infection depends on the rate of new infection and also on the *duration* of the infections. The influence of duration of infection can be illustrated with a simple example. In a herd if one new infection occurred each week and each infection lasted one week then at any one time one infection would be present: but if infections lasted 40 weeks then at any one time 40 infections would be present.

These factors affecting the prevalence of infection in a herd during a year can be equated in simple terms as follows:

$$a = b \times c \times \frac{1}{100}$$

| average prevalence of infection (as % cows infected) | total cows infected during a year (as % of total cows) | average duration of infection (as % of year that infected cows are infected) |

In the NIRD field experiment lasting one year with over 700 cows in 14 herds the average proportion of cows infected (a) was about 60%, and about 80% of all cows were infected at some time in the year (b). Therefore by calculation the duration of infections of the infected cows must have averaged 75% of the year (c). Similar high figures have been found in other large experiments. It can be seen from the equation that the average prevalence of infection (a) could be halved *either* by halving the new infection rate *or* by halving the average duration of infection (c). But if *both* the total cows infected and the duration of infection are halved then the average level of infection will fall to a quarter. In this way a marked reduction in the prevalence of infection could be attained by modest reductions in the new infection rate and the duration of infection.

Infections at start of year

759

150 in cows entering

993 now in lactation

212 new in dry cows

2114

Infections during year

* Infections not persisting to end of year

Infections at end

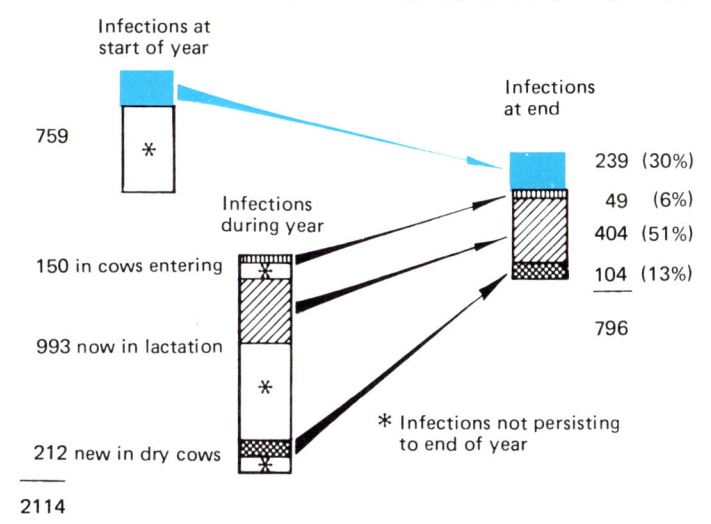

239 (30%)
49 (6%)
404 (51%)
104 (13%)

796

Fig IX 15. A pictorial balance sheet of subclinical infection during 12 months. Data from 721 cows in 14 herds[5]. Note the similarity in numbers of infections present at the start and end of the year; the large number of new infections; the infections that did not persist until the end of the year and the origins of infections present at the end of the year.

So far this analysis has not considered the time required for the prevalence of infection to fall with control systems. Even with a system based on reducing new infection the main factor governing the rate of fall in levels of disease will be the duration of infections. Taking an extreme case of the introduction of a scheme that prevents all new infections, the rate of fall in level would depend on the persistence of infections present at the start — the longer they last the slower the rate of fall. Obviously control based on eliminating infections is likely to give a rapid fall in the level of infection. Under most conditions levels of infection result from relatively low new infection rates and long durations of infection. Herds with over 50% of cows infected at any time may have only about 2 new infections/cow per year but the average duration of each infection will be several months.

These considerations make it possible to understand why in the first field experiment a reduction of over 50% in new infection rate reduced level of infection by only about 12% in a year. The persistence of infections maintained the level. In fact over 30% of the infections found at the end of the year had been present in the same quarters at the start of the year (Fig IX 15). For effective control it is necessary to devise better methods of eliminating infections.

ELIMINATION OF INFECTIONS – ANTIBIOTIC THERAPY

Infections that do not disappear spontaneously are eventually eliminated from a herd by antibiotic therapy or when cows are culled. In the 12 month field experiment 2114 infections were detected by laboratory tests on milk samples. Nineteen per cent of these disappeared spontaneously, and 7% were eliminated by the sale of cows from the herd and only 29% were eliminated by antibiotic therapy given to quarters found by the stockmen to have obvious clinical mastitis (Fig IX 16). To reduce the duration of infections considerably it is necessary either to increase the sale of infected cows or to improve the effectiveness of antibiotic therapy.

Economics limit the number of cows that can be sold from a herd, and udder disease is only one factor to be taken into account. Without information on subclinical disease it may be difficult to select particular cows to be culled to aid mastitis control. Nevertheless increasing the sale of cows that show frequent signs of clinical mastitis would be helpful. In one experiment the 10% of cows that had 5 or more clinical attacks in a year were responsible for 50% of all clinical mastitis[49].

The main reason why antibiotic therapy given to cows detected with clinical mastitis is relatively ineffective as a method of eliminating infection is not that antibiotics do not cure infections. For example in the same experiment the use of antibiotics to treat clinical mastitis eliminated 72% of the 843 infections treated once or more[48]. The more important limitation was that stockmen using foremilk cups detected less than half of the infections present in the year. Nevertheless the effectiveness of antibiotic therapy is important and in practice depends to some degree upon the type of antibiotic used, the method of formulation, the course of therapy given and the length of time that the antibiotic persists in the udder after infusion. The result obtained when about 3000 clinical cases of mastitis were treated with various penicillin, streptomycin and cloxacillin preparations illustrate the marked differences in response of infections caused by the major pathogens. Twenty-eight per cent of *Staph. aureus* treated were eliminated whereas 82% of *Str. agalactiae*, 87% of *Str. dysgalactiae*, 72% of *Str. uberis* and 75% of all other infections were eliminated[5]. Even if research improves the effectiveness of antibiotic preparations for the treatment of clinical mastitis it will not bring about a major reduction in the average duration of infections because less than half of infections are detected from symptoms of clinical mastitis.

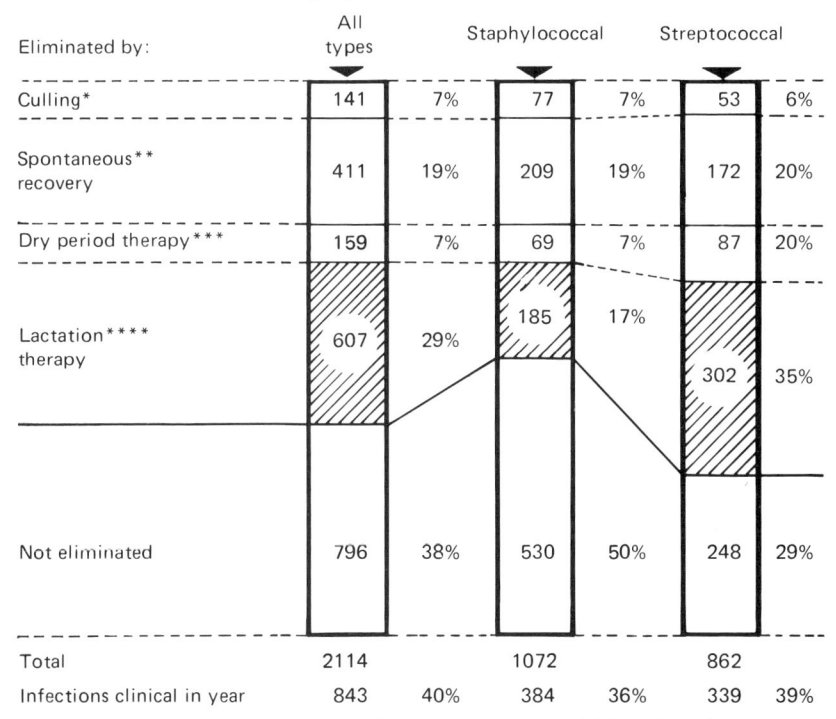

Eliminated by:	All types		Staphylococcal		Streptococcal	
Culling*	141	7%	77	7%	53	6%
Spontaneous** recovery	411	19%	209	19%	172	20%
Dry period therapy***	159	7%	69	7%	87	20%
Lactation**** therapy	607	29%	185	17%	302	35%
Not eliminated	796	38%	530	50%	248	29%
Total	2114		1072		862	
Infections clinical in year	843	40%	384	36%	339	39%

Fig IX 16. The relative importance of the various ways in which udder infections are eliminated from herds. Data from 721 cows during a year[5].
Note that only 40% of infections showed sufficient signs of clinical mastitis to be detected by stockmen and receive antibiotic therapy.

    * cows culled for any reason
   ** infections disappearing from quarters without antibiotic therapy
  *** antibiotic therapy given after the last milking of lactation to alternate cows in each herd. The antibiotic preparation used had not been formulated specifically for dry period use
 **** therapy given to lactating cows only when quarters were found clinical

If infections are to be eliminated much more quickly by antibiotic therapy then not only *clinical* but also *subclinical* infections must be treated. There are a number of possible ways in which the treatment of subclinical infections could be organized but if a control scheme is to be practical and economic then detecting subclinical infections by laboratory tests cannot be a routine method. An alternative is to treat *infections at drying off* using suitable antibiotic preparations. This can be done by treating selected quarters found infected using simple indirect tests or alternatively by treating *all cows* at drying off.

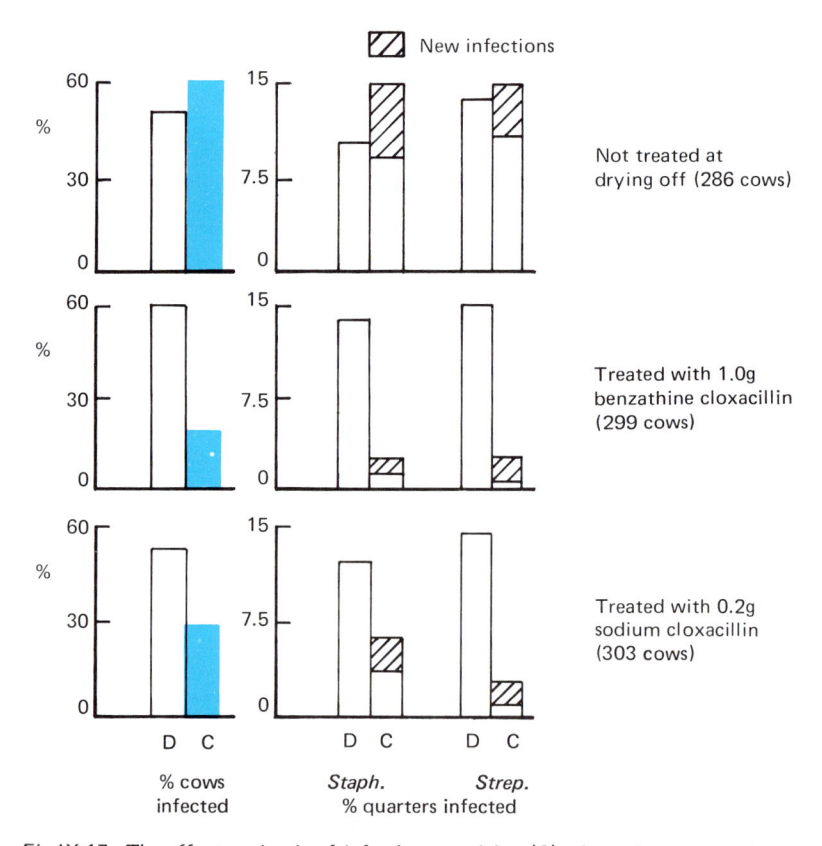

Fig IX 17. The effect on levels of infection at calving (C) of treating cows at drying off (D) with cloxacillin preparations[50].

Treatment of all cows at drying off has marked advantages because all persisting infections are treated, most of the new infections that otherwise occur in the dry period are prevented, milk does not have to be discarded because of antibiotic contamination and laboratory or other tests to find infections are not needed[50,51]. In addition with suitable products therapy particularly for staphylococcal infections is more effective when given at drying off than in lactation. Such timing for treatment ensures that most quarters are free from infection during the dry period allowing the regeneration of secretory tissue damaged by previous infections[15]. Effective products designed for use at drying off eliminate up to 70% of staphylococcal infections, over 90% of streptococcal infections and about 70% of infections caused by all other pathogens. At the same time new infections in the early dry period are reduced by about 75% (Fig IX 17), and infection with *Corynebacterium pyogenes* is usually almost completely prevented.

Another method of reducing the duration of infections would be to treat all new infections during lactation. This might be practical if an indirect test could be built into the milking machine for automatic detection of infections. It has been known for many years that a simple indirect test for infection is the electrical conductivity of milk[52] and recently there has been increased interest in automating this test procedure into the milking equipment[53]. Treating with antibiotics those 'new infections' that were detected by the conductivity test would be an alternative to drying off therapy. Its advantage would be an earlier treatment of most new infections and it is possible that the rates of elimination might be improved. The main disadvantages are economic, not the least cost being for reliable easily used conductivity devices with their associated data handling and monitoring equipment. The costs of the increased treatment during lactation would be high because of the milk that would have to be discarded when contaminated with antibiotics. This cost would be increased because the conductivity test would also detect minor infections not normally treated and also other physiological abnormalities. A scheme of this type has potential but will require much development work. Probably the simplest first stage would be to use the technique to detect infections approaching the clinical state as a guide for treatment. If this was successful the threshold of the test could be reduced to detect infections that gave a less inflammatory reaction. By trial and error the level could be found that detected infections as early as possible without an excessive number of false positives.

A BASIC MASTITIS CONTROL SYSTEM

There is good evidence that a simple routine based on teat dipping reduces new udder infection by about half under a wide range of management conditions, but the effect on levels of infection are inevitably slow. Equally convincing is the demonstration that antibiotic therapy given to all cows at drying off will reduce infection at calving to low levels and substantially reduces the duration of infections. These two techniques applied together provide a basic control system reducing levels of infection by at least half within a period of about a year. The value of such a system has been demonstrated in field experiments by the National Institute for Research in Dairying and the Central Veterinary Laboratory in England[10,54] and by Cornell University in the USA[55,56] and by others. The British experiment involved the co-operation for 3 years of 30 dairy farmers using 1 500 to 2000 cows at any one time. The herds covered a wide range of management and milking systems. Results of the many bacteriological and other laboratory tests to measure the effect of applying the system were not given to the farmers, who followed the simple control routine laid down at the start of the experiment.

The experiment measured the effects of two simple hygiene systems: one was limited to teat dipping after milking using a hypochlorite teat dip (4% available chlorine), no other disinfection being carried out in any part of the milking routine; and the other was based on a similar teat dipping routine but in addition the milkers wore smooth rubber gloves and the udders and gloved hands were disinfected in a hypochlorite solution (0.06% available chlorine) and udders were dried with a separate paper towel for each cow. Efforts were made to ensure that milking equipment was in good mechanical order; clinical mastitis was detected with the foremilk cup and a standard antibiotic therapy was given to all clinical quarters. All cows were treated with a benzathine cloxacillin product at drying off and cows with persistent clinical mastitis were culled. In

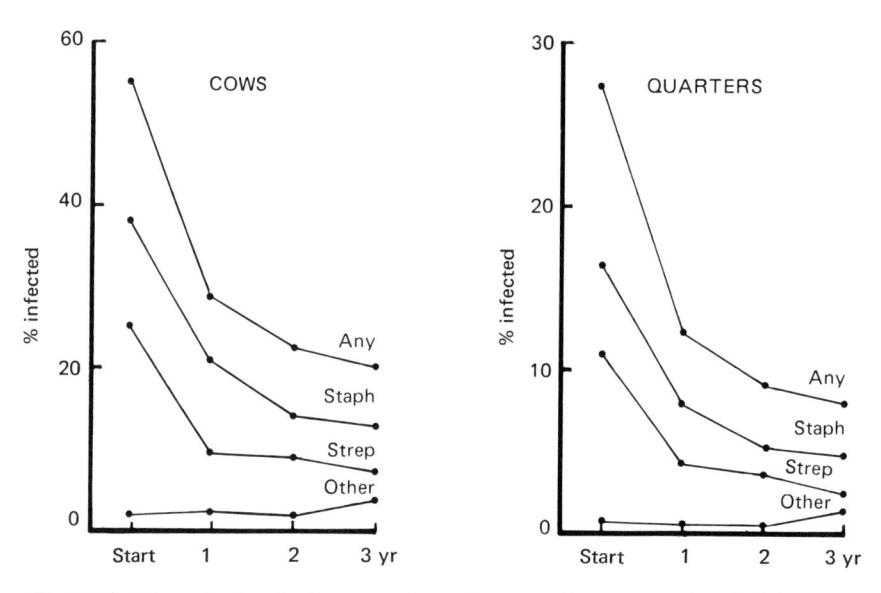

Fig IX 18. The reduction in the proportion of cows and quarters with subclinical mastitis in 30 herds in 3 years with a simple control system comprising hygiene including post milking teat dipping, and antibiotic therapy for cases of clinical mastitis and for all cows at the end of each lactation[10,54].

addition in half of the herds on each hygiene routine all subclinical infections detected at the start received antibiotic therapy[57].

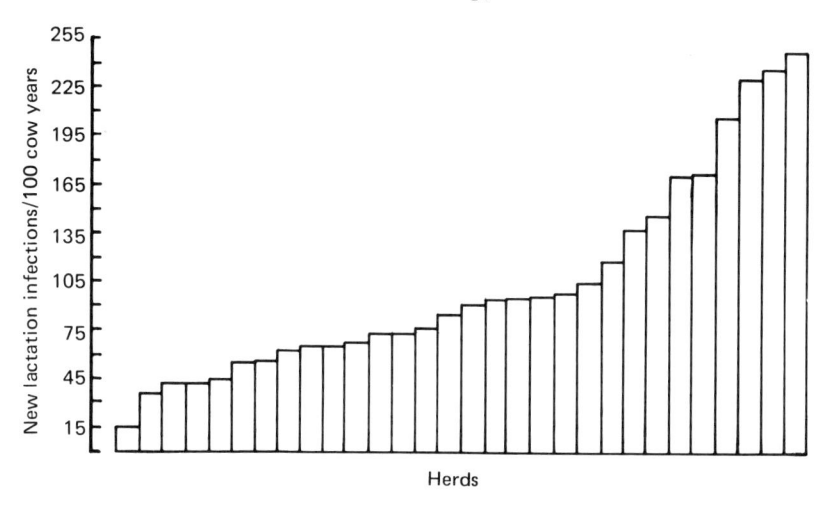

Fig IX 19. The variation between herds in the rates of new infections (all pathogens) during the three year NIRD/CVL test of a mastitis control system[5].

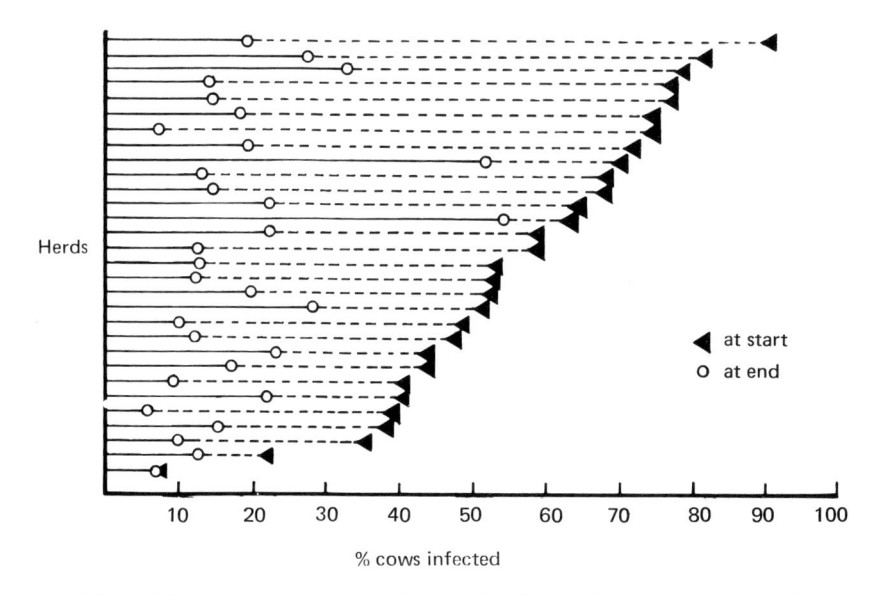

Fig IX 20. The proportion of cows with subclinical mastitis at the start and end of the 3 year NIRD/CVL test of a mastitis control system showing the differences between individual herds[54].

The reduction in subclinical mastitis in cows and quarters during the 3 years of the NIRD/CVL experiment is shown in Fig IX 18. Levels of infection fell on average from 28% to 8% of quarters and there was also a marked decline in clinical mastitis[54]. Disease declined in all herds and the system reduced both staphylococcal and streptococcal infection. However large differences were observed between herds, both in the rate of new infections (see Fig IX 19) and the response to antibiotic therapy of staphylococcal infection[5]. These influenced the extent of improvement in disease level of the individual herds (Fig IX 20).

As the experiment continued and the levels of staphylococcal and streptococcal infections declined, so did the rate of new infections with these pathogens, although this was less marked in the case of *Str. uberis* (see Fig IX 21). *Str. agalactiae* was found in 250 quarters of 16 herds at the start of

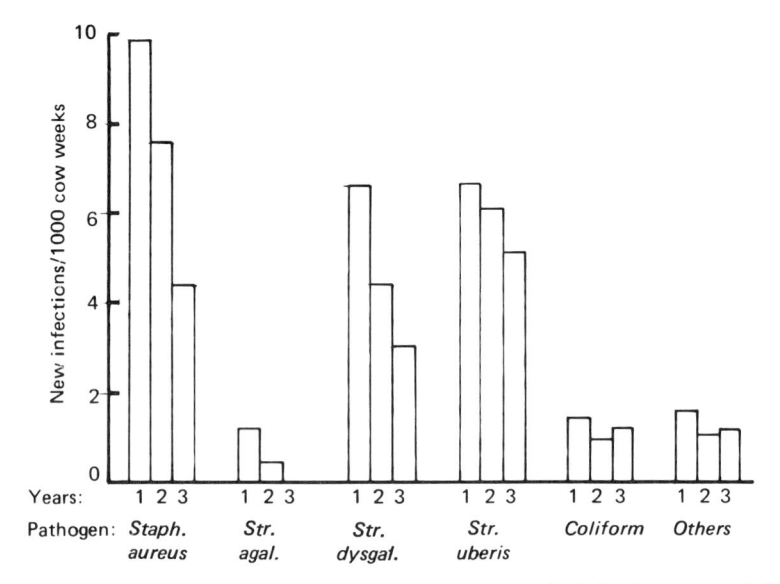

Fig IX 21. Rates of new infections during lactation in each of the three years of the NIRD/CVL test of a mastitis control system[54].

the experiment but this type of infection was eliminated from all of the herds and in spite of some re-introductions was present in only 2 quarters at the end of the 3 years. At the conclusion of the experiment 3 herds had no cows with staphylococcal infections and 13 herds were free from *Str. dysgalactiae*. Infections due to pathogens other than staphylococci and streptococci were uncommon throughout but did not show a major decline (see Fig IX 21). The herds receiving antibiotic therapy for subclinical disease at the start of the experiment showed a marked initial fall in level of infection but the levels in other herds fell to a similar value within one to two years. The results obtained in the two groups of herds using the two hygiene routines were similar. The Cornell University experiment in the USA gave an equally convincing demonstration of the value of the control system[55,56].

Summary of the control system

The essentials of the basic control system that has been developed and tested in the series of experiments described are as follows.

1. Follow good general husbandry and milking practice, including the regular maintenance of the milking machine.
2. Adopt the simple hygiene routine to reduce new udder infection, the main requirement being teat disinfection, preferably by dipping the whole teat in an appropriate hypochlorite or iodophor solution immediately after the milking machine has been removed.
3. Give antibiotic therapy under veterinary supervision:

   (a) to all cows routinely after the last milking of lactation, using effective antibiotic formulations designed specifically for infusion at drying off; and

   (b) to clinically affected quarters, keeping records of treated cases and culling cows with recurrent clinical mastitis.

A suitable hypochlorite teat dip for most herds can be made by diluting 1 part of an approved dairy hypochlorite with 3 parts of water, to give a concentration of approximately 4% available chlorine with not more than 0.05% caustic alkalinity.

Iodophor teat dip solutions containing 0.5% available iodine are much more widely used, but are more expensive and no more effective. They have the advantage, however, of being suitable for use with the addition of emollients, such as glycerol or lanolin, to promote the healing of teat lesions.

Wearing smooth rubber gloves and using clean running water, preferably containing a disinfectant, for udder washing will help to limit contamination at milking time.

BENEFITS AND COSTS OF MASTITIS CONTROL

Mastitis control is justified only if the finanical benefits clearly exceed the costs of the control methods. The main benefits of mastitis control are reduced clinical disease with a concomitant reduction in antibiotic treatment and milk discarded, and enhanced milk production. The value of the latter, following the decline in subclinical disease measured in the three year experiment quoted above, was about four times the savings from reduced clinical disease. To the farmer this benefit is less obvious because of the importance of other aspects of management on milk production.

The costs of the control system can be divided into the fixed costs incurred for teat disinfection, antibiotic therapy and routine milking machine testing, and the costs of extra food to secure extra milk production from an increasing proportion of healthy cows with a higher economic production potential.

The ratio of benefits to costs over a five year period (at 1976 prices) was about 3:1. Because it takes time to reduce the proportion of infected cows in a herd the margin of benefits over costs is slender in the first year of a control programme (see Fig IX 22) and widens steadily with continued application of the control system[58]. Other benefits that would follow are reduced teat lesions, less annual culling and improved milk composition. These benefits were not taken into account in the above study.

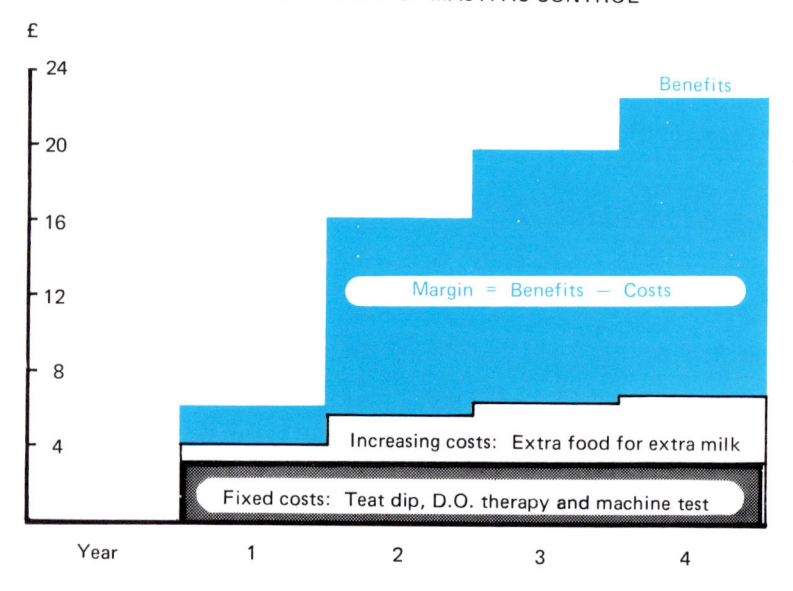

Fig IX 22. The relationship between the mean benefits and costs per cow year of mastitis control over a period of 4 years[58]. (Calculated at 1976 prices with milk at 40p/gallon.)

APPLICATION OF MASTITIS CONTROL

Although mastitis is a costly disease farmers have applied mastitis control measures erratically. Probably the main reason for this is that at any one time nearly all mastitis is subclinical and therefore unseen to the stockman.

Clinical mastitis is usually infrequent even in herds where infection is prevalent, and the symptoms usually subside following antibiotic therapy. However, the occurrence of clinical disease is an unreliable indicator of the proportion of cows affected by subclinical disease, which has the greater total effect on herd milk production. Because of the difficulty of separating the influence of disease from other factors governing milk yield this main benefit of mastitis control is rarely obvious to the farmer. Even the most effective systems are unlikely to give a real improvement in the level of disease in less than 6 months and achieving the maximum benefit takes several years. For these and other reasons it has been found that mastitis control is best encouraged in regional or national schemes of various complexity. These do not increase the benefit to those farmers who adopt the best control methods but they provide an awareness of the problem, monitor the disease level in individual or national herds and are a source of up to date information on techniques, disinfectants and other control measures.

SOMATIC CELL COUNTING IN MASTITIS CONTROL SCHEMES

The development in the last 10 years of techniques of electronically measuring the somatic cell count of milk has had a major effect on mastitis control schemes[59]. It has provided methods whereby one laboratory can measure regularly the cell counts of the bulk milk of thousands of herds. In this way the degree of success in a national or regional control scheme can be monitored and information provided to individual farmers to encourage them to adopt or continue the control methods[60].

There is no absolute cell count threshold that separates infected from non-infected quarters though quarters with milk cell count of less than 100 000 cells/ml will usually be uninfected and those with over 1 000 000 cells/ml infected. A herd's bulk milk cell count reflects the proportion of quarters infected and the severity of infections. Measures are available to control, to some degree, the former but little technical advice can be given on methods of reducing severity. It has been proposed that a cell count of 500 000 cells/ml in herd bulk milk is a useful threshold and that herds below this level have little mastitis problem. For most herds this may be a reasonable conclusion but there are some with a bulk milk cell count of under 300 000 cells/ml that have a problem with clinical mastitis caused by infections of short duration (e.g. coliform infections). These have little effect on level of

Table IX.8 Examples of the reduction of infection in a year in a group of 30 herds when only some herds are selected for mastitis control, compared with control of all herds[61]

Mean % quarters with any infection

		Mastitis control herds		Other herds		All herds	
		Start	After 1 yr.	Start	Start and after 1 yr.	Start	After 1 yr.
(1)	*Control not used in any herd*	—	—	26.5	26.5	26.5	26.5
(2)	*Control applied in only 3 herds (10% of total)*						
	At 1st herd test select the 10% of herds with either:						
	(a) The largest bulk milk cell counts	26.6	12.3	26.5	26.5	26.5	25.1
	(b) Average bulk milk cell counts	34.3	20.8	26.5	25.7	26.5	25.2
	(c) The largest % quarters infected	42.5	19.5	26.5	24.8	26.5	24.2
(3)	*Control applied in only 8 herds (27% of total)*						
	At 1st herd test select either:						
	(a) All herds (8 in number) with bulk milk cell counts >1 million cells/ml.	33.0	16.0	26.5	24.2	26.5	22.0
	(b) 8 herds with average bulk milk cell counts	27.0	15.8	26.5	26.4	26.5	23.5
	(c) 8 herds with the largest % quarters infected	38.9	20.6	26.5	22.0	26.5	21.7
(4)	*Control applied in all 30 herds*	26.5	13.1	—	—	26.5	13.1

Note: The data were obtained by notionally selecting herds from the first year's results of the NIRD/CVL mastitis field experiment[10]. The % quarters infected in a herd not selected for treatment is assumed to be unchanged after 1 year.

infection and therefore a small effect on the cell count of the herd milk. Conversely some herds will have cell counts of over a half a million and yet have only a small proportion of more severely affected quarters. In addition, the relatively large error variation in the bulk milk cell count of herds must also be recognized.

In summary, mastitis awareness campaigns or control schemes that aim to encourage the adoption of control measures require some tangible evidence of progress for the individual farmer. Bulk milk cell count provides a useful and convenient method; no better alternative is available. It is particularly useful for monitoring the progress of groups of herds and can be used for individual herds providing its limitations are recognized[61]. Schemes which confine mastitis control to a small proportion of herds selected on the basis of herd bulk milk cell count will have little impact on the level of infection of the total population, as illustrated in Table IX 8.

CONCLUSIONS

This chapter describes the complex way in which machine milking methods influence udder disease and outlines a control system based on management.

It is generally accepted that the properties of the milking machine and the way it is used are important factors that influence the prevalence of infection in herds. Avoiding mastitis was a major factor governing the early development of the milking machine. Certain milking machine factors predispose to new udder infection and clinical mastitis and these are usually associated with the effect of pulsation on the flow of milk and air within the milking machine, or with vacuum level or with the malfunction of other components. Nevertheless our understanding is incomplete. It is not known how much of the variation between herds in the prevalence of infection is due directly to milking machines or the way they are used. The current re-examination of milking machines with single chambered teatcups may advance the development of machines to reduce levels of mastitis.

In contrast with the lack of information on the importance of milking machine properties there is good evidence that the main transfer of mastitis pathogens takes place on the milking machine liners and, in preparation for milking, on the milker's hands, and in the use of common udder cloths. Furthermore, some commercial methods of cow housing and milking tend to increase chaps and teat lesions which become colonized with bacteria and therefore become an important source of pathogens. From this understanding of the epidemiology of mastitis simple hygienic methods to be used at milking time have been developed which substantially reduce the rates of new infections under most conditions. These coupled with improved methods of eliminating those infections that occur form a basic system of mastitis control that will reduce levels of infection by more than 50% within 12 months and by at least 75% over a longer period. Nearly all herds respond but there are still very large differences between the new infection rates of herds adopting the control and some types of infection (e.g. coliform) are not reduced. More research is necessary to improve this basic control that is based on improved milking and husbandry methods.

(1) Hunter, A.C. & Jeffrey, D.C. (1975) *Veterinary Record,* 96, 442.

(2) Neave, F.K. (1959) *Bulletin, Machine Milking, Ministry of Agriculture, Fisheries and Food,* No. 177, p.104. London: HMSO.

(3) Fell, L.R. (1964) *Dairy Science Abstracts,* 26, 551.

(4) Schalm, O.W., Carroll, E.J. & Jain, N.C. (1971) *Bovine Mastitis.* Philadelphia, Pa: Lea & Febiger.

(5) Dodd, F.H. & Neave, F.K. (1970) *Biennial Reviews, National Institute for Research in Dairying,* p.14.

(6) Bramley, A.J. (1975) *Bulletin, International Dairy Federation,* Document 85, p.377.

(7) Dodd, F.H. & Neave, F.K. (1951) *Journal of Dairy Research,* 18, 240.

(8) Dodd, F.H., Neave, F.K., Kingwill, R.G., Thiel, C.C. & Westgarth, D.R. (1966) *International Dairy Congress,* 17, A, p.383. Munich.

(9) Oliver, J. (1955) *Dairy Science Abstracts,* 17, 353.

(10) Kingwill, R.G., Neave, F.K., Dodd, F.H., Griffin, T.K., Westgarth, D.R. & Wilson, C.D. (1970) *Veterinary Record,* 87, 94.

(11) Brown, R.W., Morse, G.E., Newbould, F.H.S. & Slanetz, L.W. (1969) *Microbiological procedures for the diagnosis of bovine mastitis.* Washington DC: National Mastitis Council Inc.

(12) Neave, F.K. (1975) *Bulletin, International Dairy Federation,* Document 85, p.19.

(13) Wheelock, J.V., Rook, J.A.F., Neave, F.K. & Dodd, F.H. (1966) *Journal of Dairy Research,* 33, 199.

(14) Rowland, S.J., Neave, F.K., Dodd, F.H. & Oliver, J. (1959) *International Dairy Congress*, **15**, London, *1*, p.121.

(15) Smith, A. Dodd, F.H. & Neave, F.K. (1968) *Journal of Dairy Research*, **35**, 287.

(16) O'Donovan, J., Dodd, F.H. & Neave, F.K. (1960) *Journal of Dairy Research*, **27**, 115.

(17) Newbould, F.H.S. & Neave, F.K. (1965) *Journal of Dairy Research*, **32**, 171.

(18) Neave, F.K. *Unpublished.*

(19) Thomas, C.L., Neave, F.K., Dodd, F.H. & Higgs, T.M. (1972) *Journal of Dairy Research*, **39**, 113.

(20) Phillips, D.S.M., Whiteman, D.P. & Walker, H.T.M. (1969) *New Zealand Veterinary Journal*, **17**, 90.

(21) Frost, A.J. & Phillips, D.S.M. (1970) *Veterinary Record*, **86**, 592.

(22) Reiter, B. & Bramley, A.J. (1975) *Bulletin, International Dairy Federation*, Document 85, p.210.

(23) Neave, F.K. (1971) In: *The Control of Bovine Mastitis*, Dodd, F.H. & Jackson, E.R. (Editors) British Cattle Veterinary Association, p.55.

(24) Tolle, A., Zeilder, H., Worstoff, H. & Reichmuth, J. (1970) *Proceedings of the International Conference on Cattle Diseases* 6, Philadelphia, Pa: American Association of Bovine Practitioners, p.44.

(25) Neave, F.K., Dodd, F.H., Kingwill, R.G. & Westgarth, D.R. (1969) *Journal of Dairy Science*, **52**, 696.

(26) Philpot, W.N. (1975) *Bulletin, International Dairy Federation*, Document 85, p.155.

(27) Hoyle, J.B. & Dodd, F.H. (1970) *Journal of Dairy Research*, **37**, 133.

(28) Moak, H. (1916) *Cornell Veterinarian*, **6**, 36.

(29) Gordon, C. (1973) *Teat disinfection in relation to the milking routine:* Report, Agricultural Development and Advisory Service, Ministry of Agriculture, Fisheries and Food, London.

(30) Neave, F.K., Shearn, M.F.H. (1964) *Report, National Institute for Research in Dairying*, 1963, p.85.

(31) Gibbs, E.P.J., Johnson, R.H. & Voyle, C.A. (1970) *Journal of Comparative Pathology*, **80**, 455.

(32) Shearn, M.F.H., Kingwill, R.G., Griffin, T.K., Westgarth, D.R., Neave, F.K. & King, J.S. *Report, National Institute for Research in Dairying*, 1973-74, p.57.

(33) Witzel, D.A. & McDonald, J.S. (1964) *Journal of Dairy Science*, **47**, 1378.

(34) Neave, F.K., Sharpe, M.E., Oliver, J. & Dodd, F.H. (1957) *Report, National Institute for Research in Dairying, 1956*, p.69.

(35) Nyhan, J.F. (1968) *Proceedings of the Symposium on Machine Milking.* Reading: National Institute for Research in Dairying, p.71.

(36) Thiel, C.C., Cousins, C.L., Westgarth, D.R. & Neave, F.K. (1973) *Journal of Dairy Research,* **40,** 117.

(37) Thiel, C.C., Thomas, C.L., Westgarth, D.R. & Reiter, B. (1969) *Journal of Dairy Research,* **36,** 279.

(38) Cousins, C.L., Thiel, C.C. Westgarth, D.R. & Higgs, T.M. (1973) *Journal of Dairy Research,* **40,** 289.

(39) Thiel, C.C. (1974) *Biennial Reviews, National Institute for Research in Dairying,* p.35.

(40) Tolle, A. & Hanmann, J. (1975) *Bulletin, International Dairy Federation,* Document 85, p.193.

(41) Phillips, D.S.M., Woolford, M.W., Millar, P.J. & Phillips, E.M. (1975) *Bulletin, International Dairy Federation,* Document 85, p.200.

(42) Hopkirk, C.S.M. & Palmer-Jones, T. (1943) *New Zealand Journal of Science and Technology,* **A25,** 49.

(43) Petersen, K.J. (1964) *American Journal of Veterinary Research,* **25,** 1002.

(44) Stovlbaek-Pedersen, P. (1975) *Bulletin, International Dairy Federation,* Document 85, p.371.

(45) Bramley, A.J. & Neave, F.K. (1975) *British Veterinary Journal,* **131,** 160.

(46) Eberhart, R.J. (1975) *Bulletin, International Dairy Federation,* Document 85, p.371.

(47) British Standards Institution (1968) *CP3007.*

(48) Dodd, F.H., Westgarth, D.R., Neave, F.K. & Kingwill, R.G. (1969) *Journal of Dairy Science,* **52,** 689.

(49) Jackson, E.R. (1971) In: *The Control of Bovine Mastitis,* Dodd, F.H. & Jackson, E.R. (Editors) British Cattle Veterinary Association, p.25.

(50) Smith, A., Westgarth, D.R., Jones, M.R., Neave, F.K., Brander, G.C. & Dodd, F.H. (1967) *Veterinary Record,* **81,** 504.

(51) Dodd, F.H. & Griffin, T.K. (1975) *Bulletin, International Dairy Federation,* Document 85, p.282.

(52) Munch-Petersen, E. (1938) *Review No. 1, Imperial Bureau of Animal Health,* Weybridge.

(53) Linzell, J.L., Peaker, M. and Rowell, J.G. (1974) *Journal of Agricultural Science*, **83**, 309.

(54) Wilson, C.D. and Kingwill, R.G. (1975) *Proceedings Seminar on Mastitis Control*, p.422. 1040 Bruxelles: International Dairy Federation.

(55) Meek, A.M., Natzke, R.P., Everett, R.W., Roberts, S.J., Guthrie, R.S., Merrill, W.G. and Schmidt, G.H. (1970) *Proceedings of the 74th Annual Meeting US Animal Health Association*, p.41.

(56) Natzke, R.P. (1970) *Proceedings of the VIth International Conference on Cattle Diseases*, p.166. Philadelphia, USA: American Association of Bovine Practitioners.

(57) Wilson, C.D., Westgarth, D.R., Kingwill, R.G., Griffin, T.K., Neave, F.K. and Dodd, F.H. (1972) *British Veterinary Journal*, **128**, 71.

(58) Asby, C.B., Ellis, P.R., Griffin, T.K. and Kingwill, R.G. (1975) *The Benefits and Costs of a System of Mastitis Control in Individual Herds*. Study No. 17, Department of Agriculture and Horticulture, University of Reading.

(59) Heeschen, W. (1975) *Proceedings Seminar on Mastitis Control*, p.79. 1040 Bruxells: International Dairy Federation.

(60) Booth, J.M. (1972) *Journal of the Society of Dairy Technology*, **25**, 18.

(61) Westgarth, D.R. (1971) *The Control of Bovine Mastitis*, p.105. Shinfield, Reading: National Institute for Research in Dairying.

Chapter X

CLEANING AND DISINFECTION IN MILK PRODUCTION

Christina M Cousins and C H McKinnon

The minimum expectation of the consumer of milk and milk products is that cows be milked under conditions which satisfy the aesthetic requirements of any food producing process. These include active steps to keep dirt out of the milk, the use of visibly clean milking equipment and containers, and minimizing bacterial contamination from mastitic udders, teat surfaces and the equipment. There are also economic or commercial reasons for hygiene practices; bacterial contamination during production may have harmful effects on milk in its raw state, after pasteurization or during and after manufacture into milk products. The buyer normally exercises control over the hygienic quality of ex-farm milk by regularly testing consignments and producers of supplies failing the tests suffer penalties in the form of price deductions.

This chapter is concerned with principles and practices of cleaning and disinfection and descriptions of detergents and disinfectants; it includes preparation of the cow for milking. To emphasize the need for hygienic practices, sources of bacteria and visible extraneous matter (sediment or dirt) and the increase in numbers of bacteria in milk during storage at refrigeration temperatures are briefly discussed, together with problems of monitoring farm practices. Control of mastitis by hygiene is covered in Chapter IX. Public health aspects are not included.

SOURCES OF CONTAMINATION OF MILK

There are three main sources of bacteria in raw milk: the exterior of the cow's teats, the interior of the udder (i.e. udder infections) and milking equipment. The levels of contamination which can be derived from each of these justifies the various hygienic precautions and practices advocated in milk production, including milk cooling.

EXTERIOR OF THE TEATS

Between milkings cows' teats may become soiled with dung, mud and bedding materials such as straw, sawdust and sand. If not removed beforehand this dirt on the teats is washed into the milk during milking. Dirty teats are the main source of visible extraneous matter or 'sediment' in milk. It is unlikely that much dirt gains access to milk when clusters fall off udders during milking unless the teatcups fall directly into dung or bedding materials.

The large numbers of bacteria associated with the soil on teat surfaces are also washed into the milk. Numbers and types vary according to the type and degree of soiling but the milk from cows with heavily soiled unwashed teats will have an unacceptable sediment content and also bacterial counts approaching 100 000/ml. The apparently clean teats of cows kept on sawdust or sand may contribute up to 10 000/ml of milk if the bedding is heavily contaminated with bacteria (10^9 - 10^{10}/g) even though it appears relatively clean and dry. At the other extreme when cows are on pasture in dry weather and the teats do not become soiled the bacterial count of milk from unwashed teats may be less than 100/ml.

Visible dirt should be kept out of milk by cleaning teats before milking. Removing dirt by filtration is a commonly recommended practice but will not remove the bacteria, dissolved dirt or particles of the size of fat globules or smaller. A filter should be regarded only as a safeguard against accidental occurrence and as a monitor of milking hygiene. A clean filter after milking, indicative of clean teats, should be the producer's aim. Milk filters are described in Chapter III.

UDDER INFECTIONS

Milk as secreted is sterile. Udder quarters can become infected with many types of bacteria (see Chapter IX) so that it is not uncommon for 50% of quarters to be shedding bacteria in the milk. Normally the contribution of infected udders to the bacterial count of bulk herd milk is less than 10 000/ml. However milk from infected quarters at the clinical stage (clots in the foremilk) of streptococcal or coliform mastitis may contain more than 10^8 bacteria/ml. If milk from one or more such infected quarters is inadvertently included in the bulk milk of a herd the total bacterial count may be increased by 100 000 or more/ml. Such high counts of coliforms are rare but those caused by streptococci occur occasionally in milk from many herds. Thus udder infections may be responsible for test failures where the bacterial count test is used for hygienic quality control. Staphylococcal mastitis pathogens have not been implicated as the cause of excessively high counts in refrigerated bulk tank milk.

Drawing and examining foremilk enables the milker to detect and exclude abnormal milk from the herd bulk milk (Chapter IX). Often foremilk has a higher bacterial count than the remainder of the cow's total yield. However, because of the small volume (about 10 ml) of foremilk in comparison with the total milk (say 5 - 10 ℓ ; 1 - 2 gal) the effect of including foremilk is normally insignificant.

MILKING EQUIPMENT AND BULK MILK TANKS

Bacterial contamination of milk from air drawn into machines during milking is negligible in terms of numbers/ml of milk[1]. The milk contact surfaces of the equipment are therefore the only major source of bacteria in milk after it leaves the udder. In practice cleaning and disinfection do not remove from equipment all milk residues and associated bacteria which tend to accumulate daily. Except in very cold weather bacteria multiply in the equipment between milkings and their numbers may increase much more rapidly than visible residues. Surveys have shown that numbers and types of bacteria from supposedly cleaned and disinfected machines and bulk tanks vary widely from one farm to another and from time to time. Thus, as with the contribution of the teat surfaces and the udder itself to bacterial contamination of milk, that of the equipment cannot be estimated simply by inspection. However, as the following example shows, if the total bacterial count of bulk herd milk is significantly increased then the equipment must be heavily contaminated. To increase the bacterial count of 1000 ℓ(220 gal or 1 million ml) of milk by 1/ml, obviously requires 1 000 000 bacteria; thus to increase the count by 10 000/ml requires 10 000 million bacteria. This means that equipment, consisting of an average sized milking machine and bulk tank together having a milk contact surface area of 10 m^2 (100 ft^2), would need to contribute about 1000 million/m^2 or 100 million/ft^2.

Large numbers of any types of bacteria in milk are aesthetically undesirable as well as indicating faulty or careless practices. Furthermore some types, in particular those capable of multiplying at refrigeration temperatures, should for reasons of milk quality be kept to a minimum.

MULTIPLICATION OF BACTERIA IN REFRIGERATED RAW MILK

A storage temperature of 4 - 5°C (39 - 41°F) will prevent bacterial multiplication in milk for at least 24 and usually 48 h after production.

However, the presence of psychrotrophic bacteria, i.e. those able to multiply, albeit slowly, at 5°C for example, can result in marked increases in counts during longer storage periods. By using good cleaning and disinfection methods to keep psychrotrophic bacterial contamination to a minimum, and ensuring that the storage temperature does not exceed 5°C, it may take five days for numbers to reach a level of more than 10^6/ml at which taints or enzymic activities may have deleterious effects on the milk. At higher temperatures, 7 - 10°C, these psychrotrophs multiply much more rapidly.

Table X.1 Mean* total bacterial counts of ex-farm road tanker milk

Stored at:		Colony count/ml of milk after storage for:			
°C	(°F)	0 days	2 days	3 days	4 days
5	(41)	160 000	430 000	3 100 000	26 000 000
7	(45)	160 000	3 400 000	19 000 000	No result
10	(50)	160 000	68 000 000	230 000 000	No result

*Geometric mean counts of eight samples

Table X.2 Total bacterial counts of individual farm bulk tank milks stored at 5°C

	Colony count/ml of milk after storage for:			
Farm	0 days	2 days	3 days	4 days
A	5 800	3 300	7 900	14 000
B	14 000	10 000	11 000	70 000
C	14 000	10 000	710 000	15 000 000
D	28 000	83 000	2 800 000	18 000 000
E	62 000	400 000	9 500 000	41 000 000
F	170 000	110 000	110 000	130 000
G	240 000	1 800 000	8 900 000	17 000 000

These effects of increased length of storage at 5°C and storage at temperatures above 5°C on bacterial numbers are shown in Table X 1. On eight occasions a sample was taken from a bulk milk road tanker returning from daily collection from six or seven farms. On average these samples showed a marked increase in bacterial numbers after three days even at 5°C, and the effects of relatively small increases in storage temperature, to 7 or 10°C, were dramatic. The pattern of increase in bacterial numbers in stored samples from the road tanker shown in Table X 1 was reasonably consistent from time to time whereas milk samples from individual farms show a wide range of responses during storage at 5°C; examples are given in Table X 2. Some showed small changes only in total count after four days, others large changes in three days. It is also noticeable that the numbers after storage show little relationship to the initial level of contamination.

Ineffectively cleaned and disinfected milking equipment, particularly the refrigerated farm bulk tank, is the major source of psychrotrophic bacteria in milk[2]. If bulk milk after leaving the farm is to withstand handling and storage for much more than 24 h without risk of deterioration, the chief control measures available to the dairy industry are storage at or below 5°C and good farm hygiene.

PREPARATION OF THE COW FOR MILIKING

The time required to clean visibly dirty teats is a large part of the milker's work routine at milking time. Action to prevent cows' teats from regularly becoming heavily soiled between milkings is therefore an essential part of hygienic milk production. In cowsheds teats are washed using buckets and cloths and washing takes longer than in parlours where a hose can be readily available at each unit. Water supplied from the hand held hose is directed to the teats which are washed with the other hand, preferably gloved, and are then dried with a paper towel. This method, with cows that are moderately clean coming to milking takes about 15 s/cow. In principle only dirty teats need to be washed but in practice a routine of washing all teats is more satisfactory because it avoids subjective judgments of initial cleanliness.

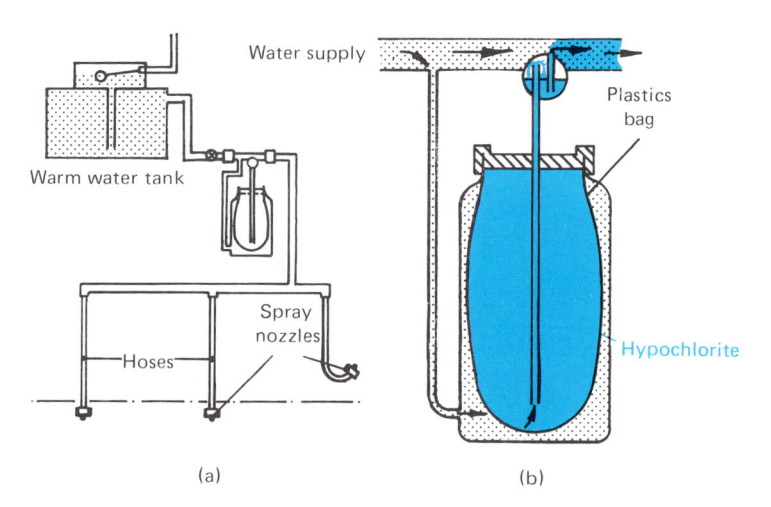

Fig X 1. An in-line chlorinator for hose washing of udders and teats: (a) position of chlorinator in the system; (b) schematic layout of the chlorinator.

Experiments have shown that if the teats are washed but not dried bacterial contamination of the milk is about the same as if teats are not washed at all. Either adding hypochlorite to the water or drying the teats are both advantageous and similar in effect but the use of hypochlorite together with drying is best of all in reducing contamination of milk with bacteria from teat surfaces[3]. Teats are the only important source of bacterial spores in raw milk before it leaves the farm and drying the teats has the additional advantage of significantly reducing their number, presumably because they are wiped off. Hose washing equipment which entrains disinfectant into the wash water is commercially available. Fig X 1 shows the principle of a system developed at NIRD[4]. Automatic udder washing is as yet rarely used and has not been evaluated.

METHODS FOR MILKING EQUIPMENT

A good cleaning and disinfecting routine in practice is one that with the minimum of time, effort and cost results in visibly clean equipment and milk consistently meeting the buyer's requirement for bacteriological quality. Officially recommended procedures are given in leaflets issued by the Ministry of Agriculture, Fisheries and Food (MAFF) for hand cleaning and sterilization[5], circulation cleaning[6] and cleaning and sterilizing refrigerated bulk milk tanks[7]. A recently issued British Standard[8] includes detailed recommendations for cleaning and sterilizing pipeline milking machines and bulk milk tanks, for checking the efficiency of these procedures and for the construction and operation of water heaters. Manufacturers of equipment and chemicals also provide instructions for all types of milking equipment. In all these publications the terms 'sterilization', 'chemical sterilizing agent' and 'detergent-sterilizer' are used incorrectly to denote respectively 'disinfection', 'disinfectant' and 'detergent-disinfectant'[9].

MILKING SYSTEMS AND CURRENT CLEANING PRACTICES

Milking systems

Methods of cleaning and disinfection are in the main related to milking systems. Table III 1 shows the distribution of herds by milking system in cowsheds and parlours in England and Wales at the end of 1974. The most striking features are that 60% of herds (but only 25% of cows) were milked in cowsheds, two-thirds with bucket machines. Other striking figures are that 55% of herds in Northern Ireland but not more than 25% in Scotland used bucket milking equipment. Only Northern Ireland had an appreciable proportion (20%) of producers still hand milking. See also Fig I 14 for numbers of milking installations and herds in the UK.

Current practices

For bucket and hand milking equipment there is no real alternative to washing by hand although the most laborious part, brushing the clusters can be partially replaced by flush washing them. Recorder milking machines and milking pipeline machines are cleaned in place, most commonly by circulation cleaning or the acidified boiling water (ABW) system. They are normally dismantled only for replacement of rubber or other worn out components. Bulk tanks are cleaned either manually of mechanically; for large tanks mechanical cleaning is essential. Most mechanical tank cleaning equipment is fully automatic (Chapter XII) but a certain amount of manual cleaning is necessary at intervals.

Disinfection is now mainly by means of chemicals used in conjunction with detergents. The use of dual purpose solutions, more convenient than cleaning first and disinfecting later, is almost universal for both manually and circulation cleaned machines except for those few, mostly in portable bails, where caustic flooding based on the immersion cleaning principle is still employed.

The use of steam for disinfection is now obsolete, but heat disinfection by the circulation of near boiling water is part of the acidified boiling water process, a rapid and convenient method which also cleans the machines effectively without the need for preliminary rinsing or conventional detergents.

Most current recommendations for in-place cleaning of milking machines call for the use of hot solutions after every milking but on many farms, after evening milking, minimal treatments using cold solutions are applied. It is to be expected that official recommendations should embody the best practices, i.e. hot cleaning after every milking. However at present there is no available evidence indicating that hot cleaning only once daily as compared with twice daily has any deleterious effect on visible cleanliness of machines or on the bacteriological quality of refrigerated bulk tank milk.

EQUIPMENT REQUIRED IN THE DAIRY OR MILKROOM

A dairy should have an impervious floor with suitable drainage and good lighting and, for all types of milking installation, should be equipped as follows.

1. Supplies of hot and cold water, preferably on tap, and a cold water hose pipe.
2. At least one wash trough of the smallest size suitable for the equipment, constructed of rubber or plastics, conveniently sited for the water taps.
3. A work surface with an impervious top.

4. One or more plastics buckets, suitable brushes, the necessary chemicals and graduated measures for them.
5. Adequate storage for supplies including chemicals.
6. A thermometer (0 - 100°C).
7. Rubber gloves and goggles to be worn when handling concentrated acids and caustic chemicals.

MANUAL CLEANING

Washing by hand is appropriate for bucket milking machines and ancillary equipment such as a receiving pan, strainer and cooler which may be used where milk is filled into cans. It is also used for some of the smaller bulk tanks.

The following additional items will be needed in the dairy.
1. A double wash trough.
2. A metal rack and hooks for storing the clean equipment.
3. An extension of the vacuum line with two or three vacuum taps for flushing clusters.

Daily cleaning routine

The routine after every milking consists of three stages: a rinse with cold or tepid water; a warm detergent-disinfectant wash; and a final rinse with clean water to which hypochlorite may be added. The cold or tepid water rinse removes residues of milk which otherwise would partially inactivate the disinfectant in the next stage. For rinsing the clusters, tepid water (38°C; 100°F) is particularly beneficial because it is more effective than cold water in removing fat and milk residues.

Immediately after milking is finished remove any dirt from the outside of units by hosing them with water and using a brush if necessary. Rinse each unit in turn by connecting the vacuum tube to a vacuum tap and drawing about 9 ℓ (2 gal) clean water through the teatcups into the bucket. Rinse the remaining utensils inside and out by hosing. If the clusters cannot be washed immediately they should be left immersed in water.

The warm detergent-disinfectant wash is the most important of the three stages. Unless the solution reaches all milk contact surfaces milk residues may remain which will protect bacteria from the disinfectant. Care must be taken to avoid air locks in the clusters.

Normally 45 ℓ (10 gal) of detergent-disinfectant solution is ample for up to three milking units and ancillary equipment. It is important to measure the volume of water used and this is best done by marking the side of the wash trough at the level of the required volume.

The wash solution may be prepared using approved hypochlorite together with a dairy detergent suitable for manual cleaning. Alternatively an approved proprietary detergent-disinfectant, also suitable for manual cleaning, may be used. Instructions for the quantities to be used will be shown on the labels on the containers. To avoid waste, measure the chemicals when adding them to the warm (45°C; 113°F) water.

Wash the clusters first. Detach the long milk tubes and immerse them in the solution. Leaving the short and long pulse tubes attached to the claw, immerse each cluster in the solution with the long pulse tube hanging over the edge of the trough, so making sure that no liquid enters the pulse tubes. Detach the short milk tubes from each claw and, using the correct brushes, scrub the teatcups, the short milk tubes, the claws and the long milk tubes. Reassemble, then transfer the clusters to the rinse trough.

Next, immerse and brush the bucket lids in the wash solution, taking care to remove the gaskets from the rims. Immerse and brush in turn the buckets and any other utensils; after allowing them at least 2 min contact with the solution transfer them to the rinse trough.

For the final clean water rinse, the rinse trough should be thoroughly clean before use. Run in cold clean water and, to prevent the risk of contamination from the trough, add approved hypochlorite at the rate of 25 ml/40 ℓ water (1 fl oz/10 gal), giving about 50 parts per million (ppm) of available chlorine.

Rinse the clusters thoroughly and hang up to drain. Rinse the other utensils in turn and stack them on the storage rack so that they drain completely before the next milking.

Simplified evening routine

Time can be saved in the evening by flush washing clusters instead of brushing them. These are the most difficult items to clean manually and where warm water is available it should be used.

After the preliminary flush with cold or tepid water, draw at least 9 ℓ (2 gal) of the warm detergent-disinfectant solution through each cluster in the same manner as for the water rinse. Then rinse by flushing with clean cold water and hang up to drain. Treat the remaining equipment as described previously.

Additional treatments

In the past, because of the difficulty of keeping clusters clean and the prevalence of milk stone, special treatments were recommended as a routine but were probably rarely applied except after complaints about the bacteriological quality of the milk. Special treatment included descaling with acid, scalding metal utensils with boiling water and, more widely used, defatting of liners and daily wet storage of clusters.

It is now generally recommended that the routine daily manual cleaning and disinfection be applied sufficiently thoroughly to avoid the need for frequent regular special treatments. However some are required when deposits are detected or in the event of a bacteriological test failure or rejected milk.

Defatting liners. Two sets of liners are required and one set, after use for a week, is soaked in cold 5% caustic soda solution while the other set is in use. The treatment which is used only with natural rubber liners helps to prolong their milking life and has a disinfecting action. Synthetic rubber liners absorb little if any fat during use; defatting is therefore unnecessary and the liners may be damaged by the caustic solution (see Chapter XI). Details of the treatment if required are given in Advisory Leaflet 422[5].

Wet storage of clusters. This treatment is still useful as a supplement to normal cleaning in hot weather where there is a shortage of cooling water for milk collected in cans. The procedure consists of suspending the clusters in a rack in such a way that they can be filled with a suitable solution, e.g. alkaline detergent-hypochlorite, and kept full between milkings. This permits prolonged contact of the solution with the interior surfaces of the cluster thus improving the disinfection process. British Standard 2756[10] should be consulted for details.

Descaling. Phosphoric acid or one of the proprietary acidic materials sold for the purpose should be used to remove any build-up of resistant residues on the equipment. Full details of the procedure are given in Advisory Leaflet 422[5]. After they have been cleaned the dismantled metal components are soaked in a descaling solution made according to instructions. They are then brushed to remove the loosened scale and rinsed. To neutralize any residual acid all the parts should then be brushed in hot detergent-disinfectant solution and finally rinsed. Tinned surfaces must not be over-exposed to acid, and aluminium equipment should be treated only with proprietary materials recommended for aluminium. There is little risk of corroding stainless steel components. *Goggles and rubber gloves should be worn when handling the acid.*

Heat treatment. In an emergency, for example if the supply of hypochlorite or detergent-disinfectant is exhausted or if the milk has been rejected, the equipment, including the clusters, may be disinfected with boiling water.

For effective heat disinfection an ample supply of boiling water is essential. A 45 ℓ (10 gal) domestic wash boiler in the dairy is the most convenient source. The boiling water is applied to the equipment as follows.

Assemble the cleaned milking units and draw at least 9 ℓ (2 gal) of water as near boiling as possible and not less than 85°C (185°F) slowly through from the teatcups into the milking bucket. Disconnect the cluster and hang up to drain and dry. Swirl the bucket, place the lid in an empty container and pour the hot water from the milking bucket over the lid. After 2 min discard the water and put the lid and the milking bucket on the rack to drain and dry. Repeat the process for the remaining units.

If the cooler, strainer and receiving pan are to be treated, pour at least 9 ℓ (2 gal) boiling water slowly over the cooler and strainer, having placed them in the receiving pan. Allow to stand for 2 min then place them on the rack to drain and dry.

Alternatively, dismantle the clusters and, together with the bucket lids, immerse them in water in the wash boiler. Bring the water to the boil and switch off the heater. Run off the water and use it immediately to treat the buckets, cooler, strainer and receiving pan.

Vacuum tubes and the air pipeline

On bucket milking installations, during milking, the rubber vacuum (stall) tubes and the air pipeline carry air only. They may however become contaminated with milk if buckets are overfilled or in the event of a split liner. Moisture from the vacuum line can pass the check valve on the bucket lid and to prevent contamination of the milk from this source the air pipeline and vacuum tubes must be kept clean.

It is sufficient to clean the vacuum tubes once a day. After all the other equipment has been treated immerse them in the warm detergent-disinfectant solution. After a short soak rinse them under the tap and hang up to drain. Unless they have become contaminated with milk they need to be brushed inside only once a week.

The air pipeline and interceptor should be washed once a week and whenever milk is known to have entered the air pipeline. If it is neglected for any length of time cleaning may be difficult.

Prepare hot detergent or detergent-disinfectant solution in 10 ℓ (2 gal) quantities for each branch of the air pipeline, using 100 g (3 oz) of chemical and water at 60 - 70°C (140 - 160°F). If an iodophor is used the temperature of the solution should not exceed 50°C (122°F).

Make sure the interceptor is empty, and using a vacuum tube, draw the solution through the pipeline or branch pipeline using the vacuum tap furthest from the interceptor. If more than 10 ℓ (2 gal) is to be drawn through, empty the interceptor after the first 10 ℓ (2 gal) to avoid overfilling, which will result in liquid being drawn into the vacuum pump. Rinse the pipeline with hot water and then inspect all drainage cocks and automatic drain valves to see that they are clean and will open as intended, thus allowing the air pipeline to drain completely.

Action in the event of a milk test failure or rejected milk

Although the immediate cause of most hygiene test failures is attributable to poor cooling and failure to keep the milk cool the underlying cause is almost always failure to clean and disinfect the milking equipment adequately. If, because of rejected milk, immediate action is necessary to prevent further rejections, treatment with boiling water as described previously will be effective until other remedial measures can be completed.

1. Old and worn rubber components and metal equipment which is rusty, pitted, corroded or with open seams should be replaced.
2. The metal equipment should be descaled and the rubberware soaked in hot detergent solution followed by thorough brushing.
3. The daily cleaning and disinfecting methods should be checked immediately with the manufacturer's instructions and with the methods described here and any faults corrected.

IN-PLACE CLEANING

There are two in-place cleaning methods widely used for milking pipeline machines in cowsheds and parlours and recorder machines in parlours. The first, known as circulation cleaning, is a three stage process consisting of a water-pre-rinse, a re-circulated hot wash with detergent-disinfectant solution and a final cold water rinse with or without added hypochlorite. It is usually said to be based on chemical disinfection but its reliability depends to a considerable extent on the use of very hot water at a minimum initial temperature of 85° (185°F). The second in-place cleaning method, referred to as acidified boiling water (ABW) cleaning, relies solely on heat disinfection. Its main characteristics are: a single stage only, near boiling water being admitted to the milk contact surfaces without preliminary rinsing; the continuous discharge of the water to waste after passing through the machine; acidification of the water for the first half of the process to prevent hard water deposits and to assist cleaning, and a temperature of at least 77°C (170°F) on all milk contact surfaces for a minimum of 2 min, achieved by means of the high inlet water temperature and the absence of recirculation. Both circulation and ABW cleaning are well adapted to parlour milking machines but circulation cleaning is generally used for milking pipeline machines in cowsheds. Both would benefit from a less expensive method of water heating than by electricity as they rely almost equally on an adequate supply of very hot water, but see the section on waste heat recovery in Chapter XII where it is suggested that savings will be small and can justify only small capital expenditure on alternatives. ABW cleaning requires up to one-third more water at a temperature 10°C (18°F) higher. However circulation cleaning uses chemicals that are energy expensive in manufacture and on a total energy basis the processes are comparable. The ABW process is completed in about 6 min whereas circulation cleaning takes at least 15 min.

There is evidence that in practice a higher standard of bacterial cleanliness in the plant is more often achieved with ABW than with circulation cleaning. However, the difference is not likely to be detected in milk quality provided both are applied equally effectively. Given that installations remain visibly clean the most likely causes of serious bacterial contamination from the machine are hidden areas and crevices where bacteria are protected from the disinfection processes and which in the normal event are unlikely to be detected. These should be borne in mind when looking for the causes of high plant rinse counts.

The recorder milking machine in its present form is peculiarly a UK development and the systems of cleaning which were evolved for these comparatively elaborate machines are also specialized. The main evolutionary changes made to pre-1957 recorder machines to aid cleaning were as follows. To enable all milk contact components to be readily connected at the end of milking into a cleaning circuit three changes were made (Fig X 3): an extra pipeline, the air pipeline (milking vacuum) was installed to connect the units to the vacuum system during milking and to cleaning liquids by means of a 3-way valve for cleaning; the clusters were connected in place into the cleaning circuit by jetters (Fig X 2), an idea imported from New Zealand; and spreaders were fitted into the top centre spigot of the recorder jars so that flowing liquid evenly covered the interior surfaces. More recently spreaders have been found unnecessary.

Fig X 2. Teatcups connected to jetter. Cleaning fluids are drawn upwards through the jetter and through the teatcups.

C.M. Cousins and C.H. McKinnon

To prevent cross contamination by liquid moving between the milk and air systems all pipelines were terminated in a sanitary trap (Fig X 3). All glass recorder jars instead of ones with lids, and the releaser milk pump (from the USA) were introduced because they were better adapted than their predecessors to circulation cleaning. Finally the use of very hot water made the cleaning and disinfection process more effective. This enabled the extensive use of rubber-to-metal and rubber-to-glass joints in units and pipelines to be retained. These are inexpensive, effective in a vacuum machine and in most instances such as the many joints in a cluster it is difficult to envisage a satisfactory alternative.

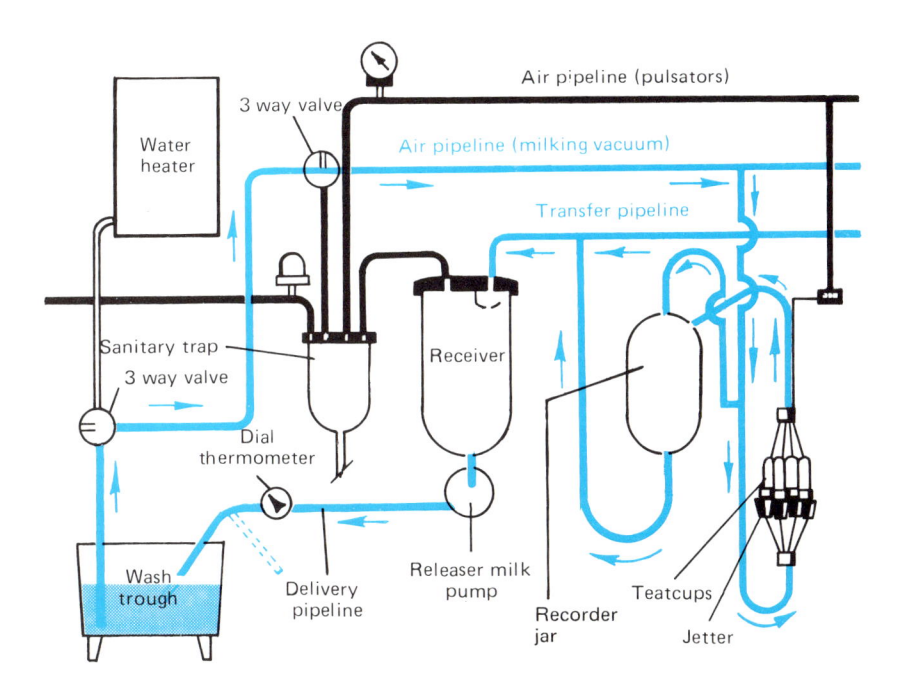

Fig X 3. Circulation cleaning circuit for recorder milking machines.

The cleaning circuit for in-place cleaning of milking pipeline machines in parlours is a simplified version of that for recorder machines as shown in Fig X 4. The circuit for milking pipeline machines in cowsheds is similar but the clusters are connected in the milk room (Fig X 5).

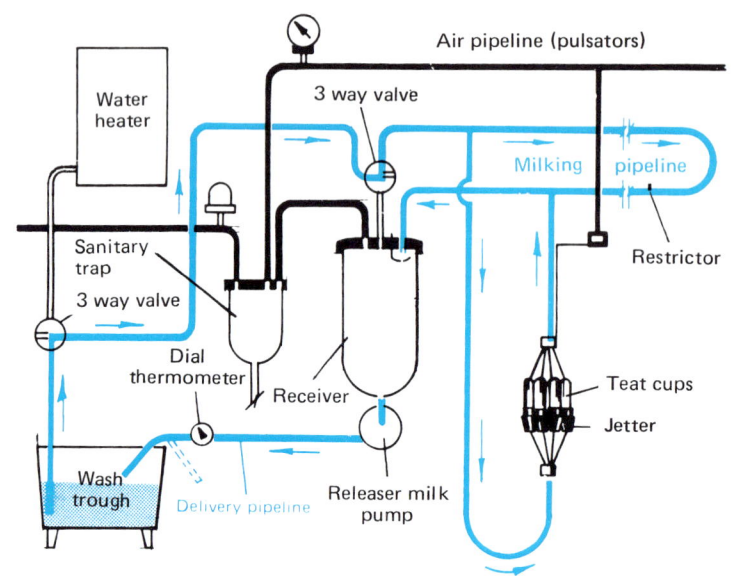

Fig X 4. Circulation cleaning circuit for parlour milking pipeline machines.

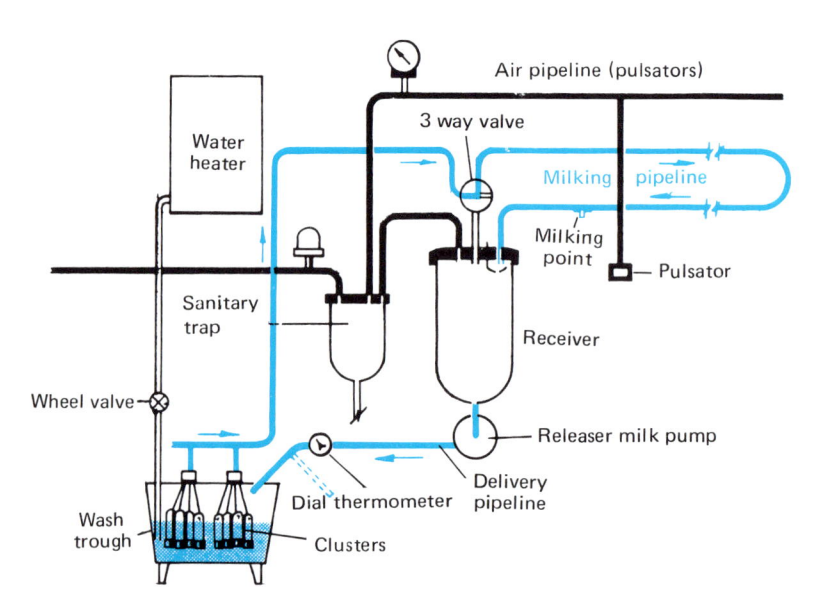

Fig X 5. Circulation cleaning circuit for cowshed milking pipeline machines, showing clusters attached to the suction end of the milking pipeline. Alternatively they may be attached to the discharge end of the delivery pipeline.

CIRCULATION CLEANING

For cleaning their particular installations most manufacturers issue instructions to be read in conjunction with a diagram of the plant. Some instructions are more detailed than others and the particular brand of detergent, hypochlorite or detergent-disinfectant may be specified; otherwise the advice given is to use suitable products (unspecified) in accordance with the chemical manufacturers' instructions.

Recorder milking machines and milking pipeline machines in parlours.

The basic routine for the procedure after each milking is as follows.

1. Start the cleaning routine immediately milking is finished, leaving the vacuum pump and pulsators running.
2. Remove all external dirt from each cluster and jetter using a brush and a detergent-disinfectant solution. (If water only is used, it will be contaminated by dirt from the teatcups and the drainings which collect at the junction of the teatcup and jetter cup may gain entry to the liner when the jetter is removed after cleaning.)
3. Fit the clusters into the jetters and set the valves at the units to the washing position.
4. Check that the water in the heater is 85°C (185°F).
5. Drain milk from the receiver and milk pump.
6. Remove the milk delivery pipe from the bulk tank and arrange for discharge to waste; remove the filter sock or re-usable filter from the in-line filter carrier.
7. Connect the air pipeline (milking vacuum) directly to the water heater, set the 3-way valve to the washing position so that the hot rinse water is drawn into the machine (Figs X 3 and 4).
8. Set the releaser milk pump to run continuously and adjust the spreader on the receiver lid to the washing position.
9. Allow hot water to pass through the machine and discharge to waste until the return water temperature reaches 65°C (150°F); then discharge the required volume of hot water into the wash trough.
10. Add a suitable detergent and sodium hypochlorite, or a combined detergent-disinfectant to the hot water. Normally the volume of solution recommended is 10 - 14 ℓ (2 - 3 gal)/milking unit. Use the chemicals at the concentrations recommended by the chemical manufacturer for the volume of solution used. Set the 3-way valve to draw solution from the wash trough through the machine. Continue circulation for 5 - 10 min. A short circulation time fits more conveniently into the work routine and there is no advantage in prolonging circulation because the temperature of the solution will fall progressively.

11. Discharge the detergent-disinfectant solution to waste by deflecting the delivery piepline.
12. Run clean cold water into the wash trough and allow it to pass through the machine to waste. Sodium hypochlorite at the rate of 25 ml/40 ℓ water (1 oz/10 gal) giving about 50 ppm available chlorine may be added to the cold water to avoid risk of contamination from supply lines and the storage tank.
13. Switch off the releaser milk pump and milking vacuum pump, drain the machine completely and set it ready for milking.

Notes:

(a) If a hypochlorite rinse is to be given immediately before milking it is not necessary to add hypochlorite to the final rinse (step 12). In this case after draining the machine, leave it set for cleaning.

Just before the next milking run clean cold water into the wash trough, add sodium hypochlorite at the rate of 25 ml/40 ℓ (1 fl oz/10 gal), circulate the rinse for 1 - 2 min and then discharge it to waste.

Take great care to drain the machine completely; disconnect the jetters from the clusters, drain the jars, transfer pipelines, the receiver, releaser milk pump and delivery pipe, and set the plant for milking.

(b) If, during milking, milk or froth passes from the receiver into the sanitary trap, this should be cleaned by flooding the receiver with detergent-disinfectant solution towards the end of the period of circulation, thus allowing cleaning solution to pass into the sanitary trap. To keep the sanitary trap clean during normal operation of the plant, this treatment will be necessary occasionally.

(c) Where a D-pan, strainer and cooler are used they should be dismantled daily for manual cleaning. Some of the detergent-disinfectant solution may be collected and used for this purpose, but rubber gloves should be worn, unless the detergent is suitable for manual cleaning.

(d) Some manufacturers provide equipment for automatically controlling the circulation cleaning cycle and dosing the cleaning solutions with detergent-disinfectant products. However, some of the operations e.g. brushing off the clusters, connecting them to the jetters, setting the machine for cleaning and milking and draining the machine still need to be done manually, so that the cost of the equipment should be considered in relation to the saving in time.

Milking pipeline machines in cowsheds

The method for these installations, except for the handling of the clusters, is identical to that for recorder and milking pipeline machines in parlours.

Immediately after milking brush external dirt from the clusters and take them to the milk room. After draining milk from the machine suspend the teatcups in the wash trough with the long milk tubes attached to a manifold either at the suction end of the air pipeline (milking vacuum) (Fig X 5) or at the discharge end of the delivery pipeline. Where the cleaning solution is drawn from the wash trough through the clusters into the cleaning circuit, care must be taken to see that none of the teatcups or clusters is starved of liquid during circulation of the solutions. Where more than four clusters are in use, special arrangements may be necessary. The volume of detergent-disinfectant solution required to maintain effective circulation is determined mainly by the length of the milk pipeline.

Milking buckets, used on occasions, must be washed by hand.

Additional treatments

Regular and frequent inspection of the installation during and after cleaning is necessary to detect and correct split liners, punctured short milk tubes, leaking rubber joints, deterioration, kinking or flattening of any rubber tubes and any malfunction resulting in impeded circulation or build up of milk residues.

Any treatment additional to the normal twice daily cleaning routine is time consuming and costly if frequently applied. Where scale is seen to build up quickly, e.g. on jars and glass liners, the suitability of the detergent or detergent-disinfectant for circulation cleaning and for the hardness of the water should be checked; if necessary, a different material should be selected. A suitable detergent-disinfectant solution should be capable of maintaining visual cleanliness for at least a month and, except where the water is exceptionally hard, for much longer. Descaling should be used at the first signs of film. If film is allowed to accumulate circulation of acid may not be completely effective and dismantling and brushing, which is even more time consuming, will be necessary to remove all residues.

Descaling. The following procedure is recommended.

1. After milking, rinse the machine thoroughly using warm water.
2. Using a proprietary milk stone remover as directed or syrupy phosphoric acid (density 1.75) at the rate of 30 ml/4.5ℓ (1 fl oz/1 gal) of warm water (45°C; 115°F), prepare sufficient solution in the wash trough to maintain circulation.
3. Circulate the solution for 5 - 10 min and discharge to waste.
4. Rinse the machine with cold water and discharge to waste.
5. Prepare and circulate a hot detergent-disinfectant solution and finally rinse with clean cold water as for the normal cleaning routine. This is required to remove any traces of acid, loosened film and associated bacteria.

Dismantling. Components in some older installations cannot be effectively cleaned in place. The circulation of solutions at high initial temperatures (80 - 85°C) may not be effective in killing bacteria associated with such components where they form a dead end and because of the poor heat conductivity in rubber-to-rubber or rubber-to-plastics joints. Fittings of the type shown in Figs X 6 and 7 need to be dismantled at least once a week after the normal cleaning routine.

The parts, including the seating of the tap or cock, should be brushed clean in detergent-disinfectant solution and replaced.

If, on inspection, milk residues are found on the rim of the receiver lid (Fig X 8) these must be dismantled and brushed as described above.

At intervals, perhaps once a year, long milk tubes and any other rubber components showing signs of deterioration or damage should be replaced.

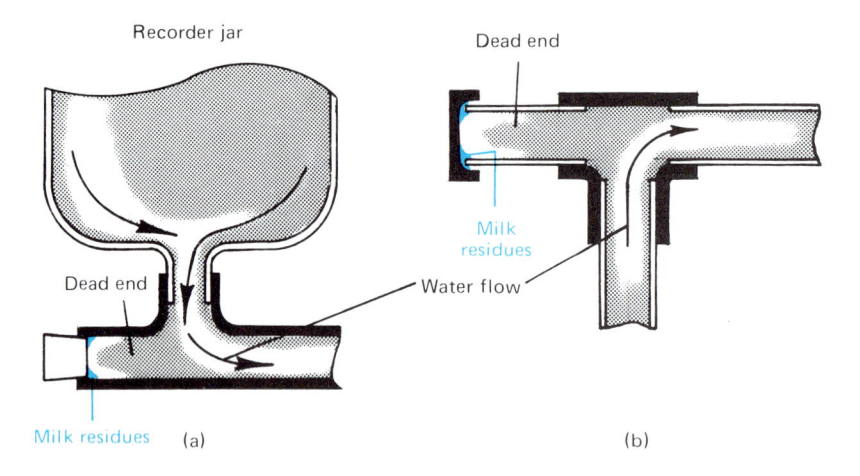

Fig X 6. Milk residues in components forming dead ends: (a) at the base of a recorder jar; (b) at the end of a transfer pipeline.

ACIDIFIED BOILING WATER CLEANING

The general layout of the pipelines and the components set for cleaning are shown in Fig X 9. Boiling water, 14 - 18 ℓ (3 - 4 gal) per unit, enters the machine directly from the water heater at the predetermined rate, is circulated over the entire milk contact surfaces and is then discharged to waste. During the first 2 - 3 min of flow, 1 ℓ (2 pints) of dilute acid is drawn into the boiling water to prevent deposition of hard water salts on the internal surfaces of the equipment. During this period all parts of the machine should reach a temperature of 77°C (170°F) and be maintained at this temperature while the remaining volume of water is drawn through the machine. The total flow time is 5 - 6 min. The labour requirement is minimal. Liner life is at least as good as with circulation cleaning[11].

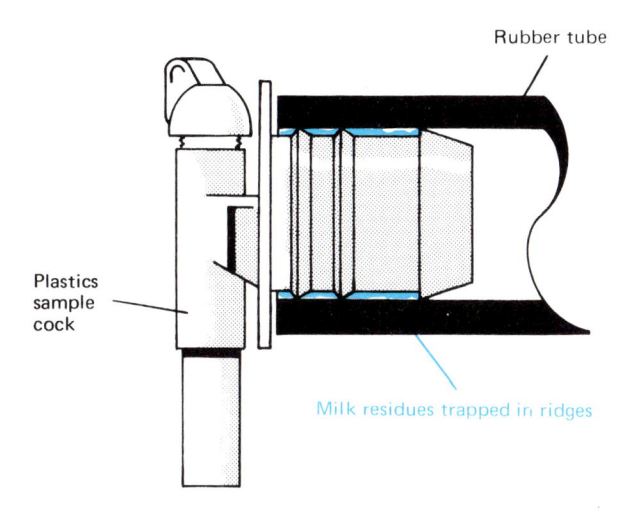

Fig X 7. Section of a plastics sample cock at the base of a recorder jar showing site of milk residues.

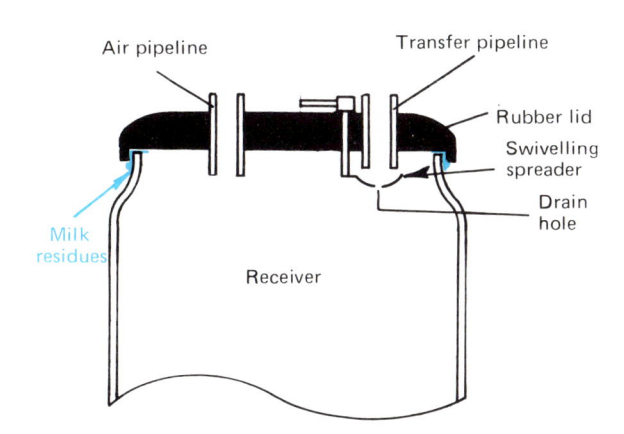

Fig X 8. Section of a receiver and lid showing site of milk residues.

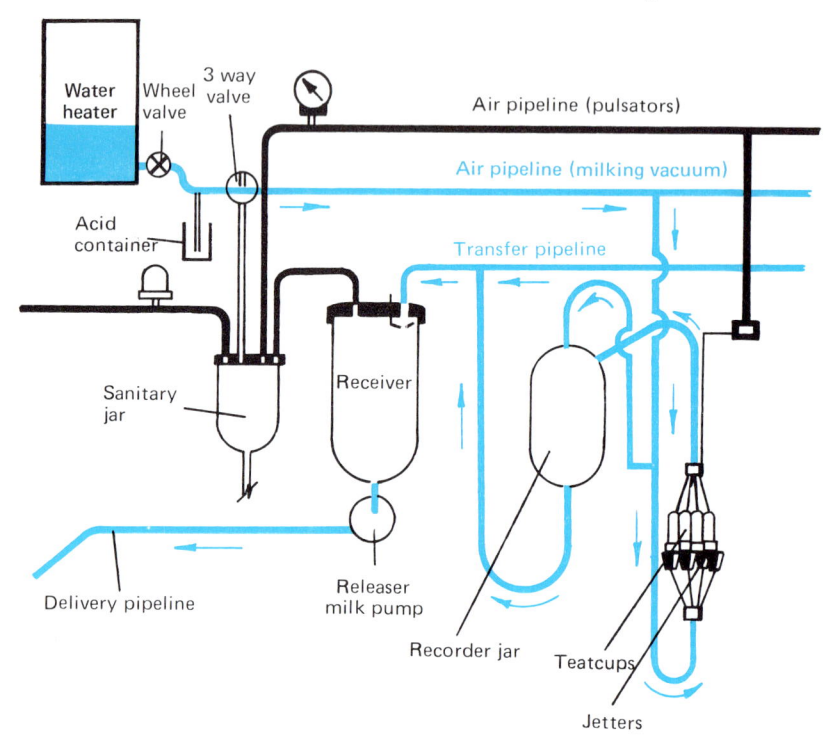

Fig X 9. Acidified boiling water cleaning circuit for recorder milking machines. Note direct connection between the water heater and the air pipeline (milking vacuum) through the 3 way valve.

Preparation of stock acid solution

Either nitric acid (70% w/w) or sulphamic acid (crystals) is used for preparing the stock solutions for various sizes of machines as follows.

To 22.5 ℓ (5 gal) of water in a plastics container add:

Plant	Nitric acid	or	Sulphamic acid
4 unit	0.75ℓ (1½ pt)		1.0 kg (2 lb)
6 unit	1.0 ℓ (2 pt)		1.5 kg (3 lb)
8 unit	1.5 ℓ (2½ pt)		2.0 kg (4 lb)
10 unit	1.75ℓ (3 pt)		2.25 kg (5 lb)

Sulphamic acid which is in the form of crystals is more easily and safely handled than concentrated nitric acid. Commercial preparations are also available and should be used according to the manufacturer's directions. Concentrated nitric acid should always be added to cold water. Sulphamic acid should be added to warm water and stirred thoroughly with a clean wooden stick until the crystals have dissolved. The greatest care should be taken when handling strong acid. *Goggles and rubber gloves should be worn when preparing the stock solution.*

Flowrate control

Control and equalization of flow of water through the machine is obtained by the use of restrictors at strategic points. The diameters suitable for most machines with a working vacuum of 51 kPa (15 inHg), are as follows.

1. At the outlet from the water heater: 6.4 - 11.1 mm (¼ - 7/16 in)
2. At the acid container outlet: 0.8 - 1.2 mm (1/32 - 3/64 in)
3. In the vacuum tube to each recorder jar: 4.8 mm (3/16 in)

In some installations there is sufficient restriction in the components used to make some restrictors unnecessary; the criterion in all cases must be a satisfactory flow so that all parts of the machine reach the minimum temperature specified.

Routine at the end of each milking

1. Brush the outside of the clusters and jetters using hypochlorite or detergent-disinfectant solution. Fit the teatcups into the jetters and open the valves at each unit.
2. Remove the milk delivery pipe from the bulk tank.
3. Check that the water temperature in the water heater is above 96°C (205°F).
4. Add 1 ℓ (2 pints) of stock acid solution to the acid container.
5. Turn the 3-way valve to the washing position and open the wheel valve controlling the flow from the water heater.
6. Adjust the spreader on the receiver lid to the washing position.
7. Set releaser milk pump to run continuously to discharge return water to waste.
8. When flow stops, stop the vacuum pump and releaser milk pump.
9. Drain the machine.

Essential features of plant design

The design details of a machine installed for or converted to ABW cleaning will vary according to the make of machine but certain features are essential to the success of the system.

1. All components should be capable of withstanding the high temperature and the acidity of the cleaning solution and, except for the rubberware, are normally of borosilicate glass, plastics, and stainless steel.
2. To avoid excessive heat loss the machine should be as compact as possible with no unnecessary runs of pipelines.
3. There should be no dead ends (Fig X 6).

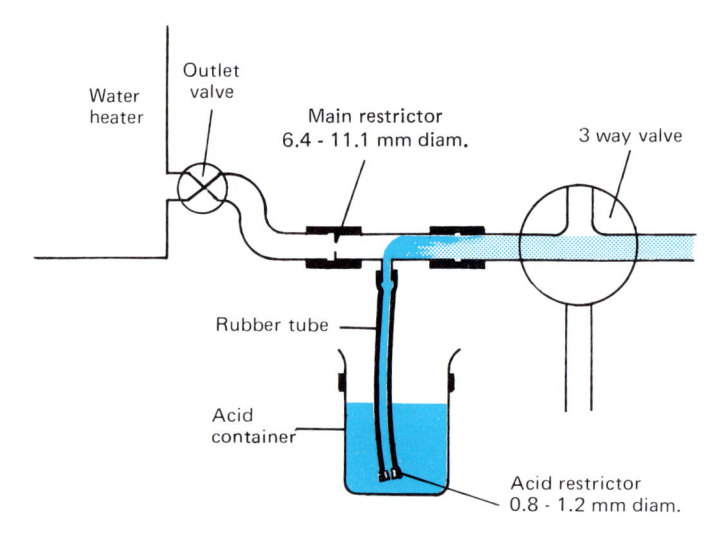

Fig X 10. Safe arrangements of the acid container and restrictor for metering acid drawn into the machine during acidified boiling water cleaning.

4. The acid container should be designed in such a way as to prevent accidental leakage of acid into the water heater tank (Fig X 10).
5. The jetter cups should be designed to prevent the liner mouths sealing tightly on to the base of the jetter cup and to allow the whole of the liner mouthpiece to be freely exposed to the cleaning solution (Fig X 11).
6. During washing the clusters should hang above the jetters so that the solutions are drawn upwards through the teatcup liners. This aids even flow through all four liners (Fig X 2).
7. A dimensionally constant air admission hole of 0.8 mm (1/32 in) in each claw prevents hold up of liquid in the jars. However this should be the only air admission point and it is essential that all joints are a satisfactory fit, and that there are no punctures in the rubber parts, in particular the short milk tubes, or leaks in valves.

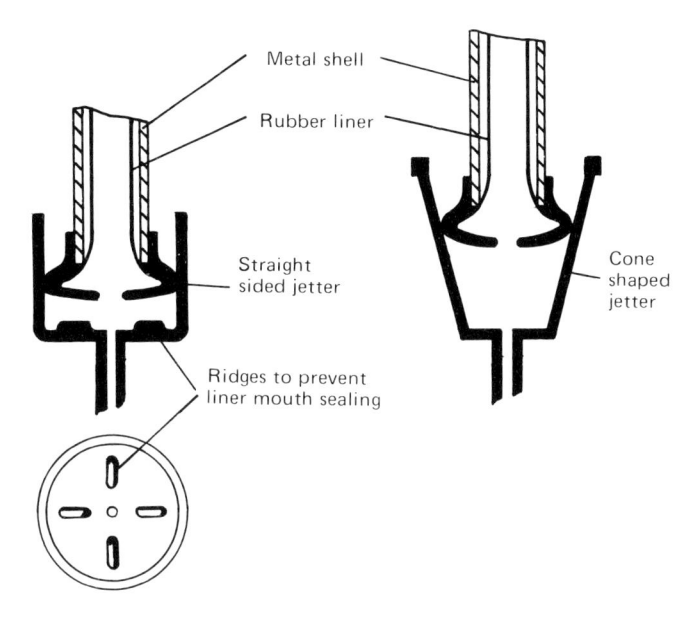

Fig X 11. Sections of jetters which have been designed to avoid compression of the liner mouthpiece against the base of the jetter cup, thus allowing cleaning of both internal and external surfaces.

The following faults have been found to be the cause of unsatisfactory plant rinse counts.

1. Unsuitable and badly designed components or layout (Figs X 6 and 7).
2. Insufficient water.
3. Water not up to temperature at the end of milking because of faulty time clock or thermostat.
4. Water flow times in excess of 6 min due to blockage by scale or incorrect restrictors.
5. Air leaks in components causing faulty distribution of hot water. The short milk tubes are particularly liable to cause trouble.
6. Imbalance between components due to incorrect restrictors or deterioration of rubber tubes resulting in kinking or flattening while under vacuum.
7. Components left out of circuit because of failure to open valves during washing.

Cleaning and Disinfection

Water requirements

The water requirements for ABW cleaning are 14 - 18 ℓ (3 - 4 gal) per unit at a minimum temperature of 96°C (205°F). The requirements for water heaters, construction, control and types, are covered in Chapter III.

Procedure for checking ABW cleaning

After installation the plant should be checked to ensure that the cleaning process is working correctly. This would normally be the responsibility of the installer. On all but the largest plants two persons are sufficient to carry out the complete check, one of whom records temperatures in the parlour and checks flowrate, while the other records the temperature in the dairy and records timings. The following equipment is required: a clip board, record sheets, stop watch, thermometer and temperature sensitive papers. These are prepared as follows: small pieces of Thermopapers (supplied by Wenz Sa RL) are cut and attached to transparent adhesive tape with the white side next to the adhesive. Each piece of tape has three Thermopapers, one changing colour at 71°C (160°F), one at 77°C (170°F) and one at 82°C (180°F). The colour change is from white to black (Fig X 12). A prepared adhesive tape is attached to each component to be checked; it is most important that the tape should be stuck firmly to a dry surface.

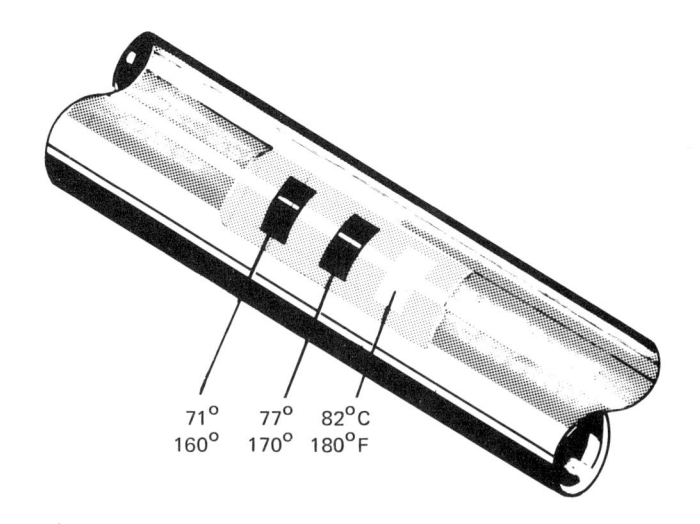

Fig X 12. The use of temperature sensitive papers for testing the surface temperature of milking machine components during cleaning. In the illustration 3 pieces of temperature sensitive paper are shown attached to a glass pipeline. After cleaning, the 2 lower temperature papers had changed colour from white to black, indicating that the surface had reached a temperature above 77°C but below 82°C.

1. Check the complete installation for air leaks and set the machine for cleaning (Fig X 9).
2. Check that the water temperature in the heater is 96°C (205°F) or above and record.
3. Attach a prepared adhesive tape to the following components: the boiling water inlet, each recorder jar, each claw, the transfer pipeline between the last unit and the receiver, the releaser milk pump and delivery pipeline.
4. Check that the vacuum level is satisfactory and the pulsation system is operating.
5. Check that all valves are in the washing position.
6. Add 1 ℓ (2 pints) acid to the acid container.
7. Start the cleaning process by opening the valve from the water heater and, at the same time, start the stop watch.
8. During the first minute of cleaning observe the flow of water to all units.
9. Record at one minute intervals the number of positive (black) temperature sensitive papers on each component.
10. Record the time that the acid container empties.
11. Record the time when water ceases to flow from the water heater.
12. If there is any doubt about the volume of water in the heater refill tank with cold water, pass it through the plant, collect and measure it.

The total flow time should be 5 - 6 min and the entry of the acid should take 2 - 3 min. All parts of the plant should reach 77°C (170°F) by the third minute. If these requirements are not met the necessary adjustment and corrections should be made, followed by a further check.

Periodic temperature checks at six-monthly intervals will ensure the continued efficiency of the system and reveal any faults that may otherwise go unnnoticed.

Additional treatment

Under certain conditions a brownish semi-transparent film may appear on the glass surfaces; it is particularly noticeable on the jars. This proteinaceous film can usually be readily removed by the following treatment with hypochlorite.

1. Set the machine for washing in the normal way.
2. Rinse out the acid container.
3. Add 500 ml (1 pint) of undiluted approved sodium hypochlorite solution to the acid container and make up to 1 ℓ (2 pints) with cold water.

4. Complete the normal washing routine.
5. At the end, thoroughly rinse the acid container to remove all traces of hypochlorite. *Care must be taken to prevent accidental mixing of acid and hypochlorite as dangerous fumes are given off.*

One such treatment with hypochlorite is usually sufficient to remove the film and it should not be necessary to treat a plant more frequently than once a month.

ABW cleaning for cowshed milking pipeline installations

The ABW system was originally designed for recorder machines and results from various investigations have been reported[12, 13]. The principle has been applied to cowshed milking pipeline installations[14]. They present their own special problems, but providing the machine is visibly clean with no apparent milk residues or scale and conditions are such that the temperature in all parts can be maintained above 77°C (170°F) for 2 min or more there is no reason why satisfactory bacteriological results should not be obtained.

LESS COMMON SYSTEMS

Immersion cleaning

This method was developed in 1956 for direct-to-can milking equipment which consists of a teatcup cluster, vacuum can lid and vacuum hose. Immersion cleaning is little used now because direct-to-can milking is almost obsolete. The equipment is cleaned and disinfected by immersion in cold 2 - 3% caustic soda solution containing a water softening additive, EDTA (section on Cleaning and Disinfecting Agents p.321). The process is cheap, simple and efficient: full details are given in Advisory Leaflet 496[15].

Caustic flooding

This system is normally only used for milking pipeline and recorder milking machines in portable bails where there is no provision for hot water. Caustic flooding is based on the principle of immersion cleaning. All the milk contact surfaces of the plant are flooded with a 3% solution of caustic soda plus EDTA for the period between milkings. A brief report[16] on the method concluded that while plant contamination tended to be greater than with circulation cleaning, nevertheless where the recommended routine was applied with due care milk could be produced which consistently satisfied all market requirements. The cost of cleaning, as might be expected, was considerably lower than for circulation cleaning. The daily operations were said to be easy and quick to apply but, with a recorder machine in particular, if the instructions for flooding twice daily and dismantling all the components at monthly intervals are followed the system is almost certainly more laborious than most circulation cleaning systems, bearing in mind also the preparations required before milking.

Clear and explicit instructions are issued by Hosier Equipment (Collingbourne) Ltd who developed and introduced the system.

Vacuum-pressure (V-P) system

A milking machine incorporating movement of milk from the recorder jar by compressed air is marketed by Weycroft-Macford Ltd and is briefly described in Chapter III. It can be cleaned and disinfected by an ABW system. Near-boiling water is moved by compressed air from the water heater along the compressed air line, where acid is entrained, into the recorder jars, out through a non-return valve and back through the milk transfer pipeline to the milk outlet. The water then passes to a manifold, to which the long milk tubes of the clusters are connected, and out to waste through the teatcups. The V-P system has the advantage that much higher temperatures are achieved than when using vacuum for circulating hot water. Jetters cannot conveniently be used to bring the clusters into the cleaning circuit because the system is under pressure and the teatcups may be forced out of the jetter cups. The non-return valve at the base of each recorder jar hinders effective cleaning of the outlet of the vessel and milk stone tends to build up at this point.

Reverse-flow cleaning

This in-place cleaning system was developed in New Zealand for milking pipeline machines with 17 or more units in parlours. The clusters do not have to be connected to jetters, which saves the cost of jetters and the labour of attaching and detaching them. Cleaning solutions are pumped to the receiver, pass into the milking pipeline and through the clusters to waste. In New Zealand the recommended cleaning procedure consists of five stages; the machine is rinsed first with cold water and then with a cold solution of non-ionic wetting agent. A hot detergent solution at 88°C (190°F) is followed by a hot water rinse at the same temperature, and the process is completed at the next milking by a pre-milking cold iodophor rinse. It requires a great deal of both cold and very hot water and, in the UK, would be costly.

CLEANING FARM BULK MILK TANKS

Refrigerated farm tanks cannot be cleaned and disinfected using hot solutions without increasing the refrigeration load and risk of damage to some components. The temperature of the inner surface of cold wall tanks cannot of

course be much increased by the application of hot solutions. All current cleaning routines consist of three stages.
1. Cold water rinse; this is the responsibility of the tanker driver.
2. Cold or warm (30 - 37°C; 86 - 98°F) detergent-disinfectant treatment.
3. Final cold clean water rinse.

Manual cleaning

The two methods involving brushing are unsuitable for larger tanks. They can be sprayed by hand but mechanical cleaning is preferable.

Brushing with solution. Prepare the solution by adding to the required volume of warm water (37°C; 98°F) the recommended quantity of approved detergent-disinfectant. Remove the dipstick and outlet plug from the rinsed tank and brush them in the solution. Close the tank outlet with the blanking nut. Using some of the solution (contained in a plastics or rubber bucket) and a long handled brush, scrub thoroughly all the interior surfaces of the tank, including the covers, underside of the bridge, agitator, thermometer and thermostat if these are of the immersion type. Remove the blanking nut and, as the solution runs to waste, brush the outlet and nut. Place the dipstick and plug in the tank and rinse the tank and its fittings by hosing with clean cold water. Allow the tank to drain completely, insert the plug and replace the blanking nut. The tank is then ready to receive milk.

Brushing with powder. Special detergent-disinfectant powders are available for this procedure which is also known as the sprinkle paste method. The manufacturer's instructions specify the quantity of powder appropriate to the size of the tank. To the wet surfaces of the rinsed tank apply the powder evenly; a container similar to a large pepperpot is convenient for this purpose. Using a long handled brush work the powder into a paste covering all the surfaces and fittings. After a contact time of at least 2 min rinse off the paste by hosing with clean cold water. Brush the outlet and blanking nut as the liquid runs to waste. Allow the tank to drain completely.

Spraying by hand. A hand operated spraying device, which entrains concentrated iodophor into the stream of water from a cold water hose, is required. The apparatus should be capable of delivering iodophor solution at the correct concentration throughout the spraying period.

After the tank has been rinsed, replace the blanking nut on the tank outlet and spray the tank thoroughly. The larger the tank the longer the spraying time required; this should be at least 5 min for a 4000 ℓ (900 gal) capacity tank. Brush the dipstick, plug and any areas showing film using iodophor solution. Remove the blanking nut and brush the outlet and nut as the solution runs to waste. Rinse the tank by hosing with clean cold water and allow to drain completely.

Mechanical cleaning

The equipment and the procedure are described in detail in Chapter XII.

General comments on cleaning bulk milk tanks

To date iodophors are used almost exclusively both in the mechanical and hand spraying equipments, one reason being that a liquid detergent-disinfectant is easily dispensed or entrained. The success of iodophor solutions in spray cleaning and disinfection of ice cold surfaces compared with the failure of liquid chlorine based detergent-disinfectants, for example, is not fully understood. The wetting agents in iodophors presumably assist in fat removal and their acidity counteracts the deposition of mineral salts. Under the extremely cold and therefore difficult conditions for both cleaning and disinfection without the aid of brushing a contact time of at least 10 min has been found necessary to achieve a reasonable degree of disinfection. This is probably because of the adverse effect of the low temperature on the disinfection process. It also permits the iodine to penetrate any imperceptible residues on the tank surface. After about a month of mechanical cleaning, film, made more apparent because it is stained yellow by the iodine, often develops on tank surfaces and therefore brushing at monthly intervals is required. Either of the manual brushing methods may be used. The success of cold disinfection for bulk tanks compared to its failure for recorder milking machines, for example, is attributed to the absence from bulk tanks of the types of joints and crevices present in milking machines.

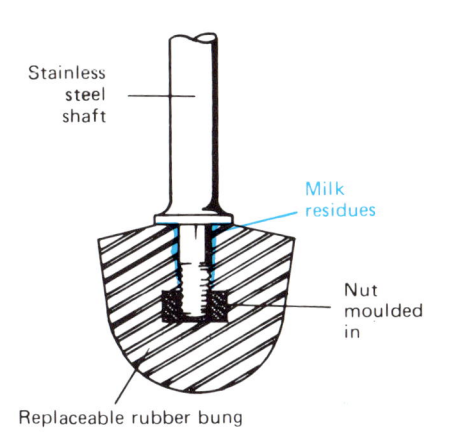

Fig X 13. Section of a bulk tank outlet plug, showing site of milk residues.

Outlet plugs sometimes become heavily contaminated with bacteria, although they appear to be clean, because of the unsatisfactory way in which the rubber bung is attached to the metal shaft (Fig X 13). Once a month the bung should be immersed in boiling water for 2 min; conduction of heat to the junction of the shaft and the bung where bacteria may accumulate effectively disinfects the plug. The trough under the tank outlet in which the plug, dipstick and outlet nut are immersed for cleaning (Chapter XII) should be cleaned if, as sometimes happens, a greasy film builds up on the sides of the trough.

Frequent inspection of the interior of the tank and its fittings is necessary. Also the operation of the mechanical equipment should be checked to ensure the iodophor is entrained and that, with fully automatic equipment, the cleaning cycle has been completed.

COSTS OF CLEANING AND DISINFECTION

It is of some interest to determine what the cost should be of different cleaning and disinfection systems, e.g. ABW and circulation cleaning. By making assumptions about herd size and milk yields the cost of cleaning as a percentage of the total cost of milk production can be arrived at. Some years ago a figure of 2%, which included labour, hot water and chemicals, was calculated for circulation cleaning of a 3-unit machine[17]. In practice both the actual cost and the percentage will vary from farm to farm, but such an estimate is a guide to the significance (or insignificance) of any savings which might be possible by using cheaper chemicals, less hot water or a different method of heating it.

In 1970 annual costs of cleaning materials and heating water for circulation and ABW cleaning of a 10-unit recorder milking machine were calculated[18]. Using electricity for heating, a saving of 45% could be achieved by using low cost as compared to high cost chemicals for circulation cleaning. It was further shown that ABW cleaning, also using electricity for heating, was cheaper than circulation cleaning using high cost chemicals. The cost advantages of using oil or gas instead of electricity for heating were considerable. The figures quoted in the paper will hardly be relevant today but they illustrate the scope for reducing costs. Up-to-date information on cleaning costs is available from the Agricultural Development and Advisory Service (ADAS).

TRENDS IN CLEANING AND DISINFECTION

Both circulation and ABW cleaning of milk installations require hot water. Water heating is energy expensive and the cost although a small part of the total costs of milk production is of concern to farmers. Cold in-place cleaning therefore seems attractive. However it has yet to be shown that twice daily cold in-place cleaning and disinfection is capable of keeping milking machines visibly clean for more than a few days except by using chemicals costing as much as heating water. Experimental work in progress suggests that the use of cold solutions of relatively cheap chemicals reinforced at weekly intervals by the use of very hot solutions will be found to maintain visible cleanliness and acceptable levels of bacteriological rinse counts. This would of course result in a saving in cost compared with current recommendations for both circulation and ABW cleaning systems. In the meantime there is already experimental evidence that cold hypochlorite solution can replace one of the two daily ABW treatments without detriment to visible cleanliness and bacterial counts in the machine provided the remaining ABW treatment is applied effectively.

Finding suitable chemicals and routines completely effective in the cold thus obviating the need for hot solutions is the ultimate aim of experimental work. The use of component designs better adapted to cold cleaning (rigid pipeline couplings in place of rubber sleeves for example) could be expected to make some contribution. However, the present successful combinations of relatively cheap forms of construction and methods of cleaning and disinfection using very hot water present a severe challenge to innovation. For the time being a water heater is still essential because the circulation of very hot solutions of suitable chemicals is the only means of removing any deposits or milk residues which might build up in machines which are cleaned in place. The alternative of dismantling a large recorder milking machine for manual cleaning is quite impracticable.

MONITORING AND ASSESSMENT OF CLEANING AND DISINFECTION

Means of assessing hygiene in production include inspection of equipment and methods, bacteriological rinses and swabs of milking equipment, and testing milk supplies. Frequent farm inspections and routine rinsing and swabbing of equipment are too costly for regularly monitoring hygiene. Current monitoring is by means of routine bacteriological tests and, less commonly, sediment tests on milk from individual farms but these tests are of limited value for measuring production hygiene.

TESTING MILK

Sediment tests

Tests to determine the amount of sediment (visible extraneous matter) in milk[19] are sometimes used, a test failure indicating that cows' teats were dirty when milked. However there is no certainty that milk which passes the test has come from cows with clean teats; most of the dirt may have been removed by filtering the milk on the farm. Sediment test results do not therefore reflect the care taken to prevent dirt and associated bacteria getting into milk.

Bacteriological tests

Some bacteriological tests, e.g. determination of the total bacterial count of the milk, are better than others for detecting unhygienic production of milk, in particular, refrigerated bulk tank milk. A dye reduction test, the 2 h resazurin test is the current Hygiene Test in England and Wales. However, the results correlate poorly with total bacterial counts and bulk tank milk very seldom fails the test even with bacterial counts greatly in excess of 100 000/ml, a level which is clear evidence of faulty production methods. The test is not therefore an effective means of monitoring hygiene. Nevertheless if a producer's milk consistently passes the current test it is reasonable to assume that his production methods are adequate to meet the buyer's requirements for hygienic quality of milk. In the future there may be changes in tests and standards. Most producers should have little difficulty in meeting a higher standard of hygienic quality should this be necessary.

Determination of the total bacterial count of the milk is increasingly used in other countries for quality control and payment schemes for farm refrigerated milk. Farmers can better relate unhygienic practices to high bacterial counts than to results of a dye reduction test. The total count does not distinguish between bacterial contamination from udder infections, dirty teats or milking equipment. However this is not a serious weakness provided that, in the event of a test failure, the producer is fully aware of the possible sources of trouble and can therefore take the appropriate action.

TESTING MILKING EQUIPMENT

Bacteriological rinses and swabs of milking equipment are the most sensitive means of determining the efficacy of disinfection. They are normally used only for advisory or experimental work.

If it should be necessary to evaluate a new method of disinfecting bulk tanks or to compare the effectiveness of different cleaning and disinfecting systems for milking machines then bacteriological rinses using a standardized technique[20] are the most satisfactory means available. Bacteriological rinses do not and should not be expected to recover from a tank or a milking machine all the bacteria remaining after a cleaning and disinfecting procedure. However under controlled experimental conditions a system consistently producing rinse counts of less than $5 \times 10^5/m^2$ (approximately $5 \times 10^4/ft^2$) of the milk contact surface of equipment could justifiably be regarded as more effective than one resulting in counts of $10^7 - 10^9/m^2$ ($10^6 - 10^8/ft^2$), although it may be possible to produce milk of acceptable quality using the latter system. When developing a new method it is advisable to aim for higher disinfectant efficiency, (i.e. lower rinse counts) than is strictly necessary because experience has shown that results are not always as good when new methods are used more widely under practical farm conditions.

CLEANING AND DISINFECTING AGENTS

WATER

A clean, wholesome supply of water is required in the farm dairy for rinsing equipment, preparing cleaning and disinfecting solutions, providing hot or near boiling water and for udder and teat washing. For these purposes the bacteriological safety and purity of the supply and its chemical quality are important. Equipment for heating water is described in Chapter III.

Bacteriological quality

The purity of a water supply direct from the mains is assured but bacterial contamination can occur in storage tanks which are not properly protected from rodents, birds, insects and dust; bacteria can also be introduced from dirty wash troughs, carrying buckets and hoses. Private water supplies are tested for bacteriological purity and if there should be any risk of pollution, suitable treatment usually with sodium hypochlorite is required before the water is acceptable for dairy purposes.

Warm water for hose washing udders and teats is usually supplied from a water tank controlled at about 37°C (98°F). Bacteria gaining access to an inadequately protected tank can multiply in the warm water and a heavy build up may occur. Such water has been implicated in outbreaks of mastitis caused by *Pseudomonas* and coliform bacteria.

Water from a small capacity (e.g. 10 ℓ; 2 gal) properly protected water heater, described in Chapter III, is unlikely to become heavily contaminated with bacteria. The rate of supply of warm water is adequate for a medium sized herd. The entrainment of disinfectant into the hose water, as well as being beneficial for washing teats, is a further assurance that the water is free from harmful bacteria.

Chemical quality

With a mains supply information on hardness or the presence of other dissolved salts may be obtained from the local water board; with a private supply having an analysis made is sometimes worthwhile particularly for very hard or very soft water. This information is useful in choosing detergents and in determining whether there are risks of corrosion, excessive scale in water heaters, or deposits on milking equipment.

DETERGENTS

Legally, all milking equipment must be cleaned and disinfected after each time it is used. Detergents, combined detergent-disinfectants and disinfectants are available for these purposes. Cleaning is a most important part of the whole process, and if milking equipment is not kept visually clean by the correct application of suitable cleaning solutions, chemical disinfectants are unlikely to be effective.

After use, milking equipment is soiled with residues of liquid milk and air dried films of milk consisting mainly of fat, protein and mineral matter. In addition soil from the environment of the cowshed or parlour, mud, dung, particles from bedding, dust and also milk residues, may be present on the outside of the machine. This soil should be removed before the normal cleaning routines are begun.

In time, teatcup liners made of natural rubber will absorb fat, and milk stone of variable composition but consisting mainly of tenacious mineral and protein residues can accumulate on all components.

The addition of suitable detergents to the water used for daily cleaning assists removal of residues by wetting the surface to be cleaned, displacing the milk soil, emulsifying fat, dissolving protein, holding soil in suspension, softening water and preventing redeposition during rinsing. With these properties a detergent solution, provided it is correctly used, will prevent build up of deposits over a long period of time. No one chemical has all these necessary properties and therefore proprietary detergents contain a blend of ingredients. The choice and proportion of different ingredients will be influenced by cost, adverse effects on hands or equipment and the method of use.

Ingredients of detergents

Manufacturers of dairy detergents provide a variety of formulations suitable for different methods of application, types of equipment and the chemical composition (or hardness) of the water supply. It is more convenient to use these proprietary products and it is not usually worthwhile for the user to obtain and mix the basic ingredients. Types of ingredients used include the following.

1. Inorganic alkalis, e.g. sodium carbonate (soda ash, washing soda), sodium bicarbonate, sodium hydroxide (caustic soda), tri-sodium phosphate and sodium silicates.
2. Surface active agents, e.g. anionic and non-ionic wetting agents.

3. Sequestering agents (water softeners), e.g. polyphosphates, ethylenediaminetetra-acetic acid (EDTA).
4. Acids, e.g. phosphoric, nitric and sulphamic acids.

Detergents for washing by hand consist largely of mild alkalis with small proportions of wetting agents and water softeners; or the formulations may be liquid and consist almost entirely of wetting agents. For in-place cleaning stronger alkalis are considered necessary to compensate for lack of brushing; water softening agents are usually included but high foaming wetting agents are omitted because excessive foam hinders effective circulation. All proprietary farm detergents as far as is known are compatible with sodium hypochlorite at normal use concentrations. The use of sodium carbonate at a concentration of 0.25% with the addition of about 0.05% of polyphosphate (Calgon; Albright & Wilson (Mfg) Ltd) has been found satisfactory for circulation cleaning. Where the water supply is exceptionally hard, extra polyphosphate is recommended.

Acids, sometimes containing wetting agents, are normally used only occasionally to prevent or remove build up of milk stone, but sulphamic and nitric acids are used regularly in the ABW cleaning system. Care is needed in the use of acids for descaling to avoid corrosion of aluminium, tinned and galvanized metal equipment, but there is less risk with stainless steel which has almost entirely replaced other metals in milking equipment in the UK. Rubber and most plastics materials may safely be treated.

Detergents may be formulated with disinfectants and sold as combined products; the basic detergent ingredients are the same as those described above and these products will be discussed in more detail later.

Principles of application

Within limits, increasing the time of contact, the temperature of application and the concentration of most detergents will improve the efficiency of the cleaning solution. However, removing residues such as those found on milking equipment after use should not normally require strong solutions at a high temperature for a long time. Solutions for cleaning equipment manually are effective when used at a temperature of 44 - 46°C (110 - 115°F) for 2 min with a detergent concentration of about 0.25%, assuming that all parts of the equipment are properly brushed. In the case of circulation cleaning, keeping all the surfaces, glass, metal and rubber, visibly clean should not require more than about 5 min contact time, a temperature in excess of 54°C (130°F) or a concentration of more than 0.5% of a well formulated detergent, assuming that all parts of the plant are adequately exposed to the flowing solution. For bacteriological reasons, however, which have already been explained, an initial temperature of 82°C (180°F) for the circulation solution, to ensure a temperature of at least 65°C (145°F) during circulation is officially recommended.

DISINFECTANTS

Chemical disinfectants are used for milking machines on the majority of dairy farms and steam sterilization is now obsolete.

Heat

The efficacy of disinfection by means of heat using steam or boiling water depends on raising the temperature of all parts of the equipment for a sufficient length of time to kill all except the most heat resistant microorganisms and bacterial spores.

Steam. Modern milking equipment is rarely if ever steamed because of the high capital cost of steam raising equipment, and the process does not conveniently fit into present day cleaning routines.

Boiling or near-boiling water. In the UK scalding with boiling water is acceptable under the milk regulations for disinfection of milking equipment, but until recently it has been regarded mainly as a method for use in an emergency such as having a consignment of milk rejected. Boiling water is however used effectively for the ABW process of in-place cleaning of milking machines; ordinary circulation cleaning requires very hot water and appears to be a combination of heat and chemical disinfection.

Chemicals

These may be disinfectants alone or disinfectants combined with detergents. Some form of control is generally considered desirable to prevent the use of chemicals which may taint milk or give rise to harmful residues in milk. Tests to ensure that products are effective disinfectants under normal conditions of use protect the user against worthless materials.

In the UK each product submitted for approval is examined, and tested if necessary, to ensure bactericidal efficiency. Only those disinfectants and combined detergent-disinfectants (chemical agents) approved under the appropriate regulations may be used as alternatives to boiling water or steam. A list of all these chemical agents under their trade names, classified according to type, and giving the approved use concentration is issued periodically by the MAAF for England and Wales; the list is also applicable to Northern Ireland. Information on products approved for use in Scotland is available from the Department of Agriculture and Fisheries for Scotland.

Disinfectants alone (sodium hypochlorite). For practical purposes, on dairy farms in the UK, sodium hypochlorite is the only available disinfectant for milking equipment. Other types of chemical disinfectant, e.g. quaternary ammonium compounds (QAC) have been approved but are used only in processing dairies. This is because QAC may be inactivated by ingredients of detergents used in farm dairies. QAC should therefore be used only for disinfection of previously cleaned equipment – not a common practice on dairy farms in the UK. Approved solutions of sodium hypochlorite are sold under various trade names and must conform to certain requirements of the MAAF. These include the following. At the date of despatch by the manufacturer the solution must contain 9 - 12% by weight of available chlorine, between 0.7 and 1.5% by weight of sodium chlorate and not more than 2% by weight of free caustic alkali. The sodium chlorate is required as a tracer for detecting the addition of hypochlorite solution to milk.

The addition of sodium hypochlorite to the detergent solution so that cleaning and disinfection are combined in one operation has been officially recommended for milking equipment since 1943 when emergency regulations permitted for the first time the use of approved solutions of sodium hypochlorite for disinfecting all dairy equipment. Manufacturers' instructions for use follow the official recommendations to add to the detergent solution approved hyphoclorite at a concentration of 0.25% (4 fl oz/10 gal solution) giving between 250 and 300 ppm of available chlorine. Approved hyphochlorite at the rate of 25 ml to 40 ℓ (1 fl oz/10 gal) of water may also be used for treating the final rinse water or for a pre-milking hypochlorite rinse.

In processing dairies, if approved hypochlorite or a QAC is used for disinfection of plant, the solution is usually applied as a separate treatment. After the plant has been cleaned and just before it is to be used again, the disinfectant solution is circulated and is followed by a rinse with clean water. This method of application has never been generally advocated for milking equipment in the UK where milk producers have always been accustomed to completing the cleaning and disinfection process soon after milking. A disinfectant treatment just before milking which would need to be followed by a clean water rinse would not readily be accepted, although there is some evidence that it is beneficial. In North America and some other parts of the world, cleaning and disinfection of milking machines are separate operations and disinfection is left until just before the next milking; hypochlorites, QAC, iodophors and special acids may all be used for this purpose. Provided the equipment is well drained there is no legal obligation to rinse with clean water before using the equipment again. This is mainly because many farm water supplies in these countries are heavily contaminated with bacteria which would contaminate the equipment.

Combined detergent-disinfectants. The practice of using a detergent-hypochlorite solution and thus combining cleaning and disinfection led to the marketing of combined materials. A combined material is more convenient to handle and measure than are two separate products, and manufacturers provide different types of formulation suitable for manual use, circulation cleaning and for bulk tanks.

The concentrations at which approved products are to be used for these purposes are shown on the approved label.

The types of disinfectant used in approved detergent-disinfectants marketed for milking equipment include:

1. chlorine-releasing chemicals (hypochlorite, salts of di- and tri-chloroisocyanuric acid, dichlorodimethyl-hydantoin); products suitable for manual and in-place cleaning and for the sprinkle paste method for bulk tanks are available.
2. quaternary ammonium compounds (QAC); these preparations are suitable for manual cleaning; foam production makes them unsuitable for circulation cleaning.
3. iodophors; these preparations are all liquids formulated with acids and wetting agents. Generally they are suitable for washing by hand; most are unsuitable for circulation cleaning because the wetting agent foams during circulation; 'low foam' preparations are available, however, for this purpose. Both types are suitable for spray cleaning of bulk tanks. It is inadvisable to use iodophors at temperatures above $50°C$ ($120°F$) because of vapourization of the iodine.

The detergents used in the combined products are selected to be compatible with the disinfectant and suitable for the purposes for which the product is recommended. Treatment of milking utensils with solutions of these products must be followed by a clean water rinse.

SELECTION OF SUITABLE CLEANING AND DISINFECTING AGENTS

With the large number of approved products available the user may have difficulty in deciding not only which type of product but also which of the many brands to choose. In the first place a decision must be taken either to use two products, approved sodium hypochlorite and a separate detergent, or to use a combined detergent-disinfectant. Hypochlorite is relatively cheap and basically there is little difference between approved brands. It is often convenient to have a supply of hypochlorite available for adding to the final rinse water and for udder washing. Detergents alone are usually cheaper than combined products and using a detergent and hypochlorite can be the most economical choice. There are wide variations in costs and the user is well advised to try the cheaper brands of hypochlorite and of detergent suitable for the method of cleaning and the hardness of the water. It should be borne in mind that detergents suitable for hard water are usually more expensive because they contain a higher proportion of water softening chemicals, which are expensive.

Combined products contain the correct proportion of detergent and disinfectant for normal cleaning routines; only one material has to be measured out and there is no risk of omitting the disinfectant; thus any extra cost involved may be worthwhile. Costs of different types and brands vary widely. The quantity recommended for use, varying from 0.25 to 0.625% (4 to 10 oz in 10 gal of solution), should be taken into account when working out costs. This and the personal preference of the user are the most important factors in selection for washing by hand.

A good guide to a suitable detergent or detergent-disinfectant for circulation cleaning is the cheapest brand that will keep the pipelines, clusters and jars looking clean over a period of several weeks or months thus avoiding the need for frequent treatment with phosphoric acid or other proprietary descaling acid.

In many instances it is wise to use the product recommended by the milking machine manufacturer. If trouble should be experienced his advice can then be sought to correct the difficulty. Otherwise, if a detergent or detergent-disinfectant is apparently unsatisfactory, the user should get in touch with the manufacturer of the product concerned.

All approved products used, at the concentration shown on the label of the container, should give satisfactory results if used correctly on equipment in good condition. There are exceptions; for example, an unusually hard water supply may be the cause of rapid build up of deposits or a faulty in-place cleaning circuit may prevent adequate contact of the solution with parts of the plant.

STORAGE AND HANDLING OF CHEMICALS

Powders should be stored so that they are kept dry. Hypochlorite and combined detergent-disinfectants should be stored in the original container. Hypochlorite and hypochlorite-based liquid detergent-disinfectants lose strength more rapidly when exposed to light and at higher temperatures; they should therefore be stored in a cool place and in the dark if the container is not opaque. *All chemicals should be kept out of reach of children and animals.*

Approved chemical agents must not be used for disinfecting milking equipment after the expiry date indicated on the label. Any out-of-date material can be used for general cleaning.

All concentrated chemicals should be handled with care, and if they accidently come in contact with the skin they should be washed off immediately with plenty of cold water. Labels of approved chemicals carry suitable warnings. *Goggles and rubber gloves must be worn when handling caustic soda and concentrated acids.*

Except for adding hypochlorite to detergent solutions for preparing combined detergent-hypochlorite solutions, it may be dangerous to mix different chemicals; for example chlorine gas, which is poisonous, may be given off if hypochlorite is added to an acid. It is advisable never to mix any approved chemical agent with another as they may react chemically and their disinfectant properties will then be nullified.

LEGAL REQUIREMENTS

As well as ensuring that his milk is bacteriologically acceptable to the buyer, the producer must comply with any relevant milk regulations. The UK regulations impose certain obligations and restrictions on manufacturers of milking equipment and cleaning materials, as well as on the milk producer, and these requirements must be taken into account in all cleaning and disinfecting procedures. The legal requirements may change from time to time to take advantage of developments and improvements so that some processes not strictly within the letter of the law may, however, be acceptable in certain circumstances.

Regulations for England and Wales[21] differ slightly from those for Northern Ireland[22]. There are no equivalent regulations in Scotland, where each local authority prescribes its own byelaws based on Model Dairy Byelaws[23].

Briefly, the regulations may be interpreted as requiring that:

1. dirty udders and teats must be washed or otherwise cleaned;
2. all equipment must be constructed so that it may be readily cleaned or where necessary easily dismantled for cleaning;
3. all components, including rubber, must be maintained in good condition free from deposits and corrosion;
4. before use the equipment must be in a state of thorough cleanliness; (This state is not, however, defined and as well as being visibly clean it is generally taken to mean satisfactory bacterial counts on rinses and swabs.)
5. after use the equipment must be rinsed as soon as possible, then thoroughly washed and disinfected by scalding with boiling water or steam, or by effective treatment with an approved chemical agent (disinfectant or detergent-disinfectant);
6. only an approved chemical agent may be used as an alternative to boiling water or steam;
7. all traces of any detergent and disinfectant must be rinsed off with clean water before the equipment is used again for milk. (However, it is permissible to add approved sodium hypochlorite to the final rinse water, and in some cases where there is a risk of the water being contaminated there may be an official requirement to do so.)

The Scottish Model Dairy Byelaws[23] specify a temperature of 180°F (82°C) for steam or flowing water, but other methods of using heat or chemicals, and the time and temperature of application, are not defined. The concentration of disinfectant to be used is not specified except in the Model Dairy Byelaws. Accepted practices for steam treatment have been detailed[24] but scalding with boiling water is open to various interpretations, for example; complete immersion in water at 185°F (85°C) for two minutes or not less than 170°F (77°C) for ten minutes; utensils and rubber parts immersed for one minute in water as near boiling as possible but not less than 185°F (85°C); freshly boiling water applied to all milk contact surfaces until they are too hot to touch; hot water, 5 - 10 (1 - 2 gal)/cluster, at 160 - 170°F (71 - 77°C) with alkaline detergent used for flush washing, provided this is followed by an equal amount of hot water rinse at the same temperature[24].

The terms of sale for milk sold by milk producers to the Milk Marketing Board of England and Wales include the following requirements related to hygiene.

The producer warrants the milk to be pure new milk sweet clean and marketable with all its cream and without the addition of any preservative, and to be produced and sold in accordance with the statutory Milk and Dairies Regulations. The producer is not legally required to filter or strain milk before it leaves the farm.

The producer of bulk milk, if the vat (i.e. bulk milk tank) is refrigerated, is required to cool it to and maintain it at a temperature not higher than 40°F (4.5°C). If the farm vat is insulated but not refrigerated the milk is to be cooled to a temperature not higher than 38°F (3.3°C) and immediately be put in the vat.

The producer can obtain further advice on methods of cleaning and disinfection and on problems concerning bacteriological quality of milk from the local Dairy Husbandry Advisory Officer of ADAS and the regional office of the particular Milk Marketing Board concerned.

REFERENCES

(1)   Cousins, C.M. (1972) *Journal of the Society of Dairy Technology*, **25**, 200.

(2)   Druce, R.G. & Thomas, S.B. (1972) *Journal of Applied Bacteriology*, **35**, 253.

(3)   McKinnon, C.H., Cousins, C.M. & Fulford, R.J. (1975) *Report, National Institute for Research in Dairying, 1973-74*, p.98.

(4)   Dawkins, J. & Slade, J. (1974) *Brief Communications International Dairy Congress, 19, New Delhi*, **IE**, 4.

(5)   Ministry of Agriculture, Fisheries and Food (1972) *Advisory Leaflet*, Hand Cleaning and Sterilization of Farm Dairy Equipment Using Chemicals, No. 422.

(6)   Ministry of Agriculture, Fisheries and Food (1974) *Short Term Leaflet* Circulation Cleaning, No. 26.

(7)   Ministry of Agriculture, Fisheries and Food (1968) *Short Term Leaflet*, Cleaning and Sterilizing Farm Bulk Milk Storage Tanks, No. 17.

(8)   British Standards Institution (1975) *Recommendations for Cleaning and Sterilization of Pipeline Milking Machine Installations*, BS 5226.

(9)   British Standards Institution. (1976) *Glossary of Terms Relating to Disinfectants*. BS 5283.

(10) British Standards Institution. (1970) *Recommendations for the use of detergents in the dairying industry*.

(11) McKinnon, C.H. (1966) *Unpublished data*.

(12) Jeffrey, D.C. & Hunter, A.C. (1970) *Journal of the Society of Dairy Technology*, **23**, 70.

(13) McKinnon, C.H., Kiddle, M. & Cousins, C.M. (1970) *International Dairy Congress, 18, Sydney*, **1E**, 649.

(14) Hunter, A.C. & Jeffrey, D.C. (1970) *Journal of the Society of Dairy Technology*, **23**, 75.

(15) Ministry of Agriculture, Fisheries and Food (1971) *Advisory Leaflet*, Caustic Soda Immersion Cleaning, No. 496.

(16) Egdell, J.W. & Widdas, D.R. (1964) *NAAS Quarterly Review*, No. 63, 132.

(17) Thiel, C.C. (1962) *Journal of the Society of Dairy Technology*, **15**, 94.

(18) Parry, D.R., Williams, D.F. & Middleton, M.S. (1970) *International Dairy Congress, 18, Sydney*, **IE**, 650.

(19) British Standards Institution. (1973) *Method for the rapid determination of sediment in milk by filtration*, BS 4938.

(20) British Standards Institution. (1968) *Methods of microbiological examination for dairy purposes*, BS 4285.

(21) Statutory Instrument (1959) No. 277 Food & Drugs. *The Milk and Dairies (General) Regulations, 1959.* London: HMSO.

(22) Statutory Rules & Orders for Northern Ireland 1963, No. 44 *Milk Regulations.* Belfast: HMSO.

(23) Department of Health for Scotland (1961) *Model Dairy Byelaws.* Edinburgh: HMSO.

(24) Ministry of Agriculture, Fisheries and Food (1959) *Bulletin Machine Milking*, No. 177. London: HMSO.

CHAPTER XI

RUBBER AND THE MILKING MACHINE

J G Anderson and L J Coombs

Avon Rubber Co. Ltd., Melksham, Wiltshire

The rubber parts of a milking machine are vital to its successful operation. This chapter is intended primarily to explain the functions and properties of the different rubbers used for components but it is necessary first to give a short description of the chemistry and technology of rubber.

The word rubber is one of the most loosely used words in the English language. Most people describe anything made from rubber simply as 'rubber', when strictly they are referring to compounded and vulcanized rubber, which may be of natural or synthetic origin. The word is used to designate any product which is capable of considerable extension with a snappy return to the original length.

British Standard 3558: 1968, *Glossary of Terms used in the Rubber Industry,* defines rubber, natural and synthetic, as follows.

    Rubber. Macromolecular material which has, or can be given properties of:

        (1) at room temperature returning rapidly to the approximate shape from which it has been substantially distorted by a weak stress;

    and (2) not being easily remoulded to a permanent shape by the application of heat and pressure.

Note: This term is also applied to articles made from rubber.
Natural rubber — rubber formed in a living plant.
Synthetic rubber — rubber other than natural rubber.

This description enables a distinction to be drawn between rubber and flexible plastics, such as plasticized PVC. The latter, unlike rubber, can be remoulded by the application of heat and pressure. The relative importance of natural and synthetic rubbers can be shown by the world consumption in 1975. A total of 10 402 000 tons of all types of rubber was consumed of which 68% was synthetic.

NATURE AND PROPERTIES OF RUBBER

NATURAL RUBBER

The main source of natural rubber is the tree *Hevea brasiliensis,* indigenous to the Amazon Valley. Until the end of the last century the best natural rubber came from Brazil but other types of rubber were obtained from trees and vines of various botanical species in Africa.

In 1876 seeds of the *Hevea* plant were taken from Brazil to Kew Gardens and the young plants grown from these were sent out to Singapore and Ceylon. This was the start of the rubber plantation industry in the Far East. Rubber trees can be grown in any country where the temperature range is 21 - 35°C (70 - 95°F) and the annual rainfall is at least 2500 mm (100 in). The rate of growth of the rubber tree and the yield of rubber from it is, however, greatly influenced by its environment. Oddly enough it has never been possible to cultivate *Hevea* anywhere in South America because of prevalence of leaf disease. Serious attempts were made in the 1930s by Henry Ford to develop rubber plantations in Brazil but they all failed.

The first operation in starting a plantation is to clear the jungle thoroughly and drain the land. The young seedlings are then planted. After approximately seven years the rubber tree has grown to a size at which it is ready for tapping. This is done by carefully cutting the bark with a notched knife and collecting the rubber latex in small cups as it slowly exudes.

On larger plantations the latex is transported in tankers to a factory where it is first diluted from its original rubber content of about 40% to a standard rubber content of about 20%. It is then coagulated with acid and formed into a sheet on a two roll mill. At this stage the higher grades of rubber are sprayed with water jets to remove sand and bark. The sheets are finally dried in hot air or with wood smoke and then packed in bales for shipment. Nearly half of all natural rubber is produced by smallholders who allow the latex to coagulate naturally into lumps or slabs. This is sold to central processors who wash and dry the rubbers to produce lower grade brown crepes.

New methods of preparing and packing natural rubber are now replacing the methods described above. In these the rubber coagulum, from fresh latex or cup lump, is mechanically comminuted or crumbed, dried in hot air and then packed under pressure into small standard sized 75 lb bales. The bales are wrapped in polythene film and stacked on standard 1 ton pallets for despatch. This new type rubber is graded to a technical specification which covers cleanliness and uniformity. It is known as Standard Malaysian Rubber (SMR) or Standard Indonesian Rubber (SIR).

In recent years the rubber growing industry has developed varieties of trees which are capable of yielding rubber at very much higher rates than previously. The annual yield of rubber from the earlier unselected seedlings was about 400 kg/ha (350 lb/acre) but the new strains give about five times this yield. Chemically, natural rubber is cis-polyisoprene which can be shown as follows:

$$\cdots - CH_2 - \underset{/}{\overset{H}{\diagdown}} C = C \underset{\diagdown}{\overset{CH_3}{\diagup}} CH_2 - CH_2 - \underset{/}{\overset{H}{\diagdown}} C = C \underset{\diagdown}{\overset{CH_3}{\diagup}} CH_2 - \cdots$$

The molecules of isoprene in the rubber form extremely long chains containing several thousand carbon atoms. These chains have a wide variety of irregular shapes, the most probable being that of a coiled up mass. In the case of unvulcanized rubber there are no strong bonds between the chains and in consequence, when the rubber is stretched, the chains are easily separated and do not return to their original form on release of the stretching force.

When rubber is vulcanized with sulphur or other agents the chains are tied together chemically at a limited number of points, known as cross links, to form a three dimensional network. In normal soft vulcanized rubber between 1 and 2% of the carbon atoms in the main chain are cross linked. When such a rubber is stretched the coiled up chains tend to straighten out but are prevented from slipping over each other by the cross links. On release of the tension they return to their original coiled up form. If the process of cross linking is taken to the extreme, it is possible to form so many cross links between chains that practically no freedom of movement remains. Rubber in this form is called ebonite which is hard and comparatively inextensible.

SYNTHETIC RUBBER

By definition, synthetic rubber is any rubber other than natural rubber. Since there are many entirely different types of synthetic rubber commercially available today, the term requires qualification. Some synthetic rubbers are very similar to the natural product in all respects, whilst others differ widely in chemical and physical properties. They are usually divided into two main groups, general purpose and special purpose rubbers.

General purpose rubbers

General purpose synthetic rubbers are rubbers which have been developed primarily to replace the natural material wholly or partially in tyres and other products which use large quantities of rubber. They are in competition with natural rubber and normally are cheaper. The main types of general purpose synthetic rubbers available commercially are as follows. All of them can be readily blended with each other and also with natural rubber to give optimum properties for particular applications and to reduce costs.

Butadiene styrene copolymer. This material is generally known as SBR (styrene butadiene rubber). It was originally developed in Germany in the 1930s under the name Buna S and its production was rapidly expanded in the USA in 1942 when supplies of natural rubber were curtailed. It was then given the name GR-S (Government rubber styrene). The original GR-S was inferior to natural rubber in every respect but improvements in the manufacturing process in the last decade make present day SBR superior to natural rubber in certain respects, particularly resistance to abrasion. For this reason SBR is used very widely in tyres, and is now manufactured in nearly every country in the world. Small percentages of SBR rubbers are often blended with natural rubber for teatcup liners to improve resistance to heat and oxidation. For milk tubing SBR is commonly blended with natural rubber to facilitate manufacture and reduce costs, at the same time improving oxidation resistance.

Polyisoprene rubber (IR). Although many attempts have been made in the past to produce a synthetic polyisoprene with good rubber like properties, they have failed because of inability to control the molecular configuration during polymerisation. Now by using alkali metal or Ziegler catalysts it is possible to produce a polyisoprene with 96-98% cis configuration which is almost identical with natural rubber. Such rubbers are commercially available today and compete with natural rubber on price and quality. They can be used as a direct replacement for natural rubber in milking machine rubbers without any significant change in properties of the end product.

Polybutadiene rubber (BR). By a similar polymerization process a very good synthetic rubber can be made from butadiene, which is more readily available than isoprene. BR is used primarily in tyres because of its superior abrasion and flex crack resistance and it is therefore manufactured throughout the world.

There are no technical or economic reasons for using polybutadiene rubbers for milking machine components.

Special purpose rubbers

This group includes all those synthetic rubbers which have special properties not found in natural rubber or general purpose synthetic rubbers, e.g. resistance to oils and solvents, high temperature, and ozone. They are more expensive than natural rubber and are therefore used only in products which require their special properties. There are very many types available today but the following are most widely used.

Neoprene (chloroprene). This was the first synthetic rubber to be produced in the USA on a commercial scale, in 1931, and is therefore one of the oldest synthetic rubbers. It is manufactured and used widely throughout the world. It possesses good resistance to heat, oils and solvents and excellent resistance to ozone attack and all forms of weathering. It is generally inferior to natural rubber in resistance to cutting and tearing and tends to be less resistant to water absorption than natural rubber. Neoprene has been used for teatcup liners and milk tubing because of its good resistance to oxidation and improved fat resistance compared with natural rubber.

Nitrile rubber. Earlier known as Buna N or Perbunan, this is a copolymer of butadiene and acrylonitrile. It was first made in Germany in the early 1930s and is now produced widely in Europe and the USA under a variety of trade names. It is extremely resistant to swelling by fats, oils and most solvents. The percentage of acrylonitrile monomer can be varied between 25 and 50%, the higher ratio giving maximum oil and solvent resistance but at the same time less resilience. Nitrile rubbers, on their own or blended with natural rubber or SBR, are now widely used in the manufacture of teatcup liners because of their resistance to animal fats. They cost almost twice as much as natural rubber liners.

Butyl rubber. This rubber is a copolymer of isobutylene and isoprene, the latter being present in small amounts between 1.5 and 3%. It has remarkably good resistance to heat, ozone and all forms of weathering. It was developed originally because it is nearly ten times less permeable to air than natural rubber. This, with its good heat resistance, led to its general use in inner tubes for vehicle tyres. Butyl rubber is not particularly resistant to mineral oils, but it is resistant to animal and vegetable oils. It is a low resilience or lazy rubber with high hysteresis properties and can be used to make an excellent long life teatcup liner providing it is not allowed to come in contact with chlorine. The oxidative effect of chlorine, or hypochlorites, causes the surface of butyl rubber to go sticky, a weakness which has prevented its general use for milking machine components.

Silicone rubber. The silicone rubbers form a large family of which the simplest example is polymethyl silicone:

$$\left(\begin{array}{c} \mathrm{CH_3} \\ | \\ -\mathrm{Si} - \mathrm{O} - \\ | \\ \mathrm{CH_3} \end{array} \right)_n$$

These rubbers are characterized by their resistance to extremes of temperature, approximately −90 to 300°C (−130 to 550°F). Because of their very high price, about twenty times that of natural rubber, their use is largely limited to the aircraft and other specialized industries. Milk liners made from a suitable grade of silicone rubber can function satisfactorily and resist all forms of deterioration over a longer period of time than any other rubber. The excessively high cost of such liners has limited their use.

Hypalon. This rubber, which is chlorosulphonic polyethylene, possesses outstanding resistance to ozone and most chemicals. It does not have outstanding resistance to oils and fats and since it is more expensive than nitrile or neoprene rubbers it is not used for milking machine components.

Ethylene propylene rubbers (EPDM). These have recently been introduced on a full commercial scale. They have outstanding resistance to heat ageing and weathering and are therefore used for milk and air tubing. They have no special resistance to fats and are therefore not used in teatcup liners.

Polyurethane rubbers. These are formed by the reaction of long chain polyesters with a di-isocyanate to produce a very long molecule which is subsequently cross linked with an amine or diol. They are characterized by having a very high tensile strength and resistance to abrasion, many times better than that of natural rubber. They are also resistant to ozone and oils.

MANUFACTURE OF RUBBER ARTICLES

RUBBER COMPOUNDING

In their raw form natural and synthetic rubbers are not used in finished products, except for crepe soles and rubber solutions. They have to be mixed with other ingredients to make a rubber compound which in simple form contains sulphur, zinc oxide and accelerators for vulcanization, reinforcing agents to increase stiffness and toughness, plasticizers to assist mixing, extruding and moulding, and antioxidants to minimize oxidative degradation.

The use of rubber in its many applications is based on the discovery of Hancock in 1838 that when rubber is mixed with sulphur and heated it changes from a plastic to an elastic material. This process, known as vulcanization or curing, is applied to all natural and synthetic rubbers. As mentioned earlier in this chapter, it consists of forming cross linkages between the long rubber molecules. Ingredients other than sulphur can be used for cross linking rubbers, but sulphur vulcanization is still the most widely used with all rubbers, with the exception of neoprene where metallic oxides form the cross links. The amount of sulphur required for vulcanization is usually between 1 and 3% by weight for natural rubber and rather less for synthetic rubbers.

The reaction between rubber and sulphur is accelerated by the addition of metallic oxides, particularly zinc oxide, and certain classes of organic compounds, particularly thiazoles, thiurams, dithiocarbamates and certain amines. In addition to reducing the time required for vulcanization these accelerators develop stronger sulphur cross links, so producing vulcanized rubber with higher tensile strength, stiffness and resilience, and lower tension set and swelling in oils and fats. The mechanism of vulcanization accelerators is extremely complex but it is generally agreed that the zinc oxide forms a complex with the organic compound and sulphur which is highly reactive. The presence of a fatty acid is highly desirable to facilitate the reaction by the formation of a zinc salt. Some fatty acid is found in all natural rubber and some synthetic rubbers but it is usual to add 1 - 2% of stearic acid to most rubber compounds.

A compound vulcanized with suphur alone will be very soft and highly extensible, like elastic thread. It can be stiffened or made harder by the addition of mineral fillers such as whiting or clay, or carbon black. If the particle size of the filler is sufficiently small it will key into the molecular network of the rubber and reinforce it. For white or coloured natural rubbers, reinforcement can be obtained with zinc oxide, precipitated calcium carbonate, finely divided clays and other materials. In a red natural rubber teatcup liner compound there is usually about 15 - 25% of such fillers.

Carbon black is widely used to reinforce natural and synthetic rubbers. It is made by controlled combustion of mineral oils and natural gas, and can be produced in a variety of types with particle size varying from 5 to 500 nm (1 nm = 10^{-9} m or 4×10^{-8} in). The smaller the particle size the greater the reinforcing effect. Those with the smallest particle size are known as reinforcing black, with intermediate particle size, semi-reinforcing black, and with large particle size, non-reinforcing black. Nitrile and SBR synthetic rubbers must be reinforced with carbon black to develop their properties adequately but natural rubber can be reinforced very effectively with other fine fillers. Thus milking machine components made from synthetic rubber are always black, whereas those made from natural rubber may be black or coloured. The mechanism of the reinforcement of rubber by carbon black and other fine fillers has long been the study of many workers. Most consider that part of the carbon black in a rubber mix is in the form of chain aggregates and part is a rubber-carbon gel in which particles of rubber surround individual particles of carbon black. Plasticizers are added to assist in the dispersion of filler during mixing and to control the viscosity of the compound for calendering, extruding and moulding. These amount to about 1 - 5% for natural rubber, usually a mineral oil, but much larger quantities are required for synthetic rubber. For nitrile rubbers it is preferable to use plasticizers of the ester type.

Antioxidants are added to protect rubber against oxidation. These are usually fairly complex organic materials and very often amines or substitutes phenols. Certain waxes, such as paraffin or microcrystalline wax, are added to rubbers to protect them against ozone attack. This is particularly important in the case of teatcup liners and milk tubing.

The following are typical formulations in parts by weight for teatcup liners:

Black natural rubber

No. 1 grade rubber	100
Semi-reinforcing black	20
Zinc oxide	5
Stearic acid	2
Mineral oil	2
Paraffin wax	1.5
Antioxidant	1.0
Accelerator	0.75
Sulphur	2.50

Nitrile rubber

Nitrile rubber (27% nitrile)	100
Semi-reinforcing black	50
Plasticizer	30
Stearic acid	1
Zinc oxide	5
Antioxidant	1
Sulphur	2
Accelerator	1
Microcrystalline wax	1

MANUFACTURING PROCESSES

Mixing

The various ingredients of a rubber compound may be mixed on a two roll mill in which one roll revolves slightly faster than the other. This produces a shearing action at the roll nip and forces the powders into the plastic rubber. Alternatively a mixer having two internal rotors may be used. Rubber and compounds are fed in through a hopper at the top of the mixer and the mixed batch is discharged through a door at the bottom. The rate of mixing is much quicker than on a mill.

Extruding

Rubber tubing is produced on a screw type extruding machine. To make tubing in long lengths the extrusion is coiled in a circular tray and dusted inside and out with talc to prevent sticking. The tray is then placed in an autoclave which is filled with steam at a pressure of about 345 kN/m^2 (50 lb/in^2). After about 30 min the rubber is fully vulcanized.

Fig XI 1. Extruding straight liners. Operator on right is forcing the tube onto a mandrel prior to vulcanization. Avon Rubber Co. Ltd.

Fig XI 2. Moulding teatcup liners. Operator is removing liners from core pins after vulcanization. Avon Rubber Co. Ltd.

For short length tubing and straight liners, which required more accurate dimensions in the finished product, the extrusion is forced on to a mandrel and wrapped with cloth before curing (Fig XI 1). This is why most straight liners have a cloth marked finish on their outer surface.

Moulding

The common method of vulcanizing rubber is by moulding under pressure in a heated mould (Fig XI 2). Moulding temperatures are usually $138 - 177°C$ ($280 - 350°F$), and heating may be by steam or electricity. Hydraulic pressure is employed for closing the mould using a pressure of $7.5 - 15$ MN/m^2 (½ - 1 ton/in^2).

Calendering

This process is used for making thin sheets or rubber, usually in long lengths. Calenders normally have three steel rolls in vertical line. Hot vulcanized rubber is squeezed between the top pair of rolls, which are heated, to form a continuous sheet which is then passed round the cold bottom roll where it is cooled. The sheet is then wound into a spool with an interleaving cloth or film to prevent it sticking to itself. Fig XI 3 illustrates the process. After calendering the sheet rubber is rewound onto a large steel drum with an interleaving cloth under tension, and then vulcanized in a pressurized steam vessel. The combined pressure of the cloth and the steam consolidates the rubber and prevents porosity. Sheeting made in this way always has a cloth marked surface. Thick sheeting is usually vulcanized in a press and exhibits a smooth surface.

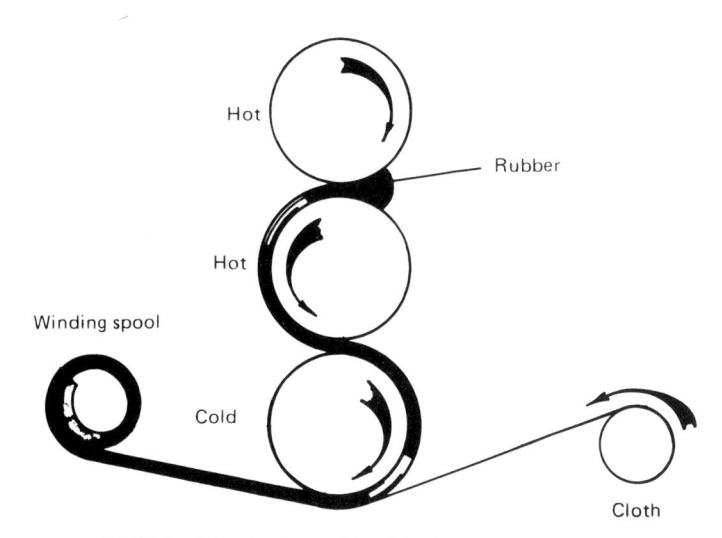

Fig XI 3. Calender for making thin sheets of rubber.

340   *Rubber*

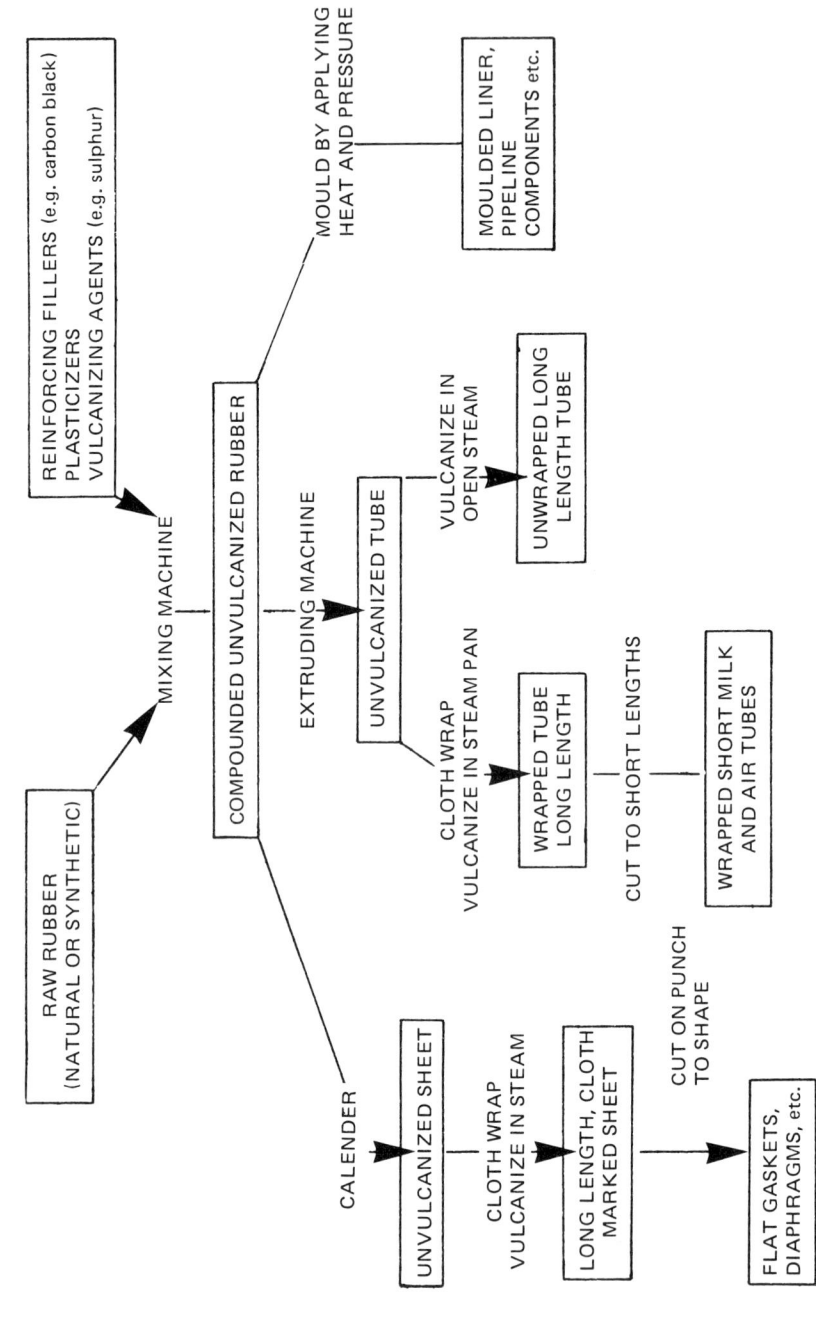

Fig XI.4 Typical production flow in manufacture of milking machine rubbers

Production of rubber components

Typical uses of the above manufacturing processes in the production of rubber milking machine components are set out in Fig XI 4.

PROPERTIES OF RUBBER COMPOUNDS

The fundamental properties common to all vulcanized rubber is elasticity, which is defined as the tendency of a body to return to its original shape and size after having been deformed. The term resilience is used to define the percentage of energy returned when rubber is made to undergo a single cycle of rapid deformation. A teatcup liner made from a high resilience natural rubber feels snappy, while one made from very low resilience butyl rubber feels lazy.

Rubber can be compounded to give a wide range of hardnesses, from soft elastic to ebonite. Hardness is measured usually on British Standard or Shore gauges which both read from 0 - 100 and give very similar figures over the range. As an indication, a pure gum rubber reads about 35 - 40 on these scales, whilst a tyre tread rubber is about 60 - 65. Most teatcup liners come within the range 45 - 55.

Tensile strength is often used as a means of defining rubber quality but this can be misleading as in practice one rarely stresses rubber to the point of producing a tensile break. A more useful property to measure for teatcup liners is modulus which is usually defined as the tensile stress required to produce a given elongation, say 300%. Hardness and modulus of rubber compounds are closely related, both giving a measure of stiffness. It is very often difficult to measure hardness accurately on a teatcup liner, but modulus can readily be determined on a sample taken from the wall of the liner.

Other important physical properties of teatcup liners are tension set (permanent set), resistance to flexing and resistance to tearing. Tension set is described as the ratio of the permanent increase in length of a test piece, after stretching and release, to the original length. Tests to measure resistance to flexing and tearing are defined in BS 903 Part A3 and A10 but very often these do not correlate with service performance. For a teatcup liner a useful indication of the flex life can be obtained by continually flexing it when assembled in its teat cup shell.

The term ageing is used to describe changes which take place in rubber with the passage of time. It is necessary to qualify this term to indicate the factor responsible, e.g. heat ageing, light ageing.

Rubber can deteriorate as a result of contact with oils and fats, solvents and other chemicals. Milking machine rubbers must have good resistance to all those materials with which they are likely to come in contact during service.

The physical characteristics of nitrile and natural rubber compounds for milking machine components can be compared in general terms, as shown in Table XI 1. Table XI 2 gives an indication of the physical properties of typical teatcup liners and milk tubing.

Table XI.1 Comparison of natural and nitrile rubbers

	NATURAL RUBBER	NITRILE RUBBER
Tensile strength	good	fair
Elongation to break	good	fair
Tear and cut resistance	good	fair
Resilience	good	fair
Resistance to butter fat	poor	excellent
Tension set	good	fair
Resistance to weathering	fair-good	poor-good
Resistance to heat (100°C)	fair-good	fair-good
Resistance to sterilizing chemicals	good	good
Resistance to flexing	good	good to excellent

Table XI.2 Physical properties of milking rubbers

		Liners		Milk tubing
		Natural rubber	Nitrile rubber	Natural rubber
Hardness (BS)	degrees	45-50	50-55	50-55
Tensile strength, minimum	MPa	17	7	10
	lb/in²	2500	1000	1500
Elongation at break, minimum	%	600	350	400
Compression set,* maximum	%	30	30	35
Swelling in butter oil, maximum) volume increase 24 h at 100°C)	%	150	5	125
After ageing 7 days at 70°C in air:				
Tensile strength retained	%	80	85	65
Hardness change	degrees	+10 − 2	+8 − 2	+10 − 2

* *Method BS 903: Part A6: 1969*

RUBBER PARTS IN MILKING MACHINES

The first part of this chapter has given a background of rubber technology which is necessary to understand more fully the functions and requirements of the various components used in milking machines. Whilst the teatcup liner is perhaps the most important part of the assembly, rubber is used for tubing, pipe line connectors, jar lids and many other components.

TEATCUP LINERS

A teatcup liner compound must meet the following requirements:

1. It must be suitable for economic manufacture to a consistently high standard of quality.
2. It must initially have the right physical properties, e.g. hardness and resilience, and must not contain ingredients which will impart odour or taste, or cause toxic hazards to milk.
3. It must have maximum resistance to the following factors causing deterioration:
   (a) milk and body fat
   (b) heat, oxygen, ozone, light and flexing
   (c) tension, cutting, tearing and abrasion
   (d) cleaning and sterilizing processes.
4. It should not be a source of bacteriological contamination of milk.

These various requirements are now considered in more detail.

1. Economic manufacture

This is obviously a primary requirement. Until about 15 years ago liner compounds had to be based on natural rubber as it was not possible to manufacture liners of consistent quality from the synthetic rubbers which were then available. With the improved quality of synthetic rubber and new manufacturing techniques, oil resisting synthetic rubber compounds can be used for the majority of moulded liners at an economic price.

2. Physical properties

Natural rubber usually has higher tensile strength than synthetic rubbers. Natural rubber also has better resilience or snap than synthetic rubbers and it is sometimes considered that because of this, natural rubber liners give more efficient milking. However, Clough, Dodd, Gardner and Cooper[1] studied the milking rate of both natural and synthetic liners by cinephotography and could find no significant difference between the two. Ward, Elsey and Jordan[2] also compared the milking efficiency of natural and synthetic liners and could find no significant difference. Cooper and Gardner[3] found that the modulus, or stiffness of a rubber, could be varied quite widely without having any marked influence on milking rate. The suggested range for modulus of elasticity, measured at a stress of 490 kN/m^2 (70 lb/in^2) is 600 kN/m^2 to 2.5 MN/m^2 (85 - 350 lb/in^2). For the particular liner studied, increasing the modulus above this range had an adverse effect on milking rate.

The rubber for liners used in tension should have low permanent set. It has been confirmed that the milking rate of a straight liner is dependent on its tension[4]. If the liner rubber has high permanent set the tension in the liner will be lost rapidly with bad effect on milking rate. As natural rubber has very much better set properties than any of the synthetics it is always preferred for liners designed to operate at high tension.

Natural and synthetic rubbers can readily be compounded so as not to impart objectionable odour or taste to milk. With synthetic rubbers it is important that the plasticizer, which might be extracted in small amounts, should be suitable in this respect. The Food, Drugs and Cosmetics Act in the USA contains a list of rubbers and compounding ingredients which are acceptable for use with foodstuffs and this now applies to milking machine rubbers used in the USA[5]. More recently the German Ministry of Health has made recommendations relating to rubber for teatcup liners and tubing for milk conveying. These recommendations list the materials which are permitted in such products and specify the limiting percentage in the finished article.[6].

3. Deterioration

(a) Fat absorption. Most oils and fats cause natural and general purpose synthetic rubbers to swell excessively, resulting in softening and general loss of physical properties. Nitrile rubbers are highly resistant to the swelling action of all mineral, animal and vegetable oils and fats (Fig XI 5). Some nitrile rubber compounds can shrink in volume when immersed in oils due to extraction of plasticizer. Neoprene is intermediate between natural and nitrile rubbers in its oil resistance, but even when swollen it maintains its original physical properties much better than natural rubber.

The rate of oil and fat absorption of articles made in natural rubber can be retarded by immersing them immediately after moulding in an aqueous solution of chlorine or bromine. A surface layer of rubber halide is formed which retards the rate of fat absorption. The process is widely used with natural rubber liners and sometimes with nitrile liners but it must be controlled carefully because excessive chlorination can cause the surface of the rubber to crack. Chlorine treatment reduces the coefficient of friction of rubber giving the article a 'silky' feel.

Fig XI 5. Swelling of natural and synthetic liners by fat. Left, natural rubber liner after 14 weeks use; right, synthetic rubber liner after 12 months use. Avon Rubber Co. Ltd.

Gardner and Berridge[7] found that natural rubber was swollen only slowly in emulsified fats, such as milk and cream, but rapidly in the fat itself. Milk was trapped between the teat and the liner during milking and the agitation arising from pulsation broke the emulsion, thus releasing fat which was absorbed into the rubber. Continuing this investigation, Cooper and Gardner[8] showed that swelling of a liner occurred mainly in the mouthpiece and in the portion of the barrel in contact with the teat. The mouthpiece was attacked mainly by body fats and sometimes udder salves, while the barrel absorbed milk fat. Rubber swollen by fat was more readily oxidized leading to loss of physical properties. Rubber under tension absorbed fat more rapidly than when relaxed and it was shown under farm conditions that liners used at 50% elongation absorbed slightly more fat than liners used at 30% elongation. There was loss of efficiency during milking in direct relation to the fat absorbed.

Cooper[9] analysed the fatty extracts from used liners and found that the fat absorbed in the mouthpiece of the liner was mainly fat from the cow's skin and hair. The fat absorbed in the barrel of the liner was a mixture of butter fat and udder fat. The percentages of each type of fat varied, but 50 - 60% of the fat absorbed by a liner during milking may be from the cow's skin and hair.

Ward, Elsey and Hurry[10] studied factors affecting blistering of the inner surface of liners made from 70% natural and 30% GRS rubber. They suggested that fatty acids of intermediate chain length may be a factor in promoting blistering. Udder salves, antibiotics and different cleaning methods had little effect, but blistering was decreased by storage in 10% lye solution. Different rates of blistering were associated with different breeds of cows and with different rations, winter rations producing more severe rates of blistering than summer rations.

Farm tests have demonstrated that teatcup liners made of nitrile and neoprene absorb much less fat and have a longer life than those made of natural rubber but they are more susceptible to mechanical damage[11,12,13].

(b) Heat, oxygen, ozone, light and flexing. These are the principal factors which cause what is known as ageing of rubber. Their mode of action in causing deterioration of milking rubbers can be briefly summarized: a more detailed treatment is given by Buist[14].

Heat alone, in the absence of oxygen, will cause deterioration of rubber compounds if the temperature is above 100°C (212°F). Under such conditions natural rubber tends to soften, but synthetic rubbers stiffen. Usually atmospheric oxygen is present and then the effect of heat is more severe. Both natural and nitrile rubbers progressively lose tensile strength when heated for prolonged periods in air at a temperature in excess of 70°C (140°F). The rate of deterioration is a function of temperature and oxygen concentration.

When heat is used for sterilizing milking components deterioration of the rubber parts is accelerated[3]. Steam sterilization at 100°C (212°F) will cause rubber to deteriorate three or four times more rapidly than hot water at 82°C (180°F). Milking machine rubbers will normally contain a heat resisting antioxidant so that they can safely withstand 72 hours steaming without undue deterioration.

One of the most common causes of deterioration of rubber is ozone, which is always present to some extent in the atmosphere. Ozone only attacks rubber which is stretched, producing surface cracks at right angles to the direction of strain (Fig XI 6). At low extension, 2%, cracks will occur very slowly or not at all, but at extensions between 10 and 20% the rate of crack development is at a maximum. The rate of deterioration is proportional to the concentration of ozone in the air, which varies considerably according to atmospheric and geographical conditions. To minimize this type of deterioration milking rubbers should as far as possible be stored free from distortion, protected from direct sunlight and with fresh air circulation reduced to a minimum.

However, milking rubbers must be stretched in use and rubbers which are compounded to give good resistance to ozone must therefore be used. This can readily be done with natural rubber by incorporating certain waxes which bloom to the surface of the rubber and form a protective layer. If the rubber is flexed the wax bloom may crack and allow ozone to attack its surface. To resist ozone under such dynamic conditions antiozonants may be added to the rubber compound. They are usually derivatives of p-phenylene diamine which function by blooming to the surface and reacting with any ozone present.

Fig XI 6. Ozone cracking of rubber long milk tube where stretched over pipe. Avon Rubber Co. Ltd.

Nitrile rubber, which is often used for milking machine components, is more prone to ozone attack than natural rubber and special compounding is necessary to overcome this weakness. Neoprene is particularly resistant to ozone. For this reason, coupled with its resistance to fats, it is sometimes used in milking rubbers.

Sunlight alone can cause an unstretched rubber to develop surface cracks. Black compounds are always more resistant to light than non-black compounds; synthetic rubbers are generally more resistant than natural rubber. A good quality natural rubber compound will withstand continuous exposure to sunlight for a long time without noticeable deterioration, so that light deterioration is not a serious problem for milking rubbers.

When rubber is subjected to prolonged flexing or bending in a limited area, cracks will appear on the surface which will eventually spread into the mass of the rubber, leading to complete breakdown. Such conditions prevail in the barrel of a teatcup liner and on the sidewall of a tyre. The cracking occurs only in the presence of air, although it will be aggravated if ozone is present as well. It is generally thought to be caused by some form of oxidative fatigue of the rubber. Certain antioxidants can be added to improve the resistance to flex cracking. Cracks can start from any imperfections in the rubber, such as badly dispersed filler. Teatcup liner compounds should be mixed to develop maximum dispersion of all compounding ingredients. In general, synthetic rubbers, or blends of synthetic and natural rubber have better resistance to flex cracking than all natural rubber compounds.

(c) Tension, cutting, tearing and abrasion. In general, natural rubber compounds are superior to synthetic rubber compounds in withstanding these damaging factors. However, modification of the design of liner can offset deficiencies of synthetic rubbers.

(d) Cleaning and sterilizing processes. The normal cleaning and sterilizing processes used for milking machines are based on aqueous solutions, hot water or steam. The rubber parts are therefore in regular contact with water. Vulcanized rubber is not completely waterproof and is affected by immersion in water, the extent of the deterioration being dependent on the composition of the rubber compound, the temperature and time of immersion.

In general, rubbers swell and soften after prolonged immersion in water and become weaker. However, the rubber compounds used in good quality liners and tubing absorb very little water. Even after 1000 hours in boiling water a good natural rubber will only swell 10%. Heat deterioration occurs in boiling water, but this need not be significant in present day rubbers. If rubber is distorted or stretched excessively during heating it will take a permanent set which could be objectionable. For example, if a liner is a bad fit in the jetter cup during circulation cleaning with hot water, permanent distortion of the orifice will occur.

Solutions of sodium hypochlorite at the strength approved for sterilization (300 ppm) do not have any serious effects on the life of natural or synthetic liners. The free chlorine in hypochlorite solution does, however, attack the surface of the rubber forming a hard layer of rubber chloride. In time this can produce a surface crazing which is readily visible under a microscope at 60 X magnification. Use of excessive concentration of hypochlorite can lead to serious degradation of the rubber surface, causing it to crumble. Both natural and nitrile synthetic rubbers are similarly affected by hypochlorite. On the other hand liners made from butyl synthetic rubber develop a sticky surface when in contact with hypochlorite. This characteristic of butyl rubber has prevented its widespread use in liners.

Alkaline detergents with 300 ppm chlorine at 70 - 82°C are used in circulation cleaning. Natural or synthetic rubbers are not noticeably affected by such treatment providing the recommended chlorine concentration is not exceeded. Similarly the ABW (acid boiling water) treatment with 0.1% nitric or sulphamic acid at pH 2 - 3 and 76 - 82°C will not harm milking machine rubbers, but if the recommended maximum temperature is exceeded, stretched rubbers will suffer from excessive permanent set and oxidative degradation may occur. Warm phosphoric acid solution at 1% concentration has no effect on rubber.

The use of caustic soda for defatting liners has been referred to in Chapter X. Solutions of caustic soda up to 5% do not have any degrading effects on natural rubber and the practice of storing such liners for periods of 7 - 14 days in caustic soda solution is highly beneficial to liner life. It is best to carry out defatting before an excessive amount of fat has been absorbed in the rubber. Although the caustic soda will remove most of the fat the structure of the rubber is weakened and does not recover fully after immersion. This can result in a roughening of the inner surface of the liners after immersion cleaning. The practice of using two sets of liners and resting each set for one week in turn is only of value if the set which is rested is defatted thoroughly. In the absence of defatting, the fat present in the liner accelerates ageing of the rubber so that it is deteriorating all the time it is not in use.

Oil resisting synthetic rubber liners do not absorb any significant amount of fat and there is no necessity for defatting these with caustic soda. In fact, it is possible that some synthetic rubber compounds may contain ester type plasticizers which would be broken down by caustic soda, causing the liner to harden and crack.

The choice between natural and synthetic rubber for teatcup liners is not easy. The oil resisting synthetic rubbers do not swell in fat, are easier to keep clean and have a long life. On the other hand they are less able to withstand stretching and rough treatment, and cost more. The old established straight liner which operates at very high stretch is best made from natural rubber as opposed to an oil resisting synthetic rubber. Probably the best all round moulded liner is one designed for and made in nitrile rubber, or a blend of rubbers with nitrile as the major component.

TUBING

Rubber tubing is used extensively in milking machines for conveying milk. The rubber compound used for tubing must have good resistance to ozone and weathering, hot water and steam, and the usual sterilizing agents. It must be soft and flexible enough to permit the use of pinch clips for closing the hose when necessary, yet it must be firm enough to resist collapse under vacuum. To maintain these properties it is necessary to keep within close limits the hardness of the rubber compound and the dimensions of the finished tubes. In order to maintain initial tension on metal nipples the rubber must not take an excessive permanent set.

The short rubber tubes connected to the teatcup cluster are liable to suffer from impact fractures and therefore the rubber compound for these must have good resistance to cutting and tearing.

There is no problem with fat absorption in milk tubing since the butter fat in the milk remains emulsified. It is therefore not necessary to use an oil resisting synthetic rubber. In fact natural rubber, or blends of natural rubber and SBR are superior because of their better impact resistance and superior flexibility.

As an alternative to rubber, translucent PVC tubing is sometimes used. This has the advantage of enabling the milk flow to be seen, but in time it becomes opaque and this advantage is lost. PVC, being a thermoplast, tends to soften in hot water and become rigid at low temperatures. Some grades of PVC tubing may lose plasticizers in use and this may result in surface cracking.

PIPELINE COMPONENTS

In circulation cleaning systems, components such as rubber connectors and elbows are widely used. These have to withstand atmospheric ageing, and heat and chemical sterilization processes. They are not, however, subjected to extreme flexing, nor are they liable to accidental damage. In order to make them easier to clean these parts should be moulded with a very smooth finish. Natural or synthetic rubber may equally well be used for these components.

Rubber pipeline parts must grip firmly the pipe or ferrule to which they connect. It is recommended that the inside diameter of the rubber part be 10 - 15% less than the outside diameter of the connecting pipe. With small diameter rubber tubing the difference will have to be increased to 25%. Higher stretch will make fitting difficult and cause ozone cracking of the extended rubber.

RUBBER LIDS

These are frequently used in the UK for glass jars of many types, as well as for buckets and cans. They are required to have excellent resistance to all forms of ageing and to cleaning and sterilizing processes. They can be expected to have a very long life.

It is essential that lids do not exhibit any significant distortion when in use under vacuum. Any slight distortion which may occur should be recovered immediately vacuum is released. As the force on the lid under vacuum is considerable, a very hard rubber compound must be used if the thickness of the lid is to be kept down to reasonable limits. To meet these requirements a rubber compound with a hardness of 80 - 90° BS is almost exclusively used.

MISCELLANEOUS RUBBER ITEMS

In addition to those rubber parts described above, there are many other miscellaneous items such as diaphragms, jetter-cups, and pulsation parts. In general it will be necessary to select a suitable rubber compound for each particular part.

ADHESION OF RUBBER TUBING TO GLASS AND METAL

This phenomenon is one which is often experienced in milking plants and is caused mainly by the rubber flowing under pressure into the microscopic irregularities on the surface of the glass or metal. Under such conditions perfect contact can be obtained between the surfaces involved and physical adhesion results. Any factor which facilitates the flow of the rubber will increase the physical adhesion between the surfaces. In milking plants the aqueous detergent solution is absorbed slightly by the rubber and can plasticize it. The interference fit of the rubber on the glass tube puts the rubber surface under pressure. The tighter the fit of the rubber tube the more likely adhesion will occur. Physical adhesion can occur with natural and synthetic rubber compounds, and also with plastics tubing. Harder rubbers are less likely to stick than softer rubbers because they will flow less. However, too hard a rubber tube is difficult to assemble. A rubber compound with low water absorption and good resistance to oxidation should minimize sticking.

In addition to this physical adhesion some chemical adhesion between rubber and glass can occur due to oxidation of wet rubber producing hydroxyl radicals which may combine with hydroxyl radicals on the glass. Since oxidation is accelerated by elevated temperature this chemical adhesion is more likely to occur when joints are subjected to steam sterilization.

CONCLUSION

This chapter will have demonstrated that the development and manufacture of rubbers for milking machines, particularly teatcup liners, is a complex operation calling for specialized knowledge of materials, processes and the operation of milking machines. For this reason the production of nearly all milking rubbers is carried out by only a few rubber manufacturers throughout the world. A very high standard of quality control of rubber parts is necessary to enable the modern milking machine to operate efficiently. Close co-operation between rubber manufacturers, milking machine manufacturers and users ensure that the latest developments in the rubber and polymer industry are made available for the improvement of the milking machine.

REFERENCES

(1) Clough, P.A., Dodd, F.H., Gardner, E.R. & Cooper, J.H. (1958) *Report, National Institute for Research in Dairying 1957*, p.26.

(2) Ward, G.M., Elsey, V.R. & Jordan, H. (1961) *Journal of Dairy Science*, **44**, 947.

(3) Cooper, J.H. & Gardner, E.R. (1955) *IRI Proceedings*, **2**, 194.

(4) Clough, P.A., Dodd, F.H., Cooper, J.H. & Gardner, E.R. (1954). *Report, National Institute for Research in Dairying 1953*, p.28.

(5) Sub. Part F, Section 121, 2562 (1976) Food Additives Amendment of the Federal Food, Drug and Cosmetic Act, USA.

(6) Anon. (1973) Recommendation XXXVIII: Pipeline for milk and suction lines for attachment to udders, made of synthetic materials and rubber *Federal German Health Gazette*. No. 3, p.44.

(7) Gardner, E.R. & Berridge, N.J. (1952) *Journal of Dairy Research*, **19**, 31.

(8) Cooper, J.H. & Gardner, E.R. (1953) *Journal of Dairy Research*, **20**, 340.

(9) Cooper, J.H. (1955) *Journal of Dairy Research*, **22**, 138.

(10) Ward, G.M., Elsey, V.R. & Hurry, J.A. (1957) *Journal of Milk and Food Technology*, **20**, 312.

(11) White, J.C. & Folds, G.R. (1954) *Journal of Milk and Food Technology*, **17**, 256.

(12) Clarke, P.M., Berridge, N.J. & Gardner, E.R. (1955) *Journal of Dairy Research*, **22**, 144.

(13) Major, W.C.T. (1958) *Queensland Agricultural Journal*, **84**, 407.

(14) Buist, J.M. (1956) *Ageing and Weathering of Rubber*. Cambridge: W. Heffer.

CHAPTER XII

MILK COOLING EQUIPMENT

J B Hoyle

In this chapter the principles are outlined of equipment for cooling milk, particularly bulk milk tanks for refrigerated cooling and storage on farms. Devices for automatic cleaning of refrigerated farm bulk tanks, heat recovery from refrigeration units to provide hot water, and current trends affecting milk cooling are also discussed.

The purpose of cooling milk soon after production is to keep it in satisfactory bacteriological condition until it is processed for liquid consumption or is manufactured into dairy products. Transport distances and storage times of raw milk have increased so that now the time delay between production and processing may be considerably more than two days on occasions. A wide range of cooling equipment, with and without refrigeration, is in use at present on UK farms although the intention is that by the end of 1979 all milk except in Northern Ireland will be refrigerated and collected in bulk. Therefore in this chapter there is only a brief mention of cooling equipment associated with milk transport in cans.

THE BACKGROUND OF BULK COLLECTION

The performance characteristics of present day cooling and storage equipment is probably more influenced by considerations relating to bulk collection than to the technicalities of cooling. For this reason a brief account is given of the transition, still taking place, from can collection to bulk collection and the way in which the requirements of bulk collection have influenced the design and performance characteristics of refrigerated farm bulk tanks.

COOLING MILK FOR CAN COLLECTION

Many methods have been used to cool milk to within 3°C (5°F) of the temperature of the cooling water available, but the commonest of the successful devices have been corrugated surface coolers and turbine coolers. These are briefly described in the legends to Figs XII 1 and 2. They can be coupled to mains or well water supplies or equally easily to refrigerated chilled water units. The cans of milk may be enclosed in insulated cabinets as in Fig XII 3 to minimize reheating during storage on farms. Cans of milk are traditionally taken to the roadside to await collection.

Fig XII 1. Lister corrugated surface cooler for flowrates of 134 - 450 ℓ/h (30 - 100 gal/h). Milk is distributed evenly by the top trough and collected in the bottom trough for distribution into milk cans. Cooling water enters at the bottom and leaves at the top, the open angled tube on the right limiting back pressure. The gap between the two corrugated tinned copper or stainless steel plates forming the water way may be as little as 0.6 mm (0.025 in).

The Milk and Dairies Regulations (1959)* require milk to be cooled without delay to 10°C (50°F) or to not more than 2.8°C (5°F) above the temperature of the cooling water available. Consequently, when natural and mains cooling waters reach their summer peak temperatures, milk may not be cooled below 21°C (70°F). However, even this is some help as shown in Table XII 1.

* Statutory Instrument 1959 No. 277. The Milk and Dairies (General) Regulations 1959. London: HMSO.

Table XII.1 *An example of the effect of temperature of storage on bacterial growth in farm raw milk.*

Milk held for 24 h at: °C	°F	Plate count, colonies/ml
0	32	2400
4	39	2500
5	41	2600
6	43	3100
10	50	11 600
13	55	18 800
16	61	180 000
20	68	450 000
30	86	1 400 000 000
35	95	25 000 000 000

Reproduced from: Davies, J.G. (1955) *A Dictionary of Dairying*, 2nd edition, p. 294. London: Leonard Hill.

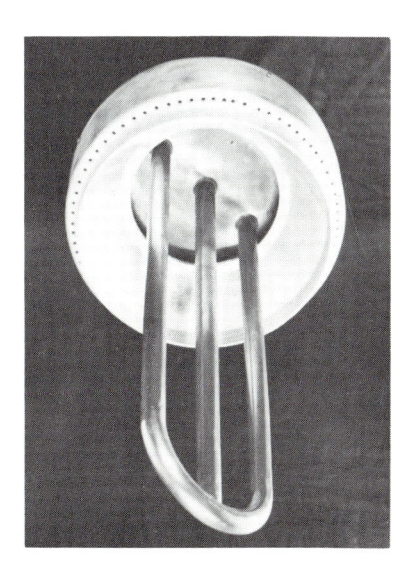

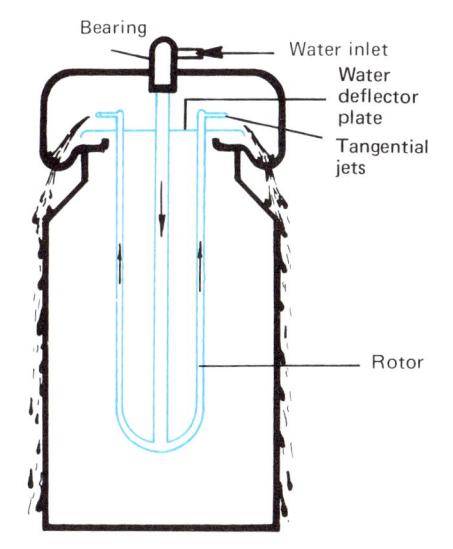

Fig XII 2. Blow in-can turbine milk cooler viewed from below, and, diagrammatically, in position on a milk can: the moving parts are in blue and the water flow paths arrowed. These coolers are capable of cooling 45 ℓ (10 gal) of milk from 35°C (95°F) to within 3°C (5°F) of the mains water temperature in about 12 min, using 150 ℓ (33 gal) of water, or using chilled water. Close fitting sleeve bearings guide the centre assembly and form a channel through which water passes to the stirrer tubes; it leaves through tangential jets at the top, which cause the rotor to turn at 50 - 120 rev/min.

Fig XII 3. A Fullwood refrigerated ice building chilled water unit for cooling milk in cans. The chilled water is circulated by means of a pump, passes through the turbine coolers and falls over the sides of the cans on to the ice bank. The insulated cabinet minimizes overnight rise in milk temperature.

The only incentives to use refrigerated cooling for milk dispatched in cans are shortage of cooling water and the risk of having milk rejected in hot weather. It was not until bulk collection started (in 1954 in Scotland and 1955 in England and Wales) that a premium was paid for refrigerated milk, and then only if stringent rules were complied with. The temperature requirements of the Milk and Dairies Regulations were not relevant to cooling for bulk collection, for which much lower temperatures were demanded. However, the Regulations have been retained for milk delivered in cans during the transition period.

GENERAL ASPECTS OF BULK COLLECTION

To the producer the main advantages of farm refrigeration and bulk collection are less risk of rejection of milk due to bacterial deterioration, freedom from handling cans and a premium on selling price to finance the buying and operation of the extra equipment required. To the buyer the advantages are more reliable bacteriological quality and considerable savings in cost of cans and milk cooling, and cost of milk intake. There are also savings in transport costs — for example a road tanker collecting from farms can take about twice as much milk per load as a lorry collecting in cans. The five UK Milk Marketing Boards (North of Scotland, Aberdeen and District, and Scottish Milk Marketing Boards; and Milk Marketing Boards for England and Wales, and for Northern Ireland) are responsible for milk transport and are the selling agents of the producers. The MMBs therefore played a central part in negotiating a scheme to share the benefits of bulk collection equitably between producers and buyers, and are still in Northern Ireland and England and Wales engaged in the difficult transition to universal bulk collection. Small average herd size has a constant retarding effect. For example in 1975 about 26% of herds in England and Wales and 65% in Northern Ireland had 19 cows or less: amounting to 6% and 30% of the cow populations.

When bulk collection is substituted for can collection, one of the effects is that the point of sale of milk is transferred from the receiving dairy to the farm. This is thought by many farmers to be an advantage[1], but the driver of the road tanker has the considerable responsibility of accepting the milk on inspection and measuring its quantity. He may be an employee of a milk marketing board, a dairy or an independent haulier. Boards are responsible for haulage but much of the work is done on contract.

Temperature of refrigerated bulk milk

From the beginning of bulk collection in the UK, it was specified that milk be cooled to and stored at a temperature not exceeding 4.4°C (40°F). This is in line with practice in most other countries. Psychrotrophs (bacteria which can grow at low temperature) are generally slow growing at 4.4°C[2] and surveys in 1957 and 1971 showed the benefit of cooling to this figure[3,4]. It is pointed out in Chapter X that comparatively small increases in temperature above 5°C (41°F) can substantially increase growth of psychrotrophs.

Surface temperatures in farm bulk tanks will generally be higher than the mixed temperature except in tanks fully jacketed with chilled water. In addition it is recognised that some rise in temperature will occur during transit, and surface temperatures can be higher still. The time-temperature history of raw milk after it leaves the farm is therefore also important in relation to damaging numbers of psychrotrophs because it can be considerably longer than the storage period on the farm.

BULK HANDLING REQUIREMENTS INFLUENCING FARM TANK DESIGN

The requirements of a bulk handling system for refrigerated milk influence farm tank design in other ways than the temperature to which milk is cooled after production. In the UK three other factors have a considerable influence. These are: the ratio of morning to evening herd yield; the time allowed for cooling the milk load in the tank to the specified temperature; and the accuracy required of dipstick measurement of milk volume. These all affect the design problem of extracting heat fast enough from milk in the tank.

Morning to evening yield, and time allowed for cooling

On average the preceding milking intervals are 14.5 h for morning milking and 9.5 h for evening milking. Consequently, yields at morning and evening milking are on average in the ratio of about 1.5 to 1. Haulage contractors and buyers of milk are anxious that milk collection from farms should start reasonably early in the morning, and, to enable collecting routes to be re-arranged easily when changing circumstances demand, that milk be available at all farms at much the same time in the morning. A high rate of heat extraction to cool the large morning yield of milk in a short time is obviously an advantage in these circumstances. The compromise arrived at was that the bulk tank should be capable of cooling a full morning load of milk, added to a full evening load in the tank, to the specified temperature within half an hour of the end of a 2 h milking period.

Effect of dipstick accuracy

Measurement of milk quantity in a bulk tank is most commonly by means of a dipstick and a calibration chart. Probably most measurements of 450 - 900 ℓ (100 - 200 gal) quantities of milk are accurate within the range of ± 0.5%. This high level of accuracy is partly achieved by restricting the cross sectional area of tanks so that up to 900 ℓ (200 gal) capacity they contain no more than 1.4 ℓ/mm (8 gal/in). A resulting effect is to restrict the most advantageous cooling area of the tank, which is the bottom.

IMPLICATIONS FOR REFRIGERATED FARM BULK TANK DESIGN

The effect of a long milking interval on morning yield and short time allowed to cool to the specified temperature of 4.4°C (40°F) is that heat extraction rate must be about 1.4 times the rate required in USA[5] and Europe. This can be achieved with ice building chilled water bulk tanks by careful attention to ice bank shape to allow consistently high rates of melting and by using all the surface coming in contact with milk for heat transfer as the tank is filled. It can be achieved by direct expansion (d.e.) tanks having very good control systems, and with cooling surfaces up the sides as well as across the bottom. However, in the early days manufacturers found that the milk froze on the side walls of the tank when milk contact was intermittent; using only the permitted surface area of the bottom they were in practice limited to 700 ℓ (150 gal) capacity if they wished to achieve the required cooling rate. Consequently almost all refrigerated farm bulk milk tanks in use in the UK at present are chilled water cooled using the surface of the tank for the purpose, with provision for an ice bank within the same structure which contains the tank.

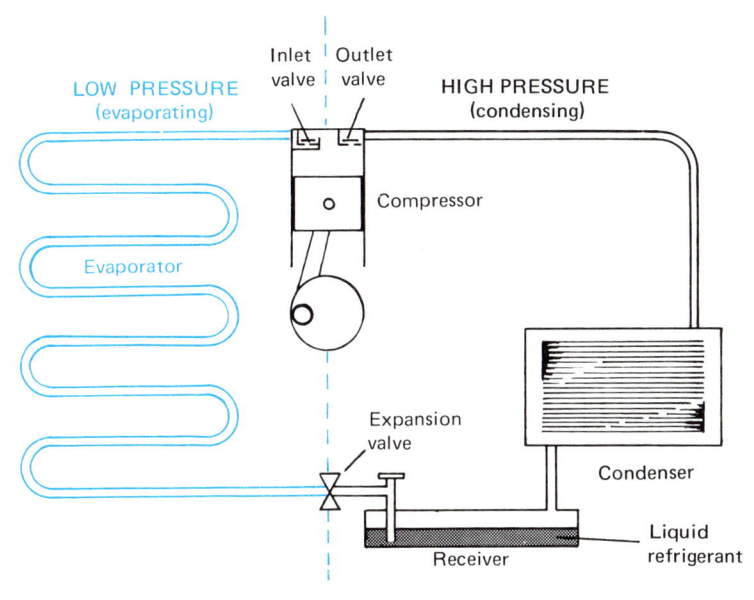

Fig XII 4. Simplified diagram of a refrigeration unit. The broken line shows the division between high and low pressure.

REFRIGERATED FARM MILK TANKS

The refrigeration cycle

The refrigeration system used in bulk tanks is the same as that most commonly used in domestic refrigerators and deep freeze units. A simplified diagram of the system is shown in Fig XII 4.

The closed refrigeration circuit is filled with a working fluid or refrigerant which changes from liquid to gas at suitable combinations of pressure and temperature. The refrigerant used is normally dichlorodifluoromethane, known alternatively as R12, Freon 12 or Arcton 6.

The compressor, which is a high pressure pump, is generally driven by an electric motor. When running, the compressor maintains a high pressure in the system between its delivery port and the expansion valve, and a lower pressure, but still above atmospheric, in the evaporator. Flow of liquid refrigerant into the low pressure side is controlled by the expansion valve. Being at low pressure the liquid refrigerant evaporates at low temperature and in doing so extracts heat from the evaporator coil and its immediate surroundings. The vapour drawn into the compressor is compressed to a high pressure and consequently will condense at a high temperature. The hot vapour passes to an air or water cooled condenser where heat is extracted which allows the refrigerant to condense back to liquid. The liquid refrigerant returns to the receiver from which it restarts the circulation cycle. Thus heat is extracted from the vicinity of the evaporator, and heats either air or water in the condenser. The compressor, condenser and receiver are generally mounted on one plate and collectively known as the condensing unit.

In direct expansion milk cooling systems, the extraction of heat from the evaporator is used to cool the milk directly, either because the evaporator coil is immersed in milk or because it is thermally an integral part of the wall of the milk vessel. For direct expansion cooling the refrigeration unit obviously must be capable of extracting the full heat load during the allowed milk cooling period.

An alternative is for the evaporator coil to be immersed in a water bath, and for a bank of ice to be built up on the coil over a relatively long period. This ice acts as a reserve which can be melted over a short period as milk is cooled. Since the period during which ice may be made extends far beyond the period of milk cooling, a smaller refrigeration unit is required than for an equivalent direct expansion machine.

When applied to a farm bulk tank extra components are introduced into the refrigeration unit to improve performance. For instance the flow of liquid refrigerant is adjusted so that evaporation takes place for nearly the full length of the evaporator coil. A heat exchanger is used to reduce the temperature of the liquid fed to the expansion valve, which at the same time ensures that the gas returning to the compressor contains no droplets of liquid which could damage the compressor. A solenoid shut off valve is fitted which closes when the compressor stops. This prevents liquid flowing from the receiver to the evaporator, which could also result in damage to the compressor when next started. The system must be fitted with controls so that the condensing unit will stop when sufficient ice is built, or when the milk has been cooled, and restart when necessary. Protective pressure switches are also fitted to prevent abnormally high and low pressures developing in the system.

UK Federation of MMBs specification

The UK Federation of Milk Marketing Boards has specified the constructional and performance requirements of farm bulk milk tanks regarded as satisfactory for UK bulk collection schemes. The present specification (BS 56)* covers the points summarized as follows [7].

* For tanks between 270 ℓ (60 gal) and 4000 ℓ (900 gal) but in principle applicable to all tanks.

1. The refrigeration unit should be capable of cooling daily the tank's nominal capacity of milk from 35°C (95°F) to 4.4°C (40°F) in an ambient temperature of 32.2°C (90°F).

2. If a morning load of 60% of nominal capacity is added in 2 h to 40% of capacity cooled the previous evening, the temperature should be reduced to 4.4°C (40°F) within 0.5 h of the end of filling. This test is made in an ambient temperature of 32.2°C (90°F).

3. The insulation should be effective in keeping the milk temperature, or the weighted mean of the milk and chilled water jacket temperatures, within 1.7°C (3°F) of the initial temperature of 4.4°C (40°F), when the tank stands for 8 h in an ambient temperature of 32.2°C (90°F). For tanks of less than 900 ℓ (200 gal) a rise of 2.2°C (4°F) is permitted.

4. The agitator should mix the tank's nominal capacity of milk of 4.5% fat content in 2 min after the milk has been left undisturbed for 6 h; samples taken from the surface and the outlet have then to agree within ± 0.05 units of butterfat percentage. Agitation tests are also carred out at 10, 25 and 40% of the tank's nominal capacity.

5. The tank has to comply with a number of structural requirements to facilitate cleaning and access, and also has to be sufficiently rigid for use as a milk measure. On this account, also, the cross-sectional area of the tank is restricted to a specific capacity of not more than 1400 ml/mm depth (8 gal/in) up to 900 ℓ (200 gal), and 700 ml/mm (4 gal/in) per 450 ℓ (100 gal) capacity thereafter.

6. The thermometer, thermostat, and ice bank controller where applicable, must be robust and reliable whilst maintaining accuracy when type tested *in situ* and also independently to ensure proper functioning at low ambient temperatures.

EQUIPMENT FOR 670 - 4000 ℓ (150 - 900 gal) PER DAY

Types of tank

Over 90% of the milk produced in the UK can be collected from tanks in this range and so this has been the main area of equipment development. The resulting types are shown diagrammatically in Fig XII 5 and the temperature of milk cooled and held in such tanks is generally as shown in Fig XII 6.

Ice bank arrangement

In order to make best use of an ice bank to give low temperature chilled water, experience shows that the ice should be formed in vertical banks with air admission from distributors beneath the evaporator coils to stir the water

Cooling Equipment

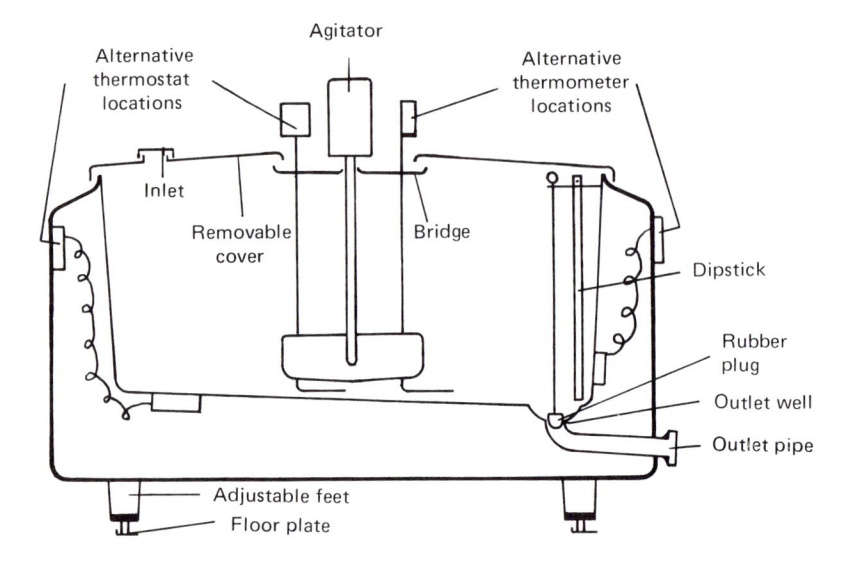

(a) (above) Typical tank showing features common to all

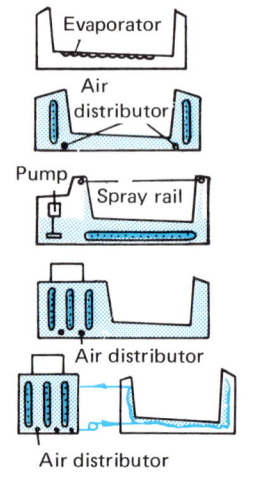

(b) Direct expansion: refrigerant evaporator bonded to bottom of tank. It may also extend up the sides

(c) Fully jacketed: ice bank surrounds the tank

(d) Sump and spray: ice below tank, pump and spray rail for cooling

(e) End ice bank: similar to (c), refrigeration equipment includes evaporator removable as single item.

(f) Remote ice bank: chilled water circulating through cooling plates.

Fig XII 5. Types of refrigerated milk tanks.

surrounding the ice banks. It is then possible to promote heat exchange over a large surface area and by this means the area of ice remains much the same until it is nearly all melted. With about 0.045 m³/min of air per m² of ice surface (0.14 ft³/min of air per ft² of ice) the water temperature can be maintained at 1°C (33.8°F) or below. This principle has been adopted by makers of fully jacketed, end ice bank, and remote ice bank tanks to achieve rapid heat transfer (Fig XII 5c, e, f). Horizontal ice banks beneath the tanks (Fig XII 5d) tend to close into a solid block, thereby reducing the surface area available for heat transfer between water and ice. The chilled water temperature is then higher with a consequent increase in cooling time. For the larger sizes of tank the ratio of milk cooling surface area to milk volume necessarily diminishes and so there can be difficulty in meeting the cooling performance required by the UK Federation of MMBs specification[6].

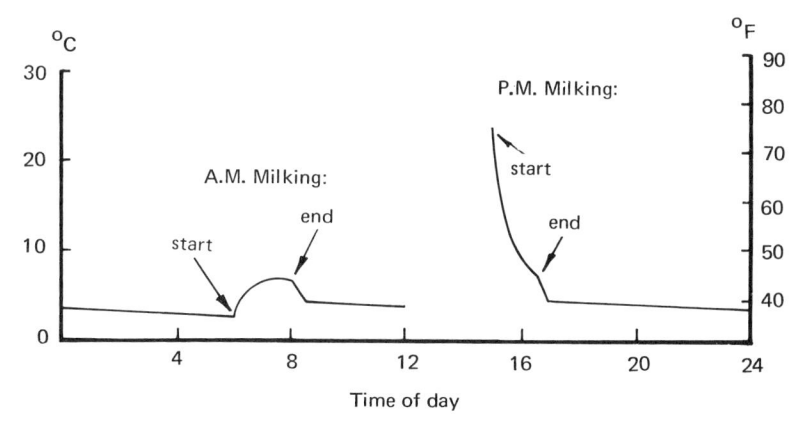

Fig XII 6. Daily cycle of milk temperature in a farm tank with ice reserve. Milk is cooled rapidly from 35°C (95°F) to 4.5°C (40°F) in the evening and then remains in the tank, usually getting cooler by conduction and convection overnight. The addition of milk in the morning brings the 'blend' temperature up to no more than 7.8°C (46°F). The bulk temperature is reduced to 4.5°C (40°F) within 0.5 h of the end of milk addition, and remains at this temperature or below until the milk is collected.

Ice bank control

Every ice-building system needs a control device to ensure that a sufficient but not excessive ice bank is built between milkings twice daily year in year out. Although many methods have been tried, only three types are in common use: thermostatic, change of state, and electrical probe. These devices sense temperature or change of state at one point only and locating them satisfactorily is a matter of experience.

With thermostatic controls the degree of heat conduction from the sensitive element to the evaporator coil and the temperature difference between them are important. The thermostat is normally set to cut out at about −2.2°C (28°F) and to cut in a little below 0°C (32°F). If set to cut in above this all the ice could be melted before the compressor cut in. If set much below 0°C (32°F) the unit will restart too soon after cut out owing to conduction of heat from the surrounding water, and the ice bank will continue to grow. Fortunately milking is a regular event, and there must be many thermostats which are not ideally set yet start the condensing unit because of the movement of warmer water in their vicinity when milk cooling commences and still give a reasonable ice bank before next milking. The farmer can tell whether the normal quantity of ice is being built by observation.

Change of state instruments depend on the expansion of water when it freezes in the sensitive element and the change in volume is transmitted to a sensitive switch through a capillary filled with a different liquid. The water in the sensitive element normally freezes progressively as the ice bank envelops it, but on the first time of operation it may be necessary for the temperature to go well below freezing point to induce the first crystals of ice to form. Therefore positioning of the sensitive element cannot be assessed until there has been at least one cycle of operation.

Electrical probe instruments depend on the change of resistance caused by ice as compared with water either between two probes, or between a probe and metal of the bulk tank. A transformer and low voltage circuit, and perhaps additional electronic components, must then be used between the probe and the compressor motor control circuit. This is in contrast to thermostatic and change of state instruments in which the switch can operate at mains voltage.

In all cases the control of ice bank size depends on a combination of reliable control during ice building with even melting of the ice during milk cooling. For instance, if the air distribution pipes beneath part of the ice bank of a fully jacketed tank (Fig XII 5c) become partially blocked, the ice will not melt on that side. If the sensitive element is on the other side, over building of ice may occur which probably first shows as ice or unusual condensation in the milk tank.

Over building of ice in fully jacketed tanks has been known to cause distortion of the tanks, with loss of accuracy of milk measurement, but intelligent observation can avoid the trouble. It is also a wise precaution if the tank is not to be used for a period to switch off altogether or, if lightly loaded, to reduce the quantity of ice built where this is possible. Over building of ice under a sump and spray tank (Fig XII 5d) can occur and may cause some milk to freeze on the bottom, but the tank itself should not suffer any distortion.

Another factor in obtaining good ice bank control is a correct and consistent shape of the ice bank, which depends on the expansion valve setting in the refrigeration unit and the refrigerant charge. For instance, loss of refrigerant charge causes less ice to build at the outlet end of the evaporator coil compared with what is built near the expansion valve. If the ice bank controller is towards the outlet end of the evaporator and controls to the normal ice thickness, very much more ice than usual is built on the first half of the coil.

When air stirring is used to agitate the water round vertical ice banks, the air bubbles expand and accelerate as they rise: the scrubbing action is therefore greater over the upper levels of ice than the lower. It is usual to try to build a slightly wedge shaped ice bank, with the thicker part at the top, by top feeding of the evaporator. Although this practice is not usually favoured by refrigeration engineers because of the danger of taking slugs of liquid or oil back to the compressor, it has nevertheless proved satisfactory in this application. If the refrigerant is fed to the bottom coil first there is a distinct possibility of over building ice on the lower coils, preventing proper agitation and eventually closing the water way. The objection to top feed can be overcome by correct evaporator design, and the use of a solenoid valve to prevent liquid refrigerant flow into the evaporator when the compressor is switched off.

In sump and spray tanks it may be necessary to maintain an open ice bank to attain the cooling rates required on test. This can usually be achieved, if the condensing unit has adequate capacity, by raising the evaporating temperature, i.e. opening the expansion valve to let more liquid through. Ice can then be formed having almost uniform diameter along the whole length of the coil so that, if the coil is evenly wound, the ice bank control can be arranged to stop the compressor whilst there is still some water between the coils and a maximum of ice surface is available for heat exchange. The condensing unit should in this case be protected from carry over of liquid by a heat exchanger.

Control of milk temperature

The behaviour of water and milk in the temperature range 0 - 10°C (32 - 50°F) is important in temperature control. Water has a maximum density at 4°C (39°F) and so temperatures invert below this point the coldest water then being on the surface and the warmer at the bottom; whilst in milk the lowest temperature always remains at the bottom. Above 4°C (39°F) both liquids behave the same in that the top surface is the warmest and may well be 5°C (9°F) warmer than liquid only a few inches below it. Furthermore, the fat in milk also rises and tends to concentrate bacteria in the cream layer which forms. The UK Federation of MMBs specification[6] lays down that whereas the mean temperature of the milk shall not rise above 4.5°C (40°F) during storage after cooling, the surface temperature shall not exceed 9.0°C (48°F). As discussed in Chapter X, this is a large rise in temperature in respect to promoting growth of psychrotrophic bacteria.

With any tank, the cooling should terminate at a bulk milk temperature between 4.5 and 2.8°C (40 and 37°F) to comply with the specifications[6], and both control and indication of temperature have to be effective with a milk volume of only 10% of tank capacity. Two types of instrument are available: (i) immersed instruments with the sensitive element encased in a stainless steel sheath suspended in the milk, and (ii) pocketed instruments with the sensitive element placed in a pocket and having good contact with the outside of the tank wall.

Both immersed and pocketed instruments can be used for any type of tank, although it is rather more difficult to insulate a pocketed instrument from the effect of chilled water temperature with fully jacketed tanks. Generally speaking, direct expansion tanks have pocketed instruments unless there are difficulties in insulating them from the effects of ambient temperature, and fully jacketed tanks have immersion instruments. Sump and spray tanks have either type.

Milk temperature control in d.e. tanks is straightforward, because as the surface temperature rises so the temperature at the bottom of the milk also rises and the thermostat can sense this change and restart the condensing unit when required. Nevertheless in some types the agitator is restarted by a timer; this reduces the surface temperature.

In a fully jacketed tank, the milk level at rated capacity will generally be above that of the surrounding chilled water; it should not be more than about 50 mm (2 in) above, or temperature stratification on the surface will still occur. With ice in the jacket, the water surface temperature will be about $0°C$ $(32°F)$ and the milk temperature will tend to be reduced by conduction after the main cooling period is finished. The thermostat should therefore never need to re-start the milk cooling system until the next load of milk begins to enter the tank.

The case with sump and spray tanks is somewhat different. The bottom of the milk tank is usually partly immersed in chilled water and so milk in that region continues to cool below $4.4°C$ $(40°F)$ after the circulating pump is switched off. Meanwhile the upper layers of the milk become warmer due to radiation from the tank covers and conduction from the rim and upper parts of the tank. Although overnight storage may exceed 12 h the milk surface is at that time low in the tank. Consequently the most adverse conditions for surface temperature occur during daytime storage, when the milk surface is only a little below the covers. To prevent the surface temperature exceeding $9.0°C$ $(48°F)$ during this time extra insulation may be fitted or the depth of chilled water increased, but usually a time switch is fitted to run the milk agitator briefly at intervals to prevent excessive temperature stratification.

Milk agitation

Rapid mixing is specified so that the tanker driver will obtain a fair sample of mixed milk 2 min after the agitator is started. This implies vigorous agitation. The agitator evolved has generally had a relatively large blade driven at low speed through gearing. Since the agitator rotates at the same speed whilst the milk is cooling, heat transfer from the milk to the tank walls is rapid and always adequate. It was thought that buttering might occur, but by and large there appears to have been little difficulty with this or with lipolysis. Lipolysis appears to be more troublesome in USA than the UK. There is an impression that this difference may be related to a combination of temperature and cooling rate, with cow management factors such as feeding playing a part.

Bail milking

No very good solution has yet been found to the problem of milk cooling for farms practising bail milking either throughout the year or in summer only. The latter is the more usual case and, because a normal bulk tank is installed at the farm for winter use, these farms have generally used a relatively cheap uninsulated plastics tank for transporting milk from the bail during summer. No cooling apart from that due to radiation losses and air convection then occurs at the bail but the milk is brought to the farm immediately after milking and transferred to the bulk tank for cooling.

On farms where road tankers cannot reach the site of the refrigerated bulk tank, it may be necessary to have a well insulated mobile tank to take the milk to the roadside. At present this equipment has also to have provision for dipstick measurement. The general use of meters on tankers could make collection from such farms easier and more flexible.

Various mobile refrigerated tanks have been constructed complete with their own condensing units and motive power, but being custom built were expensive. The use of a normal bulk tank as the prime cooling unit has generally provided the most economical solution.

EQUIPMENT FOR QUANTITIES EXCEEDING 4000 ℓ (900 gal) PER DAY

The equipment for farms producing over 4000 ℓ (900 gal) of milk per day is outside the detailed scope of the UK Federation of MMBs specification[6] and the individual Boards consider each case on its merits, though the general principles of the specification continue to apply. In some cases several refrigerated bulk milk tanks from the range up to 4000 ℓ (900 gal) are installed to cool and keep the output from different sections of the herd separate either for accounting, disease, or quality control.

If the milk is all to be bulked and if milking is extended over, say, ten or more hours per day it may well be worth considering the advantages of cooling through a heat exchanger and then storing in a well insulated tank until milk collection. The size of the cooler has to be matched to throughput, and the milk must be accumulated in a balance tank prior to pumping through the heat exchanger at a constant rate. With a milking routine operating over eighteen hours per day, and breaks for milk pipeline and tank cleaning, cooling can be spread over a long period.

If mains water is available at 13 - 15°C (55 - 60°F) and is from an assured source the milk can be cooled to 15 - 18°C (60 - 65°F) in the first part of a heat exchanger, and the water retained for yard washing or for the cattle to drink. The milk can then be cooled by chilled water to 4.4°C (40°F) and discharged into a suitable well-insulated tank. Chilled water can be produced in an ice-bank machine or by direct expansion cooling, depending whether milking time is short or long.

EQUIPMENT FOR SMALL AND INACCESSIBLE FARMS

The rental scheme of the Milk Marketing Board for England and Wales (MMBEW) is designed to bring all farms into bulk collection and to eliminate the ever increasing cost per farm of can collection over the same routes.

The requirement for additional equipment for small farms and for inaccessible farms was not envisaged when the UK Federation of MMBs specification[6] was written: but in dealing with the needs of these farms the principle that the mean milk temperature should be at or below 4.4°C (40°F) at the time of collection has been maintained. The scheme enables the MMBEW to introduce 100% bulk collection in an area without involving the individual farms in major capital outlay. The rental is paid out of the premium paid for the bulk milk and after three years reduces to a nominal sum. Tanks are bought in quantity by tender as favourable prices, and further, by owning the tanks, the MMBEW is able to obtain an EEC grant which is only available to co-operatives. All makes and sizes of tank are now rentable but clearly there is a price advantage in having one of those chosen by the MMBEW for bulk purchase.

Various ideas were proposed to solve the problem of those farms which could not be reached by the usual 8100 ℓ (1800 gal) farm collection tankers if the bulk tank was placed close to the milking parlour. These included a smaller road tanker of 5850 ℓ (1300 gal) capacity, and mobile bulk tanks but those proposed by manufacturers in 1973 were judged to be too expensive.

An in-line cooler developed at NIRD[9] had proved cleanable and was clearly adaptable for cooling milk before placing it in a transportable insulated tank. The MMBEW then developed the cooling and transport equipment shown in Fig XII 7. The cooler comprises a coil of stainless steel tube about 23 m (60 ft) long and 13 mm (½ in) outside diameter which is placed in the centre of a vertical cylindrical ice bank formed in a chilled water tank. Milk is drawn through the coil by a low vacuum at a pre-set flowrate so that when the water is air agitated, the milk outlet temperature is not above 2.8°C (37°F) at a flowrate

J.B. Hoyle 369

of 150 ℓ/h (33 gal/h). The coil can be raised out of the chilled water for cleaning by circulation.

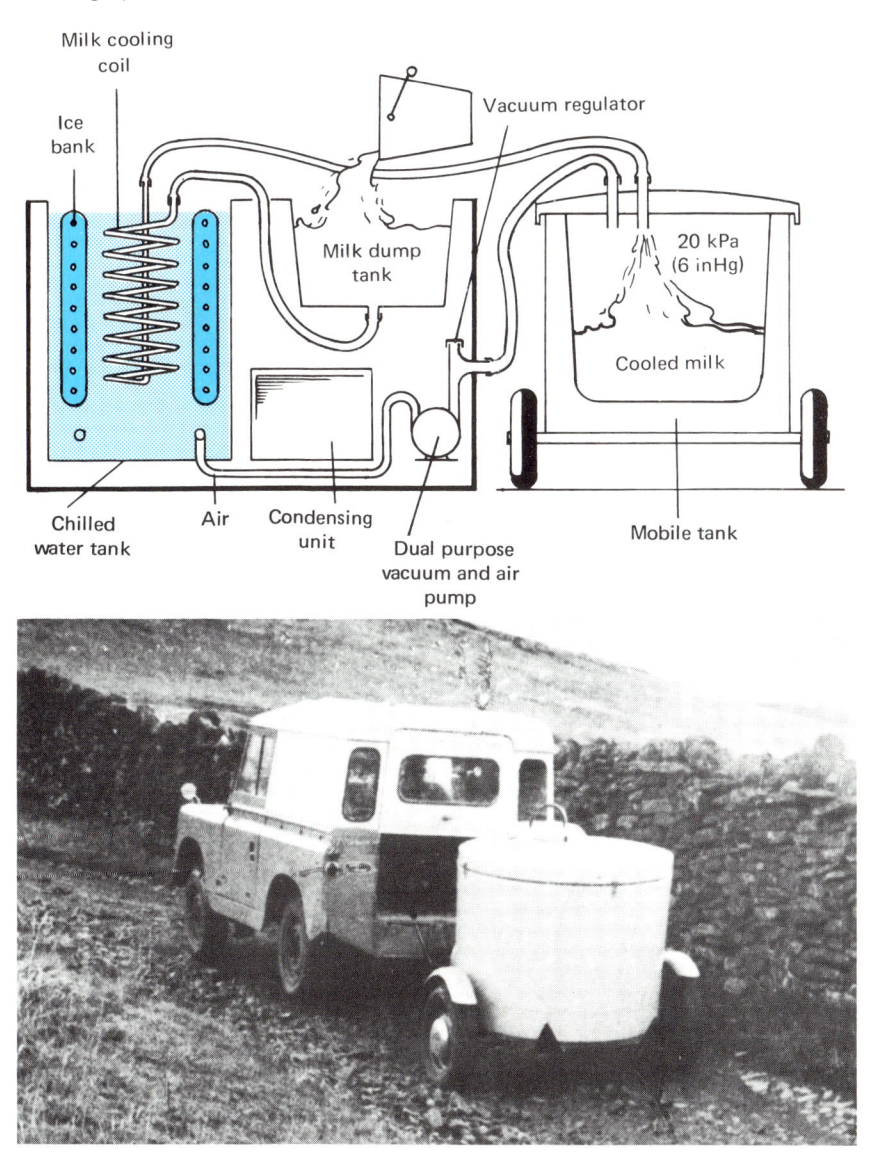

Fig XII 7. MMB chilled water in-line cooler with 500 ℓ (110 gal) mobile tank for roadside collection. The diagram shows the cooling equipment and the photograph the mobile tank being taken to the roadside on a farm in Cumbria.

370 Cooling Equipment

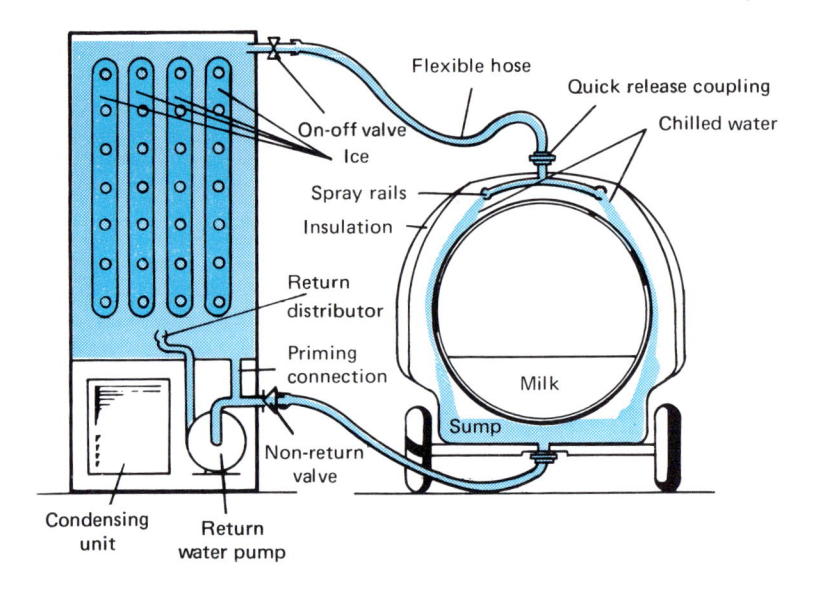

Fig XII 8. Alfa-Laval 500 ℓ (110 gal) chilled water cooled mobile tank. The diagram shows the remote ice bank cooling system and the photograph the mobile tank at a farm.

Alfa-Laval have also produced a mobile equipment which in principle is a conventional refrigerated bulk milk tank with remote ice bank (see Fig XII 5f), with the tank mounted on a chassis and having detachable hose connections (Fig XII 8). It is provided with a 12 V battery to operate the milk agitator motor at the roadside and to drive a submersible pump for spray cleaning. A battery charger is fitted beneath the chilled water tank alongside the condensing unit, which is mains operated.

Both the Alfa-Laval and the MMBEW equipment can be made to hold and cool either 500 ℓ (110 gal) or 270 ℓ (60 gal) of milk.

TANK CLEANING

General

The effects of a hand cleaning mentality on bulk tank construction have been varied and frequently unfortunate. When tanks were first introduced it was normal for dairy equipment to be scalded or steamed at least daily and chemical disinfectants were not allowed in Scotland. However, the Scottish Milk Marketing Board was able to demonstrate in a pilot scheme that chemical disinfection of direct expansion tanks was satisfactory.

Steaming of tanks was specifically disallowed in the UK Federation of Milk Marketing Boards' first joint Specification for Refrigerated Farm Tanks DE 10 issued in 1957. The tanker driver will normally use the hose provided to rinse the tank when it is empty: the tank can then be cleaned at any time before the next milking. Alternatively the tanker driver initiates automatic cleaning equipment daily and the hose is used by farmstaff every month when the tank is scrubbed by hand.

attached have always been demanded to allow easy cleaning. Fortunately for UK constructors the outlet pipe does not have to be straight, though some States of USA insist on this so that all surfaces can be inspected by eye.

The clauses in the UK Federation of MMBs specification related to hand cleaning which caused difficulties were those that required the dipstick, plug, agitator and covers to be readily removable, and also — if of the suspended type — the thermometer and thermostat. Fortunately, after some initial breakage of thermometers and thermostats due to immersion in very hot water or to rough handling, it became generally accepted that these components should be left in place for cleaning. However, an unsatisfactory consequence was that the mountings, which had not been designed for that purpose, were often difficult to clean. Similarly the agitator was not easy to remove and therefore generally stayed in place even though the coupling had not been designed for in-place cleaning. Ideally the agitator coupling should be above both the bridge and the deflector shield so that only a plain shaft has to be cleaned.

Until 1969 the width of the bridge was limited to 685 mm (27 in) for tanks up to 2250 ℓ (500 gal) capacity, with the object of facilitating cleaning by a man reaching beneath it from outside. However, it is doubtful whether the bridge on any but the smallest tanks can be thoroughly cleaned by hand unless the operator gets into the tank. Since the adoption of mechanized cleaning, wider bridges have been allowed and the hygienic state of the underside of bridges and other formerly inaccessible parts has improved.

Mechanized bulk tank cleaning equipment

The need for this equipment was seen in the early 1960s, by which time milking machines could be satisfactorily cleaned in place, but the tank remained as an item for hand cleaning. The main advantages foreseen were reduced labour and improved tank hygiene because mechanized cleaning was expected to be more consistent.

Development. The process and equipment evolved at the NIRD was designed to suit any make of bulk tank and was based on cold chemical cleaning and disinfection. The intention was that the tanker driver should switch on the mechanized cleaning unit instead of hand rinsing the bulk tank after emptying. In common with similar chemical cleaning and disinfection processes applied warm or hot to milking equipment, the cold tank cleaning process consists of a pre-rinse, a detergent-disinfectant wash using iodophor because of its effectiveness at low temperature, a contact period for continued disinfection, and a final water rinse. A minimum contact time of 10 min was then considered necessary but clear advantages could be seen for delaying the final rinse until shortly before the next milking. First, the farm staff could confirm that the iodophor treatment had taken place by observing that all surfaces were coloured and also observe surfaces having accumulated residues because these stain more deeply. Secondly, leaving the iodophor solution in contact for the maximum time available was expected to give maximum opportunity for effective disinfection.

Details of the equipment and process were published in 1966 [9]. Very few modifications have since been found necessary, although improved Mexican hat distributor spinners are now used. These spray the inner surfaces of the bulk tank more searchingly because one of the pivots is floating rather than fixed, giving a more random spray pattern.

Description of the spray cleaning unit. The equipment comprises a 135 ℓ (30 gal) water container fed through a constant flowrate valve and a ball valve at about 450 ℓ/h (100 gal/h) with a measure at one side for approved iodophor (Fig XII 9). An electric-motor-driven centrifulgal pump beneath the tank discharges at about 58 ℓ/min (13 gal/min) through an overhead pipe connected to a tee mounted over the bridge of the bulk tank. This divides the supply to two spinners, one attached to each cover of the tank about 200 mm (8 in) below the rim on the centre line of the tank and about a quarter of its

J.B. Hoyle

length from each end wall. These spinners distribute the water and iodophor over all the tank surfaces. A trough is positioned under the tank outlet so that the dipstick, plug and outlet cap can be soaked in iodophor solution draining from the tank.

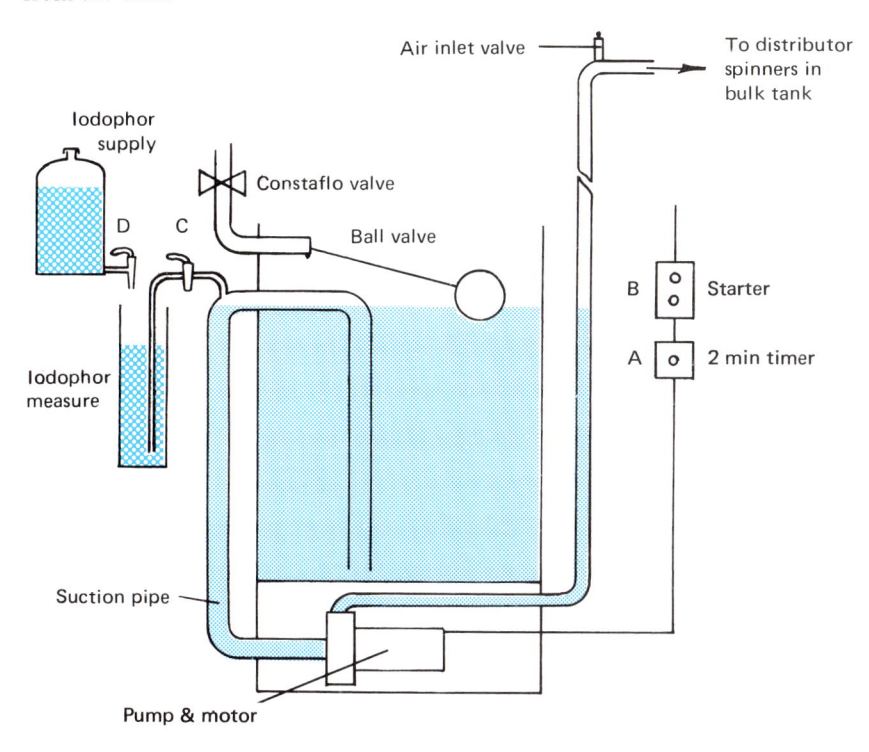

Fig XII 9. A mechanized spray cleaning unit for bulk tanks.

The pump suction is connected into the 135 ℓ (30 gal) container through an inverted U, and the iodophor measure is connected to the top of this inverted U.

The tanker driver collects the milk in the normal way but does not hand rinse the tank. Instead, he places the plug and the outlet cap in the trough beneath the outlet and closes the tank cover. He then starts the cleaning process by pressing 2 min timer A and the shielded motor starter button B. He opens the iodophor tap C and then proceeds on his route. The pump draws water through the inverted U tube and when the water level has fallen about 150 mm (6 in) there is sufficient suction at C to entrain iodophor from the measure. When it has all been entrained air follows, airlocking the centrifugal pump so that flow ceases, leaving iodophor on all the tank surfaces. The pump subsequently cuts out on the time switch, and water in the delivery pipe up to the air inlet valve at its highest point is accommodated in the pump and left leg of the U without mixing into the clean water filling the tank from the ball valve.

Before the bulk tank is next used, the cowman closes off the iodophor connection and restarts the pump on the starter and time switch to give the final cold water rinse. Before the milk is again collected the cowman recharges the iodophor measure from the adjacent supply through valve D: this is usually done while the final cold water rinse is in progress.

Hypochlorite can be included in the final rinse to comply with local requirements, but experience and research show that the tank appearance then deteriorates because of build-up of residues from chemical interaction of chlorine and milk protein, and these can be very difficult to remove [10]

The quantities of iodophor used are 113 ml (4 oz) for a 900 ℓ (200 gal) tank; 170 ml (6 oz) for a 1800 ℓ (400 gal) tank; and 226 ml (8 oz) for tanks of 2250 ℓ (500 gal) and over.

The NIRD equipment was made as simple as possible to give reliability. The only electrical components are the pump motor, the starter, and a two minute timer of a type which is also used on bulk tanks and has been proved dependable. It is a package unit although the electrics are wall mounted.

Commercial equipment. Gascoignes followed the NIRD pattern but with the iodophor transit container kept on a table instead of on a bracket on the equipment. These spray cleaning units were shown to be dependable [11].

Other manufacturers tended to complicate the equipment by introducing the concept of fully automatic cleaning. This meant that the whole process was completed under automatic control once the tanker driver had operated a switch. It could not be achieved without the introduction of a process timer and additional electromechanical components. The need to include hypochlorite in the final rinse in some areas of the UK led to further complications.

Manufacturers have recently introduced simpler equipment nearer to the original pattern but chemical is drawn directly from transit containers placed on the floor. Some at least have reverted to a two-part process, the second initiated by the cowman, and it is probably true that the cowman's interest is better maintained if he takes an active part in the tank cleaning process.

ENERGY SAVING

Electrical energy used by refrigerated milk coolers may be reduced by means of pre-coolers and heat recovery units may be used to produce hot water from the heat rejected by the bulk tank refrigeration unit and so save the energy normally required by electric immersion heaters.

Pre-coolers

At present the only pre-coolers available are plate heat exchangers similar in principle to those used for pasteurizing milk. The economics of the operation depends on using water already required for other purposes, such as yard washing or for the cattle to drink.

At present over the whole range of bulk tanks about 45 ℓ (10 gal) of milk are cooled for each unit (kWh) of electricity: at a cost of 2p per unit it therefore costs the farmer 20p per day or £73 per year per 450 ℓ (100 gal) cooled. If the milk can be pre-cooled to 20°C (68°F) about half this money for energy can be saved. The saving is probably therefore only worthwhile if the pre-cooler can be obtained very cheaply and if the average daily gallonage exceeds 450 ℓ (100 gal per day.

Heat recovery units

These units should be installed by competent refrigeration engineers because it is necessary to connect the heat recovery unit between the refrigeration compressor and the air cooled condenser. One method now in use is shown diagrammatically in Fig XII 10.

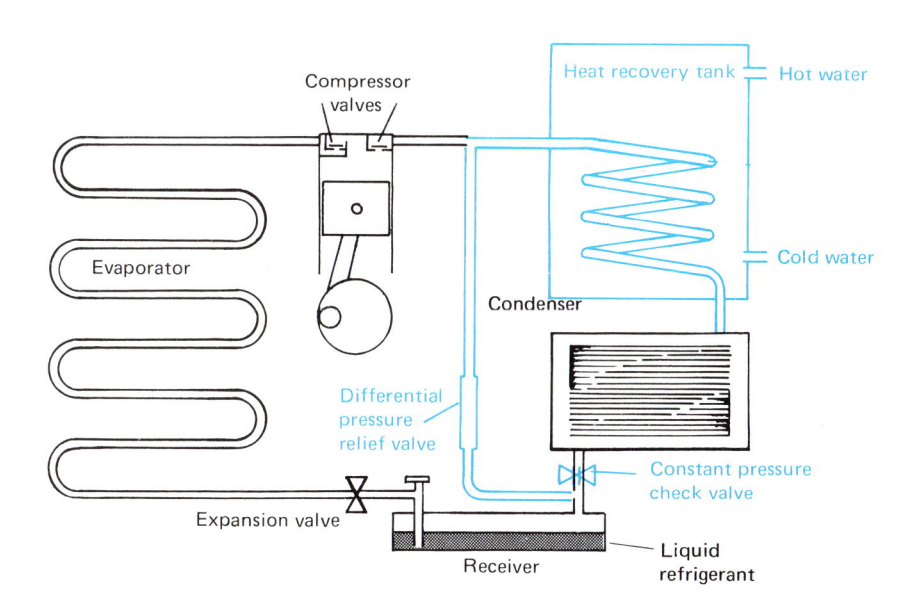

Fig XII 10. Schematic layout of heat recovery unit. The items which have been added to a normal condensing unit (see Fig XII 4) are shown in blue.

The hot gas from the compressor goes through the water heating coil first and most of it condenses and gives up its heat to the water when the water is cold. As the water temperature rises less of the heat from the condensing gas flows into the water and more is extracted by the air cooled condenser. Finally, when the water and gas temperatures are equal, all the heat is extracted by the air cooled condenser. The check valve ensures that the condensing pressure and temperature are high enough to be useful in heating the water, but also ensures that the pressure does not exceed that which the compressor reached on test at an air temperature of 32°C (90°F), i.e. about 1 MPa (150 lb/in^2).

If too much liquid were to condense in the heat recovery section due to very low incoming water temperature the evaporator could be starved of liquid due to loss of pressure. A relief valve is therefore fitted so that if the pressure in the receiver falls below 500 - 700 kPa (75 - 100 lbs/in^2) gas can pass straight through from the compressor to the receiver.

Whether heat recovery units will be worthwhile depends on their initial cost, the quantity of hot water required for all purposes, and the quantity of milk to be cooled. For a 60 cow herd with a 1100 ℓ (250 gal) tank, 66% filled on average, the quantity of hot water available would be about 135 ℓ (30 gal) at 60°C (140°F) at every milking. Thus the value of the heat recovered, in terms of electrical energy to heat the same quantity of water is about £60/year at 1976 prices (equivalent to the selling price of about 2 ℓ (0.5 gal) of milk a day). Thus ample heat is available for water heating but its value justifies only small additional capital outlay. Also, it is a disadvantage that hot water temperatures much above 60°C (140°F) are not readily produced.

If more milk is cooled, more heat is available. There may be farms producing an average of 1350 ℓ (300 gal) of milk per day where it would be feasible to use both a pre-cooler and a heat recovery unit, because generally the quantity of hot water required is not directly proportional to the gallonage cooled, but it may not always be easy to arrange to use the water the cattle drink for precooling.

POSSIBILITIES FOR THE FUTURE

Tank design has not been influenced by the general adoption of metering of milk from the farm bulk tank into the road tanker in Scotland, though some calibration costs have been saved. In England and Wales the introduction of metering can have little influence until all the collection tankers carry meters.

But if metering becomes general radical new designs of cooling equipment are possible for the replacement market of 2000 to 3000 tanks per year. There would be no constraints as regards specific capacity or rigidity and there need be no dipstick or calibration, all of which should lead to less expensive equipment.

Theoretically all sizes of tank could have direct expansion cooling systems, subject only to the availability of an adequate electricity supply or an alternative power unit.

Because there would be no need to have the milk surface still for dipstick measurement, tanks could be provided with continuously running slow speed agitators, so that the milk would never separate by stratification of the fat during storage, and a fair sample could be taken by the tanker driver immediately or from the pipeline during transfer, thus saving time. Furthermore as there would be no cream layer to adhere to the tank walls tank cleaning would be easier. Tanks would in any case be built specifically for mechanized cleaning and could be virtually clear of internal fittings, the only necessities being an agitator with its shaft, and access. Artificial illumination could be provided through a flush mounted translucent panel.

Whatever the outcome on milk measurement it should be possible to effect economies in overall costs by the judicious use of pre-cooling, heat recovery and mechanized cleaning.

REFERENCES

(1) Cheeseman, T. (1972) *Journal of the Society of Dairy Technology,* 25, 82.

(2) Thomas, S.B. & Thomas, B.F. (1973) *Dairy Industries,* 38, 11.

(3) Marth, E.H. & Frazier, W.C. (1957) *Journal of Milk and Food Technology,* 20, 72.

(4) Thomas, S.B. & Druce, R.G. (1971) *Dairy Science Abstracts,* 33, 339.

(5) 3-A Sanitary standards for farm milk cooling and holding tanks (1975) *Journal of Milk and Food Technology,* 38, 646.

(6) Specification BC 56, 2nd edition, October 1976. *Refrigerated farm milk tanks.* UK Federation of Milk Marketing Boards, Thames Ditton, Surrey. (The performance of farm tanks is also covered by BS 3976: 1966.)

(7) Hoyle, J.B. & Belcher, J.R. (1971) *Dairy Farmer,* 18, (12), 41.

(8) Slade, J.R. (1974) *Journal of the Society of Dairy Technology,* 27, 98.

(9) Cousins, C.M., Dawkins, J. & Hoyle, J.B. (1966) *Farm Mechanization,* 18, (204), 16.

(10) Jensen, J.M. (1970) *Journal of Dairy Science,* 53, 248.

(11) McKinnon, C.H. & Cousins, C.M. (1969) *Journal of the Society of Dairy Technology,* 22, 227.

TERMINOLOGY AND DEFINITIONS
(Index p. 383)

Extract from October 1975 draft International Standard ISO/DIS 3918.
Geneva: International Organisation for Standardisation.

The terms defined in the October 1975 ISO draft vocabulary for milking machine installations have been used throughout this book. This standard has not yet been confirmed by member countries so some terms or definitions may still be modified. In the following list the definitions of the draft document are reproduced in English only but the terms are also given in French, the other ISO Official language. In addition a German version of the terms is included, which was kindly supplied by Professor A. Tolle and Mr M. Helmrich. The English index of the ISO document is reproduced at the end of the list of terms and definitions.

Many of the terms are used in the illustrations in Fig III 1 at the beginning of Chapter III, Description and Performance of Components. The list of terms is by no means exhaustive but its systematic use should prevent many misunderstandings.

1 GENERAL TERMS (TERMES GÉNÉRAUX; ALLGEMEINE BEGRIFFE)

1.1 milking machine (machine à traire; Melkanlage): A complete machine installation for milking, usually comprising vacuum and pulsation systems, one or more clusters and other components.

1.2 unit (poste de traite; Melkeinheit): That assembly of milking machine components which is replicated in an installation so that more than one animal may be milked at one time.

1.3 Types of milking machines (Types de machines à traire; Melkanlagentypen).

1.3.1 bucket milking machine (machine à traire avec pot trayeur; Eimermelkanlage): A milking machine in which milk flows from the cluster into a portable milk receiving bucket connected to the vacuum system (see Fig III 1a).

1.3.2 direct-to-can milking machine (machine à traire directement en bidon (cruche); Kannenmelkanlage): A milking machine in which milk flows from the cluster into the transport can, which is connected to the vacuum system.

1.3.3 milking pipeline machine (machine à traire avec lactoduc de traite; Rohrmelkanlage): A milking machine in which milk flows from the cluster into a pipeline that has the dual function of providing milking vacuum and conveying milk to a milk receiver (see Fig III 1b).

1.3.4 recorder milking machine (machine à traire avec récipient de contrôle; Messbehaltermelkanlage): A milking machine in which milk flows from the cluster into a recorder jar under vacuum from an air vacuum pipeline. Milk is discharged when required from the recorder jar into a transfer pipeline to a milk receiver or collecting vessels under vacuum (see Fig III 1c).

1.3.5 independent air and milk transport milking machine (machine à traire avec circuit indépendant; Melkanlage mit getrenntem Luft- und Milchtransport): A milking machine in which air and milk are separated immediately below the teatcups and then transported in separate pipelines.

2 CLUSTER ASSEMBLY (FAISCEAU TRAYEUR; MELKZEUG-ZUSAMMENSTELLUNG)

2.1 cluster (faisceau trayeur; Melkzeug): An assembly comprising teatcups, claw, long milk tube and long pulse tube.

2.2 teatcup (gobelet trayeur; Zitzenbecher): An assembly consisting of a rigid shell (or case) with a short pulse tube and a liner.

2.3 liner (inflation) (manchon-trayeur; Zitzengummi): A flexible sleeve having a mouthpiece, a barrel and an integral or separate short milk tube.

2.4 pulsation chamber (chambre de pulsation; Pulsraum): The annular space between the liner and the shell.

2.5 short milk tube (tuyau court à lait; kurzer Milchschlauch): The connecting tube between the interior of the liner and the claw milk nipple.

2.6 short pulse tube (tuyau court de pulsation; kurzer Pulsschlauch): The connecting tube between the pulsation chamber and the claw air nipple.

2.7 claw (griffe; Sammelstück): The manifolds that space the teatcups in forming a cluster and connect them to the long milk and pulse tubes.

2.8 air admission hole (orifice d'admission d'air; Lufteinlass): An aperture in the cluster to admit air.

2.9 long milk tube (tuyau long à lait; langer Milchschlauch): The connecting tube between the claw and a milk collecting vessel or a milk pipeline.

2.10 long pulse tube (tuyau long de pulsation; langer Pulsschlauch): The connecting tube between the claw and the pulsator.

2.11 vacuum tube (tuyau à vide; Luftschlauch): The connecting tube between a milk collecting vessel and the air pipeline.

3 VACUUM SYSTEM (INSTALLATION DE VIDE; VAKUUMSYSTEM)

3.1 vacuum pump (pompe à vide; Vakuumpumpe): An air pump which produces vacuum in the system.

3.2 air pipeline (milking vacuum) (canalisation à air (vide de traite); Luftleitung (Melkvakuum)): A pipeline which forms part of the fixed installation and which carries only air during milking; the pipeline may also act as part of the cleaning circuit.

3.3 air pipeline (pulsators) (canalisation à air (pulsateurs); Luftleitung (Pulsatoren)): A pipeline which carries only air and forms part of the pulsation system.

3.4 interceptor (vacuum tank) (intercepteur; Vakuumtank): An interceptor vessel situated in the main airline immediately upstream from the vacuum pump to prevent liquid or foreign matter gaining access to the pump.

3.5 vacuum gauge (indicateur de vide; Vakuummeter): A differential pressure gauge to indicate the level of vacuum in the system.

3.6 regulator (régulateur de vide; Vakuumventil): An automatic valve designed to maintain a steady vacuum.

3.7 sanitary trap (piège sanitaire; Sicherheitsabscheider): The vessel between the milk system and the air system to prevent contamination by movement of liquid from one to the other.

4 MILKING SYSTEM (SYSTÈME DE TRAITE: MILCHSYSTEM)

4.1 milking pipeline (lactoduc de traite; Melkleitung): A pipeline which carries milk and air during milking and has the dual function of providing milking vacuum and conveying milk to a milk receiver.

4.2 transfer pipeline (lactoduc de transfert; Milchtransportleitung): A pipeline through which milk is conveyed under vacuum from the recorder jar to the milk receiver or a collecting vessel under vacuum.

4.3 delivery pipeline (lactoduc d'évacuation; Milchdruckleitung): A pipeline in which milk flows under positive pressure from a releaser milk pump or recording vessel to a storage vessel.

4.4 recorder jar (récipient de contrôle; Messbehälter): A vessel which receives, holds and allows measurement of all the milk from the individual animal.

4.5 milk meter (compteur à lait; Milchmengenmessgerät): A device between the cluster and the milking pipeline for measuring an animal's milk yield.

4.6 receiver (chambre de réception; Milchabscheider): A vessel which receives milk from one or more milk pipelines and feeds the releaser, releaser milk pump, or collecting vessel under vacuum.

4.7 releaser (extracteur de lait; Kammermilchschleuse): A mechanism for removing milk from under vacuum and discharging it to atmospheric pressure.

4.8 releaser milk pump (pompe extractrice; Pumpenmilchschleuse): A device for pumping milk out of the vacuum system.

5 PULSATION SYSTEM (SYSTÈME DE PULSATION; PULSSYSTEM)

5.1 pulsation (pulsation; Pulsierung): Cyclic opening and closing of a teatcup liner.

5.1.1 simultaneous pulsation (pulsation simultanée; simultane Pulsierung): When cyclic movement of all liners within a cluster is synchronized.

5.1.2 alternate pulsation (pulsation alternée; alternierende Pulsierung): When cyclic movement of half the number of liners in a cluster alternates with the movement of the other half.

5.2 pulsation cycle (cycle de pulsation; Pulszyklus): One complete liner movement sequence.

5.3 pulsation rate (fréquence de pulsation; Pulszahl): The number of pulsation cycles per minute.

5.4 milking ratio (rapport de traite; Milchfliessverhältnis): The percentage of the pulsation cycle during which milk flows from the teat.

5.5 pulsator ratio (rapport du pulsateur; Pulsverhältnis): The duration of the increasing vacuum phase and the maximum vacuum phase as a percentage of the complete pulsation cycle in the pulsation chamber vacuum record. The ratio is expressed by the following formula (see figure in 6.7): .

$$\frac{a+b}{a+b+c+d} \times 100\%$$

5.6 pulsator (pulsateur; Pulsator): A device for producing cyclic pressure change.

5.7 pulsator controller (générateur de pulsation; Pulssteuerung): A mechanism to operate pulsators, either integral with a single pulsator (self-contained pulsator) or a system controlling several pulsators.

6 MEASUREMENT (MESURAGE; MESSUNG)

6.1 vacuum (vide; Vakuum): Any pressure below atmosphere pressure, measured as the extent of the reduction below ambient atmospheric pressure. The site of measurement should be stated, for example: liner vacuum, pulsation chamber vacuum, claw vacuum.

6.2 free air (air libre; atmosphärische Luft): Volume of air at 20°C and 1 bar atmospheric pressure.

6.3 expanded air (air expansé; expandierte Luft): Volume of air at ambient atmospheric temperature at a given vacuum level.

6.4 vacuum pump capacity (débit de la pompe à vide; Vakuumpumpendurchfluss): The air-moving capacity of the vacuum pump when it has attained working temperature, at a specified pump speed and vacuum level at the inlet, expressed in volume of free air per minute.

6.5 effective reserve (réserve réelle; Reservedurchfluss (effectiv): Reserve vacuum pump capacity measured by admitting air near the regulator to lower the vacuum 2.0 kPa (approximately 0.6 inHg) below that existing when all units (with the liners stopped) and accessories (excluding the vacuum regulator) are operating.

6.6 manual reserve (réserve régulateur hors service; Reservedurchfluss (nominell): Reserve vacuum pump capacity measured by admitting air near the regulator to lower the vacuum 2.0 kPa (approximately 0.6 inHg) below that existing when all units (with the liners stopped) and accessories (excluding the vacuum regulator) are operating.

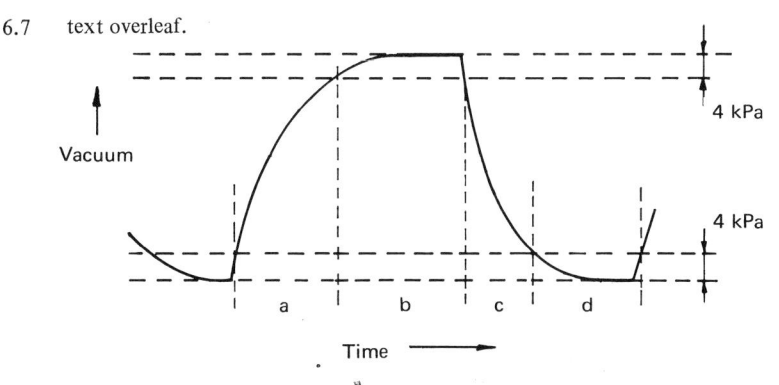

6.7 text overleaf.

Fig 6.7 Pulsation chamber vacuum record (see also 5.5, pulsator ratio)

6.7 pulsation chamber vacuum record (enregistrement du vide dans la chambre de pulsation; Pulsierungskurve (Pulsraum): Each cycle of the record of pulsation chamber vacuum is described as having four phases (see figure on page 381).

INDEX FOR TERMINOLOGY AND DEFINITIONS
(pp. 378 – 382)

A

air admission hole	2.8
air pipeline (milking vacuum)	3.2
air pipeline (pulsators)	3.3

B

bucket milking machine	1.3.1

C

claw	2.7
cluster	2.1
cluster assembly	2

D

delivery pipeline	4.3
direct-to-can milking machine	1.3.2

E

effective reserve	6.5
expanded air	6.3

F

free air	6.2

I

independent air and milk transport milking machine	1.3.5
interceptor (vacuum tank)	3.4

L

liner (inflation)	2.3
long milk tube	2.9
long pulse tube	2.10

M

manual reserve	6.6
measurement	6
milk meter	4.5
milking machine	1.1
milking machines (types of)	1.3
milking pipeline	4.1
milking pipeline machine	1.3.3
milking ratio	5.4
milking system	4

P

pulsation	5.1
pulsation chamber	2.4
pulsation chamber vacuum record	6.7
pulsation cycle	5.2
pulsation rate	5.3
pulsation system	5
pulsator	5.6
pulsator controller	5.7
pulsator ratio	5.5

R

receiver	4.6
recorder jar	4.4
recorder milking machine	1.3.4
regulator	3.6
releaser	4.7
releaser milk pump	4.8

S

sanitary trap	3.7
short milk tube	2.5
short pulse tube	2.6

T

teatcup	2.2
transfer pipeline	4.2

U

unit	1.2

V

vacuum	6.1
vacuum gauge	3.5
vacuum pump	3.1
vacuum pump capacity	6.4
vacuum system	3
vacuum tube	2.11

SOME CONVERSION FACTORS

	Imperial to metric			Metric to Imperial		
Length	1 in	=	25.4 mm	1 mm	=	0.0394 in
Area	1 ft^2	=	0.093 m^2	1 m^2	=	10.7 ft^2
Volume	1 ft^3	=	0.028 m^3	1 m^3	=	35.7 ft^3
	1 gal	=	4.546 ℓ	1 ℓ	=	0.2200 gal
Mass	1 lb	=	0.4536 kg	1 kg	=	2.205 lb
Vacuum	1 inHg	=	3.386 kPa	100 kPa	=	29.53 inHg
		=	0.03386 bar	(1 bar)		
Pressure	1 lb/in^2	=	6.89 kPa	100 kPa	=	14.49 lb/in^2
		=	0.0689 bar	(1 bar)		
Pressure, liquid columns			1 mmHg =	0.133 kPa =	1.33 mbar	
			1 inHg =	3.386 kPa =	33.86 mbar	
			1 inch water =	0.249 kPa =	2.49 mbar	
Temperature	t°F	=	$5/9$ (t−32)°C	t°C	=	$9/5$ t + 32°F
Power	1 hp	=	$3/4$ kW (approx)	1 kW	=	1$1/3$ hp (approx)
Work, energy,	1 Btu	=	1.055 kJ	1 kJ	=	0.9479 Btu
heat		=	0.293 Wh	1 Wh	=	3.413 Btu
		=	0.252 kcal	1 kcal	=	3.968 Btu
Heat transfer	1 Btu/ft^2	=	11.36 kJ/m^2	1 kJ/m^2	=	0.0880 Btu/ft^2

Volume to Weight conversions of milk (quoted in Dairy Facts and Figures 1975, Federation of UK Milk Marketing Boards)

1 Imp gal of milk = 10.32 lb (density 1.032)

1 litre of milk = (1.030 kg (density 1.030)

Names and addresses of suppliers mentioned in the text

Albright & Wilson Ltd., Detergents & Chemicals Group, P.O. Box 3, Oldbury, Warley, West Midlands B68 0NN.

Alfa-Laval Company Ltd., Oakfield Road, Cwmbran, Gwent NP4 3XE.

A.P.V. Company Ltd., P.O. Box 4, Manor Royal, Crawley, West Sussex RH10 2QB.

Blow, J.J. Ltd., Oldfield Works, Chatsworth Road, Chesterfield, Derbyshire S40 2DJ.

Froment, N.J. Ltd., Cliffe Road, Easton-on-Hill, Stamford, Lincolnshire PE9 3NP.

Fullwood, R.J. & Bland Ltd., Fullwood Works, Ellesmere, Shropshire SY12 9DG.

Gascoigne Gush & Dent (Agricultural) Ltd., Berkeley Avenue, Reading, Berskhire RG1 6JW.

G.E.C. Elliott Process Instruments Ltd., 330 Purley Way, Croydon, Surrey CR9 4PG.

Godfrey, W.M. & Partners Ltd., Brenchley, Tonbridge, Kent TN12 7LS.

Hosier Equipment (Collingbourne) Ltd., Collingbourne Ducis, Marlborough, Wiltshire SN8 3EH.

Hydrovac Products, P.O. Box 420, Glen Waverley, Victoria 3150, Australia.

Kingston Group Marketing, Upton Road, Reading, Berkshire RG3 4BJ.

Lister, R.A. & Company Ltd., Long Street, Dursley, Gloucestershire GL11 4HS.

Loheat Ltd., Everland Road, Hungerford, Berkshire RG17 0DU.

Manus (Great Britain) Ltd., Manus Works, London Road, Moreton-in-Marsh, Gloucestershire GL56 0HJ.

Precision Metal Spinnings Ltd., Masons Road, Stratford-upon-Avon, Warwickshire CV37 9NU.

Simplex of Cambridge Ltd., Sawston, Cambridge CB2 4LJ.

The Milk Marketing Board of England and Wales, Thames Ditton, Surrey KT7 0EL.

Vaccar Ltd., Southern Avenue, Leominster, Herefordshire HR6 8QH.

Wenz, Societé à Responsibilité Limitée, P.O. Box 12, Bradford, Yorkshire BD12 0PZ.

Westfalia Separator Ltd., Habig House, Old Wolverton, Milton Keynes, Buckinghamshire MK12 5PY.

Weycroft Macford Ltd., Lake Road, Hamworthy, Poole, Dorset BH15 4LG.

INDEX

Abreast layout in parlours 207-8
Acidified boiling water cleaning 297-8, 304-12
Acton 6 *see* Dichlorodifluoromethane
Adrenaline and milk-ejection reflex 175-6
Adrenocorticotrophic hormone and milk secretion 170
Ageing of rubbers 346-9
Air admission, measurement 108-9
Air admission to claw
    effect on liner vacuum 134
    and mastitis 262-3
Air admission to milk chamber 62
Air consumption and vacuum level 110
Air flowrate, measurement 106-9, 113
Air leakage past teat 142-3
Air: milk ratio 31
Alcohol in teat dips 250
Alveoli 160-3
    development 168-9
Antibiotic therapy for mastitis 270-3
Automatic cluster remover 75

Bacteria in milk
    effect of storage temperature 355-7
    multiplication in refrigerated raw milk 288-90
    sources 286-8
    tests 318
Bails 15-6, 208, 214
    milk cooling 367
Bedding and mastitis 265, 267
Blistering of liner surfaces 346
Blood supply to udder 160, 165-7
Boiling, theoretical aspects 26
Bovine herpes mammilitis virus 256
BR *see* Polybutadiene rubber
Bucket lid assemblies 65
Bucket milking
    cleaning 292-3
    equipment 63-6
    operation 28-9
Bulk milk agitation 366-7
Bulk milk collection, development 353-9
Bulk milk cooling equipment 353-77
    for large quantities 367-8
    for small quantities 368-71
    standard range 361-7
Bulk milk tanks
    cleaning 313-6, 371-4
    design 358-9
    equipment 361-71
    refrigeration cycle 359-60
    specifications 360-1
Bulk milk temperature 357, 363
    control 365-6
    refrigeration cycle 359-60
Buna-S *see* SBR
Butadiene styrene copolymer 333
Buttering in bulk tanks 367
Butyl rubber 334

Calender for sheet rubber 339
California Mastitis test 236
Can collection of milk 354-6
Cannula milking 1
Caustic flooding 312-3
Caustic soda effects on rubbers 349
Censuses of machines 21
Chemicals, storage and handling 325
Chlorhexidine as teat dip 247-53
Chloride ions in milk 172
Chlorinator, in-line, for hose-washing 290
Chloroprene *see* Neoprene
Chlorosulphonic polyethylene 335
Chute parlours 208
Cine photography of liner movements 126-7
Cine X-rays of liner movements 124-6
Circulation cleaning 214-5, 301-12
Claws
    design, and mastitis 262-3
    development 9
Cleaning dairy equipment 291-325
    costs 316
    daily routine 293-4
    manual 293
    monitoring 317-9
    selection of agents 324-5
    trends 316-7
Cleaning and sterilizing
    effects on rubbers 348-9
Cleaning units for pipelines 101
Clusters
    action during milking 116-55
    automatic removal 73-6
    cleaning 298
    connections to pipelines 76-80
    design 59-63
    maintenance 103-4

387

Clusters *Cont'd.*
   and mastitis 246, 247, 264
   and vacuum fluctuations, 147-8
   weight, and milking 122-3
   wet storage 295
Coliforms and mastitis 232, 265, 268
Coliforms in milk 287
Concentrate feeding and milk ejection 192
Contamination of milk, sources 286-8
Controllers, pulsator 57-8
Cooling equipment for milk 353-77
Cooling of vacuum pumps 43-44
Corynebacteria and mastitis 232-3, 240, 273
Cows, effects of liner design 122
Cowsheds, milking 202-4

Dairy equipment 292-3
   testing for disinfection 318
Descaling equipment 295, 303
Detectors for end of milking 74
Detergents for milking equipment 320-1
   effect on rubbers 349
Development of milking machines 1-22
Dichlorodifluoromethane
   bulk tank refrigerant 359
Dichlorophenol in teat dips 250
Dipsticks, accuracy 358
Direct-to-can units
   design 28-9
   operation 63-4
Dirt in milk, sources 290
Disinfectants for milking equipment 322-5
Disinfection of clusters 247
   of teats 245, 249
Drainage checking 103
Dry period, effect on milk yield 169
Drying off therapy 235, 272-3

Electrical conductivity of milk
   test for udder infection 273
Electrical pulse transmission 16-8
Electricity, stand-by 88-91
Endocrine control of milk secretion 170-1
Ethylene propylene rubbers 335
Exhaust systems for pumps 44

Fat absorption by rubbers 344-6
Fat secretion in milk 171
Filters 92-6
Flexing deterioration of rubbers 348

Foremilk examination 244, 288
Foremilking and mastitis 239-41
Freon 12 *see* Dichlorodifluoromethane
Friction between teat and liner 144

Gauges, vacuum 50
Glycerol for teat chaps 248, 258, 277
Growth hormone and milk secretion 170
GR-S *see* SBR

Hand milking, cleaning 292
Heat deterioration of rubbers 346
Heat disinfection of equipment 295-6
Heat recovery units
   bulk milk refrigeration 375
Herringbone layout in parlours 208
Hexachlorophane as teat dip 250
History of machine milking 1-22
Hosier bail 15-6
Housing and mastitis 264-5, 267
Hygiene, legal requirements 326-7
   machine milking 18-9, 241-58
Hypalon 335
Hypochlorite disinfectant
   bulk tanks 374
   milking equipment 322-5
   rubbers, effects 349
   teat dips 247-50, 256, 274, 277
Hypothalamus and milk ejection reflex 173

Ice banks
   arrangement 361-3
   control 363-5
Immersion cleaning 19, 214, 312
Impact effect and mastitis 148-50, 261
In-line cooler for milk 368
In-line filters 241
In-place cleaning of milking equipment 292, 297-9
Infection of udder 232-65
   subclinical 269
Interceptors 45-6
Iodophor disinfectant
   bulk tanks 372
   teat dips 247-53, 256, 277

Jetters 309

Lactation, physiology 170-7
Lactators 3
Lactose and osmotic pressure in milk 172

Lanolin for teat lesions 277
Let down 172-7
Lid assemblies 64
Lids, rubber 351
Liners
   action in relation to teat 136-45
   characteristics, effects 121-2
   defatting 295
   description 59-61
   effect on teat during pulsation cycle 124-7
   movement, cine photography 126-7
      cine X-rays 124-6
      and milk flow 127
      and teat 141-5
   vacuum effects 130-4
   wall movement
      factors affecting 129, 132-5
      rate 130-2
Lipolysis in bulk tanks 367
Liquid level controls 84-5
Liquid ring vacuum pump 42-3
Lubricating vacuum pumps 40, 42
Lymphatic system, udder 167

Maintenance of equipment 102-4
Mammary glands
   anatomy and physiology 156-78
Mamilitis 256
Manometers 27-8
Mastitis
   control 267-85
   detection filters 95-6
   effect of liner design 122
   effect of vacuum fluctuation 148
   effect of milk yield and composition 236-8
   incidence 231-67
Metering of milk
   description of meters 71-2
   effect on bulk milk cooling 376-7
Micrococci and mastitis 233
Milk cock 77
Milk composition
   effect of mastitis 236-8
Milk constituents 171
Milk contamination, sources 286-8
Milk ejection
   and machine milking 190-3
   physiology 172-7
Milk fat
   effect of milking interval on yield 181-4, 187
   secretion rate 180-3

   variation during milking 182-3
Milk filters 92-6
Milk flow indicators 69-70, 74
Milk flow from udder
   effects on liner vacuum 134
   factors affecting 117-22
   measurement during pulsation cycle 128-9
   pattern within pulsation cycle 123-9
Milk flowrate 194
   effect of streak canal bore 195
   effect on vacuum in cluster 147
   patterns during milking 117-23
   and teat movement 136-41
Milk meters 71-3
Milk pressure, intramammary 185-6
Milk pumps 81-2
Milk residues in bulk tanks 315
   in milking equipment 304-7
Milk sampling 66-7
Milk secretion
   effect of milking interval 185
   physiology 170-1
   rate 180-5
Milk stone 294
Milk testing
   bacteriological tests 318
   sediment tests 317
Milk transfer control valve 68
Milk transport systems 76-85
Milk tubes
   effects on vacuum fluctuation 147-8
Milk yield
   effect of dry period 169
   effect of mastitis 236-8
   effect of milking interval 181-7
   effect of milking rate 193-6
   graphs 118
Milking
   checks during 102-3
   delayed, effect on milk secretion 180-1
   duration 194-5
   frequency 186-8
   incomplete 188-90
   mechanics 116-55
Milking characteristics
   effect of liner design 121-2
Milking equipment
   cleaning and disinfection 291
   contamination of milk 288-9
   testing 318-9
Milking intervals 186
   effect on rate of milk secretion 185
   effect on yield 181-4, 187
Milking installations 28-33

Milking machines
  growth of use 21
  trials 12-4
  types 38
Milking performance
  work study 205, 218-27
Milking pipeline machines 69
Milking pipelines 76-80
Milking rate
  breeding to improve 195
  effect on milk yield 193-6
Milking routine 179-200
  cowsheds 203-4
  and mastitis 239-41, 263-7,
  and milk ejection 190
  parlours 206-14, 220-9
Milking stimulus 188
Milking systems, distribution 37, 291
Milking units 59-76
Mouthpiece chamber vacuum
  effect of air leakage past teat 142
Myoepithelial cells 174-5

Neoprene 334
Nerve supply to udder 107-8
NIRD experiments on mastitis control
  251, 274-7, 280
Nitrile rubber 334

Overmilking and mastitis 264
Oxytocin and milk ejection 173-4
Ozone deterioration of rubbers 346-8

Papers, temperature sensitive 310
Paravaccinia virus and teat lesions 256-7
Parlours
  high performance 228-30
  history 19
  and mastitis 252
  milking routine 206-14, 229
  selection 224-8
  types and layout 206-16
Pathogens, udder 232
  effect of milking 239
  mode of infection 238-9
  sources 240
  transfer during milking 241-3
Pipeline, air 50-1
  cleaning 296
Pipeline components, rubber 350
Pipeline milking installations 202-17
  in cowsheds 202
  development 18-9

Pipeline milking machines
  cleaning 301-3
  description 29-30, 69
  and mastitis 263, 266
Pituitary and milk secretion 170-1
Polybutadiene rubber 333
Polyisoprene rubber 333
Polyurethane rubber 335
Pre-coolers and bulk milk refrigeration 375
Pressure, discussion 23-4
  measurement 26-7
Pressure, barometric effects 33-4, 111-2
Prime movers for pumps 37-9
Prolactin and milk secretion 170
Protein secretion in milk 171
Pseudocowpox and teat lesions 256-7
Pseudomonads and mastitis 268
Pulsation
  and mastitis 261-2, 266
  milking without 141
  and teats 141-2
Pulsation chamber
  vacuum record 115
Pulsation characteristics
  and milk flow 120-1
Pulsation cycle and milk flow 128-9
Pulsation equipment 52-8
Pulsation rates 15
  and milk flow 128
Pulsation ratio and liner wall
  movement 132
Pulsation system, mechanical 58
Pulsation tests 112-3
Pulsator controllers 57-8
Pulsators 52-8
  electromagnetic relay 52-4
  maintenance 103-4
  pneumatic 54-7
  properties 58
  Shiels' patent 7
Pulse transmission, electric 16-8
Pressure-type machines 3-5
Pumps, releaser 80-4
Pumps, vacuum 33, 37-45
  maintenance 102
  testing 105-9

Quaternary ammonium compounds for
  milking equipment 322-4

R 12 see Dichlorodifluoromethane
Receivers 80
Recorder, vacuum 107, 114

Recorder jars 66
　　positioning in parlours 215-6
Recorder milking machines
　　cleaning 298, 301
　　description 29-30
Recorder unit 67
Refrigeration cycle for bulk milk tanks 359-60
Regulators, vacuum 46-9
　　maintenance 104
　　testing 110-1
Releaser milk pumps 80-4
Resazurin test 318
Residual milk 174, 180, 182
　　effect of milking routine 190
　　fat percentage 183
Reverse-flow cleaning 313
Rotary milking parlours 96, 209-14
　　routines 229
Rubber
　　adhesion to glass and metal parts 351
　　manufacture 335-42
　　natural 331-2
　　properties 343-4
　　synthetic 333-5
Rubber hand teatcup 11

Sampling of milk 66-7
Sanitary trap 46
　　see also Interceptors
SBR 333
Sediment tests 317
Shells 59
Shield for teat orifice 150
Silicone rubbers 334-5
Sodium ions in milk 172
Somatic cell count and mastitis 232, 279-81
Spit chamber releasers 83-4
Stall cocks 52
Stand-by equipment 85-92
Staphylococci and mastitis 232-76 *passim*
Staphylococci in milk 288
Steam disinfection of equipment 292
Stimulation
　　effect on udder pressure 191
Storage of milk
　　multiplication of bacteria 289
Streak canal
　　cine X-rays 124-6
　　penetration by infectious agents 238-9
　　physiology 129

Streak canal bore
　　effect on milk flowrate 194-5
Streptococci and mastitis 232-276 *passim*
Streptococci in milk 287
Stripping
　　effect of cluster weight 125
　　effect of vacuum level 119
　　and mastitis 264
Suckling 176-7
Sunlight deterioration of rubber 348

Tandem layout in parlours 208
Tanks, bulk
　　cleaning 291, 313-6
　　contamination of milk 288-9
　　with cowsheds 202
　　with parlours 214
Taps, vacuum 52-3
Teat canal 159, 164
Teat dips and mastitis 235, 250
Teat lesions 256-7
Teat sinus
　　pressure during milking 135
Teatcup liner, rubber 343-350
　　ageing 346-9
　　deterioration 344-50
　　extrusion 338
　　manufacture 343
　　moulding 338
　　properties 343-4
　　swelling by fat 344-6, 349
Teatcup liners
　　description 59-61
　　effect of design on mastitis 262, 266
　　effect of disinfection on mastitis 245
　　maintenance 109
Teatcup shells 59
Teatcups
　　development 9
　　shell movements 137
　　single chamber 141
Teats
　　anatomy 164
　　cleansing before milking 290-1
　　congestion 140-1
　　contamination of milk 287
　　dipping 245-9
　　friction from liner 143-5
　　innervation 167
　　lesions 256-8

391

Teats *Cont'd.*
  liner effects during pulsation
    cycle 124-7, 136-45
  movement and flowrate 136
  movement in liner 137-8
  pulsation effects 141-2
Temperature of air,
  effect on system 35
Testing of equipment 104-15
Tests, records 115
Theory of machine milking 19
Thyrotrophic hormone and milk
  secretion 170
Tractor PTO 86
Transport of milk to dairy 76
Tubing, rubber 350

Udder
  anatomy and physiology 156-78
  blood supply 160, 165-7
  lymphatic system 167
  nerve supply 167-8
  supporting structures 156-8
Udder infection
  diagnosis 232
  elimination methods 271
  incidence 233-5, 239, 242
  source of bacteria in milk 287-8
Udder massage and milk ejection 192
Udder pressure, factors affecting 191
Udder sinus
  pressure during milking 135
Udder washing and mastitis
  242, 245-6, 263
Unilactor 212

Vacuum
  discussion 25-6
  measurement 26-8
  regulation 35-6

Vacuum effects
  machine stripping yield 119
  milk flow from udder 120
  peak flowrate 119
Vacuum fluctuation
  effects on liner 131
  effects on mastitis 146-50,
    260-1, 266
  factors affecting 145-8
Vacuum gauges 49-50
Vacuum in cluster 129-35
  in liner 139
  in milking system, recording 115
  in mouthpiece chamber 139
  in pulsation chamber 139
  in short pulse tube 135
  in teat sinus 135
Vacuum level 14-5
  and mastitis 259, 266
  testing 105-6
Vacuum milking, development 5-12
Vacuum pumps 37-45
Vacuum regulators 46-9
Vacuum stabiliser system 79
Vacuum tank *see* Interceptors
Vacuum taps 52
Vacuum tubes, cleaning 296
Vacuum-pressure system 85
  cleaning 313
Valve
  milk transfer control 68
  selfdraining 51
Vapour pressure, discussion 26

Water for cleaning
  quality 319
  requirements 310
Water heaters 97-101
Weigh jars and circulation
  cleaning 214-6
Whiteside test 236